A. M. Ehrly

Therapeutic Hemorheology

Springer–Verlag
Berlin Heidelberg New York London
Paris Tokyo Hong Kong Barcelona

Prof. Albrecht Michael Ehrly, M.D.
Department of Internal Medicine
Division of Angiology
J. W. Goethe University Clinics
Theodor-Stern-Kai 7
D-6000 Frankfurt am Main 70

ISBN-13:978-3-642-75489-0

Library of Congress Cataloging-in-Publication Data

Ehrly, A. M.
[Therapeutische Hämorheologie. English]
Therapeutic hemorheology / A. M. Ehrly
p. cm.
Translation of: Therapeutische Hämorheologie.
Includes bibliographical references and index.
ISBN-13:978-3-642-75489-0 e-ISBN-13:978-3-642-75487-6
DOI: 10.1007/978-3-642-75487-6

1. Blood circulation disorders – Treatment. 2. Blood – Rheology.
I. Title.
[DNLM: 1. Hemodynamics – drug effects. 2. Rheology. WG 106 E33t]
RC691.4.E37 1991
616.1'306 – dc20
DNLM/DLC
for Library of Congress

2119/3140-543210 – Printed on acid-free paper

For my family

Foreword

I am pleased to have been asked by Albrecht M. Ehrly to write a foreword for this book. Hemorheology continues to be the most active branch of the science of biorheology. As an organized science, biorheology began in 1948, when I presented, at the First International Congress on Rheology, Scheveningen, Holland, a Plenary Lecture entitled "Rheological Problems in Biology," in which I introduced the term "biorheology". Biorheology comprises the study of deformation and flow of biological systems or of materials of biological significance directly derived from living organisms. It deals as well with materials of nonbiological origin, employed in a biological system or setting.

Three years later, I introduced the term "hemorheology" at the 25th Anniversary Meeting of the American Institute of Physics, Chicago, in a Plenary Lecture before the Society of Rheology, one of the Institute's Founding Member Societies.

Ehrly refers to my original definition in the introductory chapter. He also refers to the definition of hemorheology which I gave in 1981 in a Plenary Lecture on the "Future of the Science of Biorheology" at the Fourth International Congress of Biorheology, Tokyo. I postulated that "the blood together with the blood vessels, in which it circulates, constitute two markedly differing but symbiotic portions of a very special organ which is unlike any other organ." This is emphasized because, in my appraisal, we have to conceive the symbiosis of the vessels and the blood as one single sophisticated organ with its diverse populations of cells and proteins, structures, functions, its vastness, and its diverse rheological behavior.

Hemorheology, in summary, is the study of rheological aspects of the vessel-blood organ. It is the study of the deformation behavior including flow of blood and its constituents and those materials of the blood vessels and surrounding tissues with which blood or its components come into direct contact. In addition, it is also the study of the interaction of blood or its components and the vascular system with added foreign materials, such as drugs, plasma expanders, or prosthetic devices. All in all, hemorheology is the study of how the vessel-blood organ in its two portions, viz., the blood vessel and the blood, can function and interact with other organs which it penetrates, thus playing essential roles in the living organism.

It goes without saying that the above definition gives the science branch of hemorheology and its clinical applications a wider scope which, I am confident, will be dealt with in future studies including clinical hemorheological investigations.

As I emphasized in a plenary lecture of clinical hemorheology last year at the conference in Siena [1], clinical hemorheology, in my appraisal, will become one of the main fields in the practice and teaching of medicine in the not distant future. In the past 100 years many areas in medicine and surgery have developed as special fields of medicine, such as hematology, angiology, dermatology, and ophthalmology, among many others. These highly developed specialties, some of which deal with clinical hemorheology, will profit greatly from the advancement of the latter at present and in the forthcoming years. These advances will include the disciplines of diagnostic, therapeutic, and preventive hemorheology.

As I pointed out in 1981 in Tokyo, in a Plenary Lecture before the International Congress of Biorheology [2] on the future of biorheology, this modern life science of merely three decades is the missing link between all life sciences. As far as clinical hemorheology, one of the many branches of the science of biorheology, is concerned, it appears to be likewise the missing link between the different fields in the practice of medicine and surgery. As this rapidly advancing field concerns the clinical aspects of the rheology of the vessel-blood organ, therapeutic hemorheology will cover not merely what is dealt with in this book, but include such differing areas as hemorrhagic and thrombotic disorders, inflammatory processes, circulatory shock, stroke, edema, embolism – in short any clinical manifestation due to changes in the rheology of the vessel-blood organ. Therapeutic hemorheology will expand by entering most fields and specialties in the practice of medicine and surgery.

Already in 1959, at a conference called "Flow Properties of Blood and Other Biological Systems", convened jointly by the Faraday Society and the British Society of Rheology and held at the University of Oxford, certain areas in the field of clinical hemorheology were dealt with. "Clinical Hemorheology" was the title of an entire section of the First International Conference on Hemorheology, held at the University of Iceland, Reykjavik, in 1966. It is noteworthy that at that time Ehrly and other clinical investigators reported their studies in the field of clinical hemorheology. He became there a charter member of the International Society of Hemorheology, which three years later, at the Second European Conference on Hemorheology, held at the University of Heidelberg, became the International Society of Biorheology. At subsequent International Congresses of Biorheology, Ehrly took an active part in presenting studies in clinical hemorheology. Thus, as a clinical investigator and clinician, Albrecht Ehrly became a pioneer in clinical hemorheology.

Ehrly's activities in clinical hemorheology alone give him exceptional credentials for his writing of the first book on therapeutic hemorheology. These high qualifications are further augmented by his leadership in introducing new therapeutic approaches and drugs in the treatment of circulatory disorders due to pathological conditions affecting the vessel-blood organ. His clinical hemorheological findings on defibrinogenation, which are of particular importance, invite further detailed clinical investigation.

Recent advances in experimental hemorheology, including surface hemorheology, have led to new facts and insights pertaining to the vessel-blood organ [3–7]. Thus, for instance, endoendothelial fibrin lining (EEFL) as well as in vivo fibrinogenin formation, i.e., the aggregation of fibrinogen molecules and subsequent gelation without thrombin action, are expected to have a great impact on the practice of medicine and surgery.

The interface between the two portions of the vessel-blood organ is EEFL, which serves as the crucial filtration barrier for transcapillary transport between the blood and tissues. Moreover, the EEFL plays highly significant roles in the circulation of the blood. It acts as a physical anticoagulant and an antithrombotic agent, and, in the microcirculation, it aids the action of the heart by decreasing markedly the apparent viscosity of blood, known as the Copley-Scott Blair phenomenon. These roles provide the basis for new approaches in hemorheological therapy, including the selection of prosthetic devices.

It should also be noted that the recent biomicroscopic investigations pertaining to thrombogenesis by S. Witte [4, 6] (1980) and earlier findings pertaining to the initiation of hemostasis by B. W. Zweifach [4, 6] (1954) show clearly that their findings of in vivo fibrinogenin gelation open up new vistas in hemorheological therapy. Our recent studies suggest the antithrombogenic action of a number of substances, such as low molecular weight heparins, chondroitins A, B and C, among others [4, 5, 8]. These new advances in experimental surface hemorheology are expected to be developed within the next decade for clinical application. What likewise may be expected is the development of in vivo hemorheological testing, as well as hemorheological tests, which come as close as possible to the in vivo situation.

In 1982 it was pointed out by Witte [9] that the processes involved in capillary (or vascular) permeability included those which occur within the interstitial space, across the parenchymal cell membrane in the lymph. Witte proposed to call these processes the "inner circulation." I fully concur with Witte that the field of clinical hemorheology must also deal with this inner circulation. Modern clinical hemorheology will, therefore, need to be extended to include the rheology of the interstitial spaces, their contents, parenchymal cell membranes, the lymph, lymphatics, and its walls. "Perihemorheology" was introduced by me, at the Siena Conference, as a new term to designate the above extension of clinical hemorheology. As I emphasized, perihemorheology has already a long history in medicine and, as in the case of modern clinical hemorheology, modern clinical perihemorheology will play a significant role in the future. One of its important practices will be perihemorheological therapy [10, 11].

This book is a courageous effort by the author to acquaint the physician and surgeon with many drugs in the management of circulatory disorders and diseases. As a first book on the subject, it cannot yet deal with a number of aspects of clinical hemorheology pertaining to the vessel wall, referred to above. This is in great part also due to the paucity of clinical investigations and of reliable methods of testing, thus far available, of the vascular portion of the vessel-blood organ.

Therapeutic Hemorheology will be greatly appreciated by clinicians and surgeons as a guide for further exploration of the growing field of clinical hemorheology. I applaud Albrecht Ehrly in presenting his book to the practitioners and specialists in the different branches of medicine and surgery. I feel confident that the book will prove as a stimulus to continued clinical investigation and to clinical evaluations of drugs in the management of the diseases and pathological conditions affecting the vessel-blood organ.

References

1. Copley AL (1985) The history of clinical hemorheology. Clin Hemorheology 5:765–812
2. Copley AL (1982) The future of the science of biorheology. Biorheology 19:47–69
3. Copley AL (1981) Hemorheological aspects of the endoendothelial fibrin lining and of fibrinogen gel clotting. Their importance in physiology and pathiological conditions. Clin Hemorheology 1, 9–72
4. Copley AL (ed) (1983) The endoendothelial fibrin lining. Symposium of the XII. European Conference on Microcirculation, Jerusalem, Israel, 7. 9. 1982. Pergamon, New York. Second printing 1984, Thrombosis Research Suppl V
5. Copley AL, Seaman GVF (eds) (1983) Surface phenomena in hemorheology: their theoretical, experimental and clinical aspects. Ann NY Acad Sci 416:761 pp
6. Copley AL (1984) The endoendothelial fibrin lining as the crucial barrier and the role of fibrin(ogenin) gels in controlling transcapillary transport. Biorheology 21:135–153
7. Copley AL (1986) Endoendothelial fibrin(ogenin) lining: the interface between the two portions of the "vessel-blood organ." In: Seno S, Copley AL, Hamashima Y, Venkatachalan MA (eds) Satellite Symposium of the Third International Cell Biology Congress, Kurashiki, Japan, 1984. Glomerular Dysfunction and Biopathology of Vascular Wall. Academic, London
8. Copley AL, King RG, Chien S. (1983) On the antithrombogenic action of low molecular weight heparins and of chondroitins A, B, and C. Biorheology 20:697–704

9. Witte S (ed) (1982) Microcirculation, interstitium, lymph, pathophysiology and disease. Pergamon, Oxford; Clin Hemorheology 2:415–767
10. Copley AL (1987) Perihemorheology rather than parahemorheology. Biorheology 24:283–284
11. Copley AL (1987) Perihemorheology rather than parahemorheology. Clin Hemorheology 7:311–318

New York, NY (USA) *Alfred L. Copley*

Foreword

As one of the co-founders of clinical hemorheology at an international level, especially with regard to its therapeutic aspects, Albrecht Ehrly has undertaken the enormous and very successful task of bringing intellectual and factual order into this clinical area which is steadily increasing in significance. It is even more to his credit since especially the therapy of hemorheologic diseases is in a complex state of development. There are still great gaps between its theoretic basis and its practical application, even though the material available both on the research and clinical levels is immense.

It is easy to deal with terminology and facts that are already backed by a historic development of their own. However, especially in clinical hemorheology, the relationship between the clinical and therapeutic aspects is undergoing a highly active exchange. The trends are very evident but the final product cannot as yet be foreseen.

The principle of organization chosen by the author of *Therapeutic Hemorheology* was based on the fact that, with growing clinical experience, the clinicians are confronted on the one hand with a constantly increasing number of pathologic hemorheologic cases requiring therapy, yet on the other hand therapeutic measures in general and treatment with medications in particular often only indirectly show a certain logic of procedure. For here neither the understanding of pathogenesis and the cause of disease can be easily brought to a common denominator, nor can an immediate conclusion be drawn from this constellation as to the necessary medicinal effect. The declarations of drug manufacturers as to the rheologic efficacy of medications as well as to the cause of this efficacy often have only a very indirect relationship to the therapeutic effect achieved. In the extreme case, nothing of what is said to be the mode of action of a drug holds true, despite the fact that the positive effect is, as expected, perhaps even better.

While the clinical impression of a therapeutic effect is of primary importance, particular significance must be attached to a therapeutic control such as measuring the blood flow properties. This has always been a special concern on the part of the author which, especially in the Federal Republic of Germany, has given him the reputation of a critical therapist of the first order. At a time when this science is only in its adolescence, it takes creative courage to present the reader not only with informative but also with a constructive insight on the clinical aspect in particular. The incoherence still existing between theory and practice allowed Albrecht Ehrly to formulate the idea that, contrary to the customary way up to then, practice should not be primarily led by theory. He has accepted the principle of the ex juvantibus therapy deliberately. With the change-over from the inductive to the deductive mode of knowledge, he utilizes the facts of pharmacology in this science by not starting out from the effects which can be theoretically expected, but rather proceeding in a critical and comparative manner from the effects experienced in practice. This is a method that leads to a modus vivendi, i.e., a concrete therapeutic tool for the practitioner. Without this, the urgently needed expansion of an aspect of medicine that up to now has been under-represented cannot be achieved.

This book is primarily directed at an extensive target group of clinicians and practitioners. Its niveau, which is beyond that achieved within this framework up to now, also addresses the relevant specialists, the angiologists, the intensive care clinicians, the hematologists and anesthesiologists, as well as pharmacologists and pathophysiologists. They are offered a critical evaluation of all the therapeutic possibilities known within the rheologic frame up to now. In addition, each section is presented with a well-founded commentary that also makes the study of this realistic guide to hemorheologic therapeutic activity a useful and lasting enjoyment for all readers.

Kaiserslautern *Hellmut Hartert*

Preface

Therapeutic measures to improve the flow properties of blood (hemorheologic therapy) have become increasingly important since their tentative beginnings about 80 years ago. Nowadays, drugs which have or should have a beneficial effect on the flow properties of blood are being used more and more and are being administered in a large number of diseases and groups of diseases, in particular vascular disorders.

Accordingly, in the last few years reports on the research and application of such therapeutic measures have been increasing in the relevant literature and at congresses on clinical hemorheology. This book was written because, on the one hand, a detailed study of this aspect of clinical hemorheology had not existed before and, on the other hand, the interest of clinicians and general practitioners in this field has increased considerably. The author feels competent to present this study not only because of his extensive experimental research in the field of hemorheology, but also more importantly, because he is a therapist.

During the time when I was taking my first steps in the field of hemorheology in 1962 – supported by my colleagues F. Gramlich and H. E. Müller in Mainz – there was neither a "school" nor a "teacher" of hemorheology or clinical hemorheology in Germany. So I had to rely on publications by Bayliss, Copley, Dintenfass, Fahraeus, Gelin, Gregersen, Merrill, Scott-Blair, Wells, etc. – scientists whom, for the most part, I did not get to know until 1966 at the first International Congress on Biorheology at Reykjavik, Iceland and for whose fundamental studies I am greatly indebted. At about this time, I also became acquainted with two German physiologists, Peter Gaethgens und Holger Schmid-Schönbein, who were beginning to do work on experimental hemorheology. As an angiologist I greatly profited from the, in part, controversial discussions to add impetus to my more pathophysiologic and clinically oriented research. The same applies to many other colleagues of that time, such as J. P. Barras, D. Braasch, A. L. Copley, S. Chien, J. Ditzel, H. Meiselman, Y. Isogai, G. V. F. Seaman, and T. Somer, to mention only a few.

In the following years, clinical hemorheology and, as a result, hemorheologic therapy continued to develop, especially after clinical practitioners also increasingly discussed the significance of hemorheologic factors with regard to prevention, diagnosis, pathogenesis, pathophysiology, and therapy of diseases. The increasing significance of the hemorheologic therapy of diseases must, therefore, be seen as a logical consequence based on the knowledge already acquired in the fields of hemorheology, biorheology, and especially clinical hemorheology. It must be pointed out that it is not the goal of this book to compile or comment on clinical studies or individual studies on the question of the clinical efficacy of hemorheologic therapy in patients (see Sect. 2.2.1).

In this book on therapeutic hemorheology, the attempt was made to collect and comment on as high a percentage as possible of the literature of those publications in which hemorheologic changes were found within the framework of therapeutic measures and in which a connection with the therapeutic goal was assumed. I am nevertheless aware that owing to the wide distribution of scientific contributions to *Therapeutic Hemorheology,* I have inevitably left out some thematically relevant publications. I would be very grateful if these authors or the readers of this book would point out such missing contributions to me.

I would also like to take this opportunity to thank all my co-workers for their assistance with this book. Special thanks are due to Dr. Winfried Volkmann who, for the purposes of his dissertation, accomplished the great task of compiling the literature, as well as to Dr. J. Schenk and Dr. R. Dehn.

Prof. A. L. Copley and Prof. H. Hartert, both pioneers in biorheology and hemorheology, have both agreed to write a foreword to this book. I would like to take this opportunity to express my sincere gratitude to my friends Alfred and Hellmut for performing this task.

Frankfurt, September 1991 *A. M. Ehrly*

Contents

Introduction

Hemorheology, the science of the flow properties of blood, is still a young field of medical research. Only in the last 20 years has it increasingly gained entry into medical science, which is by no means lacking in specialized terminology. While at first, only the terms "blood viscosity" and "plasma viscosity" were known, nowadays other rheologic factors of the flow properties of blood such as "erythrocyte aggregation" and "erythrocyte deformability" are used as well.

After the fundamental physiologic principles were worked out for this specific branch (general hemorheology), the pathologically modified flow properties of blood in different diseases were then recognized and systematically studied. This "clinical hemorheology" may best be classified within the framework of vascular pathophysiology; there are also connections to hematology [232].

There is a whole list of books and congress reports which present an overall view of general and clinical hemorheology and in which a major part of basic research is included as well (e.g., *Replogle* and *Meiselman* 1967 [723], *Copley* 1968 [146], *Larcan* and *Stoltz* 1970 [548], *Dormandy* 1970 [197], *Dintenfass* 1971 [175], *Chien* 1972 [125], *Messmer* and *Schmid-Schönbein* 1972 [640], *Dintenfass* 1976 [178], *Stoltz* und *Drouin* 1980 [823], *Schmid-Schönbein* et al. 1981 [762], *Lowe* et al. 1981 [588], *Dintenfass* 1981 [180], *Manrique* and *Müller* 1981 [610], *Ehrly* 1983 [296], *Dormandy* 1983 [211], *Ehrly* 1984 [311], *Dintenfass* 1985 [183]. In the German speaking countries various inaugural dissertations (e.g., *Krosch* 1959 [529], *Ehrly* 1969 [234], *Gaethgens* 1970 [359], *Schmid-Schönbein* 1970 [757], *Chmiel* 1973 [128], *Heidrich* 1975 [424], *Rieger* 1976 [726], *Leonhardt* 1978 [566], *Volger* 1980 [886] *Landgraf* 1984 [537]) have been published in this field.

As knowledge regarding the pathophysiologic correlations between the impaired flow properties of blood and different groups of diseases increased, it was natural to try to draw therapeutic consequences from these (see Sect. 3.3 for details). Thus the International Congress of Hemorheology in Heidelberg in 1969 dealt mainly with clinical problems as well as with some therapeutic ones [420]. In a monograph published in 1973 [252], the current state of the art regarding a rational basis for hemorheologic and especially defibrinogenation therapy of chronic occlusive arterial disease was described.

After 1979 societies of clinical hemorheology were founded in Europe: hemorheologic societies or groups were formed not only in Germany but also in France, the United Kingdom, Italy, and Portugal. They held scientific conferences at regular intervals at which therapeutic contributions were presented as well. From 1979 on, European congresses on clinical hemorheology have been taking place at biennial intervals (First European Conference on Clinical Hemorheology in Nancy in 1979 [823] and Second European Conference on Clinical Hemorheolog n London in 1981 [335]). In 1981 the journal Clinical Hemorheology, edited by *A. L. Copley* and *S. Witte* (Pergamon) appeared for the first time. It also included therapeutic studies.

In 1983 the International Congress of Biorheology [466] and the Third European Conference on Clinical Hemorheology [336] took place in Baden-Baden. Here clinicohemorheologic topics were dealt with as well as therapeutic problems. Review papers were presented on hemodilution therapy, defibrinogenation, and therapy with rheologically active medications [388, 303, 593].

At the Second Congress of the German Society of Clinical Hemorheology in Munich in 1983 [302], hemorheologicotherapeutic themes were discussed exclusively. Such themes were, of course, also discussed at the Fourth European Conference on Clinical Hemorheology in Siena in 1985 [337], as well as at the annual conferences of the other national societies. Short reviews on hemorheologic therapy, especially concerning various drugs, may be found in journal articles [252, 444, 549, 826, 827, 212, 320, 551, 828, 872, 616, 51, 749, 500], as special chapters of books as well as in the form of congress reports [175, 638, 824, 208, 290, 589, 438, 725, 332]. To date, a complete and annotated summary of therapeutically induced hemorheologic changes in the blood of patients ("therapeutic hemorheology") has not been presented.

The term "therapeutic hemorheology" suggested by the author therefore implies, contrary to "hemorheologic therapy," that the subject of this book is not so much the difficult question of the clinical efficacy of the different hemorheologic modes of therapy, but rather the therapeutically induced changes in the flow properties of blood (see also Sect. 2.2.1).

The generally used term "hemorheologic therapy" claims that the success of treatment may, almost exclusively, be traced back to an improvement in the flow properties of blood. This conclusion is not imperative for two reasons. First, the clinical efficacy of each medication or therapeutic measure with hemorheologic effects would have to be conclusively proven. This is, in general, a difficult proposition. Secondly, one would have to assume that the therapeutic measures or medications chosen show exclusively hemorheologic effects. This is, as with many so-called vasoactive substances, not the case.

Therefore, in the author's opinion, the expression "rheotherapy" suggested by *Lowe* in 1984 [593] is just as inappropriate as the general term rheologics ("Rheologica") for hemorheologically effective substances [321, 500]. *Schmidt-Schönbein* [765] now talks about antithixotropic therapy. However, the general and well-established term "hemorheologic therapy" is not only the most widely used one, but also readily comprehensible to clinicians and indicates the treatment of patients with hemorheologically effective medications and therapeutic measures.

"Therapeutic hemorheology" is, on the other hand, the science of influencing the flow properties of blood in patients by means of certain therapeutic measures and medications which can influence the pathologic mechanisms responsible for the diseases and which affect their symptoms favorably. This includes the analogous pathophysiologic, pharmacologic and clinicotherapeutic problems. "Therapeutic hemorheology" in the strict sense, according to general opinion, is primarily concerned with the hemorheologic components of whole blood viscosity, plasma viscosity, erythrocyte aggregation, and erythrocyte deformability. Little is known about the hemorheologicoclinical significance of the other cellular components of the blood (leukocytes and thrombocytes). "Therapeutic hemorheology" in the wider sense correlates to *Copley's* definition of hemorheology [148] and therefore also includes therapeutic interactions between the blood and the blood vessels.

The flow properties of blood – for example, within the framework of microcirculatory disturbances in chronic occlusive arterial disease - must not be seen as an isolated parameter [291], but must stand in very close correlation to other factors and covariations of the circulatory system [286]. The material properties (flow properties) of the blood which have an influence on the flow resistance are therefore always to be seen as part of

a standard whole. Accordingly, a therapeutic influence on the flow properties of blood should primarily lead to an improvement in the microcirculation and thus optimize the blood supply of the tissue (see Sect. 3.3.9, Table 3).
Macrocirculation, microcirculation, hemorheology, and tissue blood supply are all cogwheels in the mechanisms of the human and the animal organism, but they can only function adequately when they interact properly with one another.

2 Aim and Structure of the Book

2.1 Aim

The aim of this book is to give a synoptic presentation and critical evaluation of those hemorheologic data which were gained from in vivo studies with patients and which were published and available for investigation by the end of 1986. Here the term "in vivo" means, in fact, "extra vivum", since the hemorheologic results, even after the therapeutic application of hemorheologically active substances, were actually measured in vitro, i.e., outside the body.

Drug therapy covers a wide realm beside such simple therapeutic procedures as venesection. Special consideration is also given to changes in the flow properties of blood caused by certain diagnostic and therapeutic measures, such as, for example, X-ray contrast media, which can lead to side effects. This book does not only deal with a simple, chronologic sequence of studies on this subject, but also places particular emphasis on critical commentary and analysis.

A critical stand and evaluation of studies from previous years can only be achieved on the basis of up-to-date knowledge and modern measuring techniques. This means that the relevance of some of the older studies available can often only be judged in the light of successive studies on the same theme. The critical evaluation of individual studies for therapy necessarily implies that the personal attitude of the author toward the problems presented will appear again and again.

An evaluation or a comparison of scientific studies is also made very difficult by the fact that the criteria for choice of patients, disease groups, and, in particular, the technical requirements are seldom comparable. The same or similar viscometers were used only in very rare circumstances. In many cases, in particular in earlier studies, only one or two factors of the flow properties of blood were measured. Nowadays, the major research groups in this field are in full agreement that a reliable statement about changes in the flow properties of blood (whole blood viscosity, plasma viscosity, erythrocyte aggregation, erythrocyte deformability) is one in which the properties are measured selectively and evaluated together (hemorheologic status [287], Aachen Test Profile [764]).

This book dispenses with an exhaustive discussion of the theoretical bases as well as an extensive presentation of the changed flow properties of blood in particular diseases, since these topics have already been dealt with by other authors (see Chap. 1). Instead, a short, synoptic description of the theoretic bases and the pathophysiologic correlations, including a summary of hemorheologic measurement techniques, and a glossary of the most important terminology have been added.

In addition to the comprehensive studies published in medical journals and specialized text books, the latest results, only available as abstracts, are also included. The latter were presented at hemorheologic congresses between 1981 and 1986 (e.g., the Second European Congress of Clinical Hemorheology in London in 1981 [335], the Fourth Annual Meeting of the Japanese Society of Biorheology in 1981 [23], the Fifth International Congress of Biorheology and the Third European Conference on Clinical Hemorheology in Baden-Baden [336] and others). The abstracts from some of these congresses were brought out as special issues of *Clinical Hemorheology* or *Biorheology*. Some of these articles were published in full in these journals later. Since the abstracts do not contain any detailed descriptions of the methods and results, these findings are not evaluated comparatively, but rather just mentioned and labeled "abstract, no details."

2.2 Exclusion Criteria

2.2.1 Clinicotherapeutic Results Within the Framework of Therapy with Rheologically Active Measures

Most publications on the therapeutic use of medications and measures that improve the flow properties of blood include information on the clinical efficacy. Some publications only present case reports, while others are more or less well structured and well-founded studies in which hemorheologic tests were also carried out. As a rule, when reporting on therapeutic improvements in the hemorheologic parameters of patients, authors also report on clinical improvements. If, on the one hand, the multitude of therapeutic measures in which a hemorheologic effect is described and, on the other hand, the discrepancy in the meaningfulness of the clinical data of these publications are considered, then it is easy to understand that a critical comparative study of the clinical efficacy of hemorheologic measures would have gone beyond the framework of this book, however significant these questions may be for practitioners. Often a quantitative assessment of the clinical efficacy is also impossible owing to missing objective parameters. For this reason, only the therapeutic effects of hemorheologic measures are described and commented on comparatively. But even this brings methodological difficulties that make it difficult to answer the question of whether there is in fact a hemorheologic effect (see Sect. 3.3.4). If one assumes that there are causal correlations between the impaired flow properties of blood and clinical symptoms and wants to draw the therapeutic consequences (see Sect. 3.2.1), then one must make sure that the individual medications and measures do actually improve the flow properties of blood. Should this not be the case or should it be improbable, then such a medication cannot be considered hemorheologically effective, and further considerations are superfluous. In fact, there are a number of therapeutic agents and measures whose hemorheologic effect has been described, but whose actual effects must be contested according to the present state of knowledge.

Therefore, when dealing with the individual hemorheologic therapeutic measures (see Chap. 4), the authors' statements as to the clinical efficacy are only mentioned if this is particularly conspicuous in either the negative or positive sense. A description of the clinical results or comments thereon were, however, not possible because of the reasons mentioned above.

2.2.2 Clinicotherapeutic Studies with Hemorheologically Active Substances and Measures Without Simultaneaous Measurement of Hemorheologic Parameters

Many scientific publications describe therapy with rheologically effective medications in connection with the most varied pathologic symptoms and discuss the mode of action in the sense of an improvement in the flow properties of blood (e.g., [95, 385, 386, 387, 204]). These studies were, however, not considered in this synopsis because the authors did not carry out their own viscosity measurements. Usually the authors draw correlations to experimental studies of other scientists or discuss a rheologic mode of action without even citing concrete results of other authors. It would go well beyond the scope of this book to include this type of publication which mentions a conceivable or probable therapeutic effect on the flow properties of blood without the authors' own measurements. These publications are therefore omitted on purpose, but with no intension of underestimating their clinical relevance.

2.2.3 Publications with In Vitro Studies and Microcirculatory Measures

Studies with healthy human subjects and results of animal and in vitro studies are not the topics of this book. They are only mentioned and cited if they appear absolutely necessary for understanding or clarification. Moreover, results which in general may be considered as "therapeutic influences on the microcirculation" are neither listed nor commented upon. The same applies to many capillary microscopic results as they were mainly published in earlier decades.

The results of such microcirculatory studies depend on multiple factors of which the flow properties of blood present only one aspect. The microscopic assessment of therapeutic measures used to be mainly subjective. Thus, for example, the reduction of the sludge phenomenon of the erythrocytes, e.g., when administering plasma substitutes, used to be interpreted as a direct disaggregating effect. It is, however, obvious nowadays that an increase in the flow velocity alone, as a result of the reduction of the whole blood viscosity, leads to a dynamic disaggregation and feigns a substance-specific disaggregating effect when the flow rate is increased [230, 761].

2.2.4 Review Articles

In some publications and congress contributions, particular areas of hemorheologic therapy (e.g., hemodilution, defibrinogenation, and plasmapheresis) were treated in the form of summaries, e.g., at the Third European Conference on Clinical Hemorheology in Baden-Baden in 1983. As a rule, these reports are neither discussed nor quoted here unless they include new, original results. Original studies with first- hand concrete rheologic data regarding medication or a specialized therapy are discussed in greater detail than subsequent studies on the same medication.

2.2.5 Publications on Hemorheologicotherapeutic Effects of Leukocytes and Thrombocytes

In recent years, leukocytes and also thrombocytes have gradually become the subject of basic rheologic research. The therapeutic principles which cause a general improvement of blood circulation in ischemic regions by influencing the flow properties or the behavior of thrombocytes or leukocytes are, however, barely known up to now (e.g., [712, 573, 126, 331, 627, 642, 214, 756]). This statement does not belittle the role which might be played by thrombocytes in the development of thrombosis or perhaps even of arteriosclerosis. Whether the rheologic properties of leukocytes and thrombocytes can be studied in vitro at all, as can those of the erythrocytes, seems to be doubtful because of the special function of these cells (adhesivity, pseudopodium formation, etc.)

2.2.6 Publications on the Prophylactic Use of Hemorheologic Measures

The subject of the prophylactic utilization of hemorheologic measures, e.g., preoperative hemodilution, has not been considered in this book since this has already been treated elsewhere [638–641, 440]. Changes in the flow properties of blood which were possibly due to the elimination of risk factors (e.g., smoking) have not been treated either.

2.2.7 Publications not Allowing a Specific Categorization

Studies which do not say which substances, for example, were used in therapy have not been considered (e.g., [707]), nor have those in which test substances with code names were used.

2.3 Organization of the Book

2.3.1 Structure

The book is divided into a general and a specific part. In the general part there is a short historic synopsis, as well as the pathophysiologic correlation between impaired flow properties of the blood and particular diseases. In addition, the validity of statistical evaluations on changes in the viscosity parameters is examined. A further part is concerned with therapeutic hemorheology. After brief comments on the historic development of therapeutic hemorheology, the therapeutically feasible modes of action of hemorheologic therapy are outlined. In addition to a critical examination of in vitro studies, animal experiments, and studies with medications on healthy subjects, the most commonly used hemorheologic measurement methods are discussed. Special importance is attached to the critical analysis of the hemorheologic data measured in connection with therapy aimed at improving the flow properties of blood.

In this general part on therapeutic hemorheology, the indications for therapy with rheologically effective medications are presented, as are the clinical aims of therapeutic hemorheology including objectivized results. After a short look at the future development of therapeutic hemorheology, a synopsis of the most important rheologico-therapeutic measures is given, listed according to their modes of action.

In the second, specific part of the book, the various hemorheologically effective measures and drugs are dealt with chronologically, including a critical commentary on each. This is followed by a glossary of terms and a comparative presentation of short forms (free names; international nonpriority name, INN) with the corresponding trade names.

2.3.2 Chronologic Order of the Publications Discussed in the Specific Part

Faced with the choice of arranging the publications with concrete, rheologic results according to groups of diseases or according to medications, the latter was chosen. It would also have been conceivable to arrange them according to the rheologic mode of action [826]. However, since the effect of many therapeutic measures is not limited to one component of the flow properties of blood and since some therapeutic measures have been recommended for a whole series of diseases, it seemed reasonable to order the publications according to the therapeutic measures. For this purpose a chronologic procedure was used, i.e., the oldest known study on influencing flow properties of blood, with, for example, infusion of dextran solutions in patients, was discussed first. This was followed by further studies on the effect of dextran solutions on the flow properties of the patients' blood right up to the end of 1987. Each discussion of the relevant individual studies on a medication (e.g., dextran solutions), is followed by a critical resumé. Substances mentioned in only one publication have not usually been given a special introduction or summary.

2.3.3 Monotherapy/Combined Therapy

In addition to various kinds of monotherapy using, for example, venesection, plasmapheresis, pentoxifylline, etc., studies which included a simultaneous application of rheologically effective measures *(simultaneous hemorheologic therapy)* were considered separately . The latter means the application of various therapeutic measures at the same time in order to improve the flow properties of blood. Most often, different principles of action such as, for example, hemodilution combined with defibrinogenation are used. These studies are relatively rare; it seemed nonetheless reasonable to cite them separate from the monotherapy with single substances or single measures (see Sect. 4.2). The assessment of rheologic changes in the blood using two or more drugs or rheologic techniques simultaneously is of course much more difficult than when using single substances. Nevertheless, from the clinical point of view, combined a hemorheologic therapy is one possibility for "therapeutic hemorheology." Finally, there is the successive combination of rheologic therapies which, however, has rarely been the subject of scientific studies.

2.3.4 Procedure for the Treatment of Publications on Hemorheologicotherapeutic Measures and Medications

The rheologic medications and measures discussed in the specific section are ordered chronologically, i.e., the substances or measures that were published first are also mentioned first. All subsequent publications on this therapeutic agent are added to the first one in order to allow a comparative commentary. The procedure in the assessment of a publication on the hemorheologically effective therapy was as follows.

Introduction
The introduction includes the name of the therapeutic agent, the main pharmacologic characteristic, the chemical or other short form, and a short description with indications.

Publications
This section includes information on patients and diagnoses, the dosage and mode of application of the medication, measurement methods, and temperature etc., and possibly details of statistics and clinicotherapeutic results inasmuch as these are apparent from the publication (see Sect. 2.2.1).

Note
This section presents the authors' remarks on the methodology.

Discussion
Here the individual studies on one type of therapy are presented in chronologic order and summarized with an evaluation and commentary.

2.3.5 Drug Terminology

Most publications on therapeutic hemorheology report on the effects of drugs. The substances are listed in the individual chapters under their international short forms (INN), preceded by a short description of the active substances. Since the drugs are often distributed by a number of firms under different names and these names are also not uniform from country to country, the tabular list of trade names (see Chap. 8) only mentions the name of the substance with which the earliest study was made. If a trade name is not discernable, then only the short form (INN) is given.

2.3.6 Sources, Literature

In our search for relevant literature, the Index Medicus and DIMDI programs like Medlars 2 etc. were especially useful (as of the end of 1987). Moreover, the literatur from the bibliographies of the available publications was used. This is particularly the case for the various review articles, books, inaugural dissertations and theses, as mentioned in Chap. 1. For many substances, the manufacturers were also asked to present literature.

3 General Part

3.1 Short Historic Presentation and Explanations of the Terms "Biorheology", "Hemorheology", "Clinical Rheology", and "Therapeutic Hemorheology"

The term "viscosity" is already quite old. It describes the internal friction of an object, in most cases a fluid. Heraclitus already spoke about *pantha rei* (everything flows), no doubt not only in connection with the transitoriness of earthly things. In the eighteenth century *Newton* (1642–1727) described the basic correlation between viscosity, shear force, and shear rate. The discovery of the laws governing the flow of liquids in tubes was made in the year 1839 by the construction engineer *Hagen* [412] in Berlin and in 1842 by the Parisian medical practitioner *Poiseuille* [702]. Further studies by *Hagenbach* [413] in 1860 led to the formulation of the law named after Hagen and Poiseuille. The Hagen-Poiseuille law states that the quantity of fluid flowing through a capillary in a certain time (flow rate) under constant pressure is directly proportional to the pressure and to the fourth power of the radius of the capillaries and indirectly proportional to the viscosity of the liquid and the capillary length. The prerequisite for the validity of the Hagen-Poiseuille law

$$\frac{V}{t} = \frac{\Delta p \times r^4 \times \pi}{8 \times \eta \times l}$$

is a laminar flow. This was proven in 1883 by *Reynolds* in the experiment for Newtonian fluids in rigid tubes [724]. In 1890 *Couette* [150] was the first to describe the flow of fluid between cylinders. In the same year *Schwedhof* described the increase in viscosity for certain fluids in decreasing velocity gradients [783].

While *Poiseuille* [702] and later *Haro* [419], *Ewald* [341], and *Levy* [572] were still experimenting with in vitro defibrinated blood, *Hürtle* [459] was the first to make his measurements with fresh, uncoagulated blood. As early as 1900, *Burton-Opitz* [100] discovered that in vitro defibrinated blood had a lower viscosity than oxalated blood. In 1904 the same author discovered that the addition of a saline solution to circulating blood reduced its viscosity [101].

In 1906 *Pre'Denning* and *Watson* [710] found that the blood viscosity was dependent on the temperature, as did *Burton-Opitz* [102] and *Hess* [446] in the same year.

"*Hirsch* and *Beck,* however, deserve the honor of having transferred the first practical applications of these findings and methods of *Hürtle* and other investigators to the requirements of clinical medicine" [449], cited in [631]. A short time later, *Bence* and *Determann* were the first to measure normal blood anticoagulated with hirudin (cited in [631]). These last mentioned teams may be considered as the real forerunners of modern clinical hemorheology.

Already at an early stage, various authors [710, 446, 103, 863] discovered that the blood viscosity level depends on the number of erythrocytes in the blood, and in 1912 *Hess* [447] found that the variability of blood viscosity depends on the flow velocity. In 1924 *Ostwald* [675] introduced the term "structural viscosity" for this phenomenon. Ever since, the increase in viscosity with decreasing tangential pressure has been called structural viscosity.

With the founding of the Society of Rheology by *E. C. Bingham* and others in Washington, DC, in 1929, the term "rheology" was coined for the field of science dealing with the deformation and the flow of matter. Since practically all materials can be deformed, the term "rheology" must be interpreted broadly (fluids, gasses, mixtures, oils, steel, bones, etc.).

In the year 1948 *Copley,* at the First International Conference of Rheology in Holland, recommended the name "biorheology" for the rheology of living systems [142]. In 1951 *Copley* named the science of the rheology of blood "hemorheology" and defined it as follows: "Hemorheology is concerned with the deformation and flow, i.e., with rheologic characteristics of cellular and plasma components of blood in macroscopic, microscopic, and submicroscopic dimensions and with the rheologic characteristics of the structure of the vessels with which the blood comes into immediate contact" [143].

As early as 1960, *Copley* [145] considered the vessels and their content as one common organ. In 1981 this was described in detail (see below). A book by *Copley* and *Stainsby* [144] concerning the "flow properties of blood" appeared in 1960. This term and the individual rheologic factors were described in detail in 1966 by *Groth* [396]. In 1961 *Gelin* [366] had already spoken about them as "viscous factors".

In the year 1966, the International Society of Hemorheology was founded by the biorheologists present on the occasion of the First International Conference on Hemorheology in Reykjavik (Iceland). In 1969 the Second International Conference on Hemorheology took place in Heidelberg with the title "Theoretical and Clinical Hemorheology". The corresponding conference volume appeared in 1971 with *Hartert* and *Copley* as editors [420]. At this conference the International Society of Hemorheology was renamed the International Society of Biorheology. A detailed treatise on the history of rheology and hemorheology has been published by *Copley* [149].

In 1969 *Ehrly* restricted the term "clinical hemorheology" to the abnormal behavior of flow properties of blood and defined it as the "science of the flow capacity or flow properties of blood under abnormal, i.e., pathologic circumstances. Nowadays clinical hemorheology must be considered mainly as a part of angiology" [232]. This definition was a much narrower interpretation than that of *Copley* [143]; at that time, as well as today, the majority of all clinical studies (pathogenesis, pathophysiology, diagnosis, therapy, and prognosis) referred only to the changed flow properties of blood.

The most comprehensive definition of the term "hemorheology" was presented by *Copley* in 1981 at the Fourth International Congress on Biorheology in Tokyo. According to this, hemorheology is "the study of the rheologic aspects of the vessel-blood organ. It is the scientific study of the deformation behavior including the flow of blood and its components and those constituents of the blood vessels and the surrounding tissue with which blood or its components come into direct contact" (synopsis in [147]; see also Foreword by Copley in this volume).

I would suggest the term "therapeutic hemorheology" for the science of therapeutic measures to improve the flow properties of blood (see also Introduction). In its widest sense, this term includes the possibilities available for improving all the factors which, according to *Copley's* definition [147], constitute hemorheology. In fact, all the relevant therapeutic measures (see Sect. 3.4) known to the author up to date are based on the improvement of flow properties of blood and their parameters such as whole blood viscosity, plasma viscosity, erythrocyte aggregation, and erythrocyte deformability.

The authors of the individual publications discussed in this book on hemorheologic therapy have attributed their clinical success to the improvement of the flow properties of blood. This does not, of course, exclude the fact that other factors also play a role (see Sect. 3.2.1.2). This book, however, is concerned exclusively with the therapeutic influence on the flow properties of blood themselves (see Sect. 2.2.1).

3.2 Clinical Hemorheology

3.2.1 Pathophysiology of Diseases with Pathologically Changed Flow Properties of Blood

Blood is the organ that is directly responsible for the transportation of materials to and from cells and tissue. Oxygen and metabolites are delivered (supply) and metabolic products are carried away (waste removal). This blood, which the heart pumps into the periphery via the lungs and the vascular system, possesses special inherent physical characteristics which are contributing factors to the supply and removal in the circulatory periphery. The flow properties of blood are one link in a chain of different factors which determine tissue supply. Whereas in most diseases with hemorheologic components, other parameters such as blood pressure, vessel morphology, and circulatory rate are altered simultaneously, there are diseases in which the impaired flow properties of blood become directly active, i.e., inherently and pathogenetically. Even if the presence of pathologically altered flow properties of blood does not allow a statement in individual cases as to whether these changed values are causal or secondary in the sense of an epiphenomenon, therapeutic consequences can, nevertheless, be drawn on the basis of the physical principles laid down in the Hagen-Poisseuille law.

3.2.1.1 Direct Systemic Deterioration of the Flow Properties of Blood

The situation can be understood relatively easily when the impaired flow properties of blood or of the erythrocytes hinder the flow directly, as, for example, in the case of erythrocyte rigidity in sickle cell crisis. The deterioration of the erythrocyte deformability leads to ischemic conditions and symptoms of various kinds without causing disturbances in the cardiovascular system, as for example, through coronary insufficiency, or by narrowing or occluding the major vessels. In this case, the clinical symptoms may be clearly traced back to the impaired flow properties of blood as the primary, pathophysiologic source. Polycythemia and polyglobulism as well as paraproteinemias with normal hematocrit values also belong to the category of primary hemorheologic changes.

3.2.1.2 Indirect Systemic Deterioration of the Flow Properties of Blood

The group of diseases with deterioration of the flow properties of blood includes those in which primarily the flow prerequisites, such as the vascular perfusion pressure in cardiac shock, are disturbed. Here the decreased pumping capacity of the heart is responsible for the drop in blood pressure, the decrease in the peripheral pressure gradients, and the reduction of the vascular diameter in the extreme periphery (critical closing pressure). In addition, owing to the reduction of the shear forces, blood viscosity increases the sense of hemorheologic vicious circles (see Sect. 3.2.1.4). This clinical situation is an example of the fact that major hemodynamic disorders can result in the secondary impairment of the flow properties of blood, and this may also contribute to the impairment of the peripheral perfusion. If the flow properties of blood were measured under physiologic flow conditions, as they exist with normal blood pressures values just before the onset of circulatory shock, then normal values could be expected. Cardiac shock is an example of how the flow properties of blood can be negatively influenced by impaired flow conditions. As Fig. 1 shows, the flow behavior of blood is the result of flow conditions (flow prerequisites) and flow properties ("viscosity") of the blood.

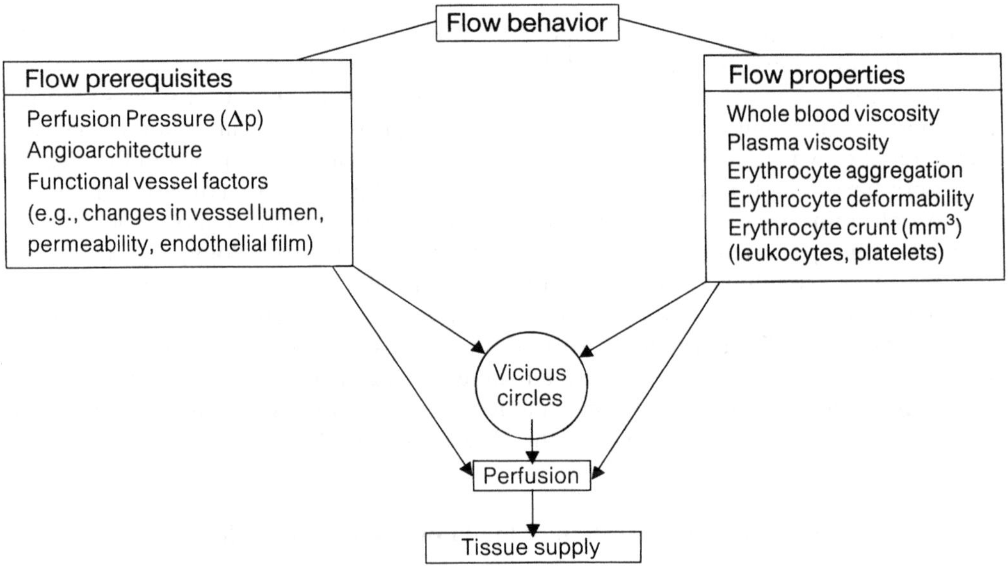

Fig. 1. Schematic presentation of the interaction of the factors flow properties and flow prerequisites (flow conditions), which together determine the flow behavior of the blood and consequently the blood flow in the microcirculatory system

3.2.1.3 Local Changes in the Flow Properties of Blood

The flow properties of blood are generally measured from mixed venous blood, normally taken from the antecubital vein. In this area, changes in, for example, the erythrocyte deformability are possibly only detectable to a minimal degree. The local changes in the flow properties of blood are greater in ischemic muscles at work than in mixed venous blood [408, 876, 354, 355]. Owing to the greatly increased metabolism in muscles at work, while at the same time the oxygen supply is decreased with a corresponding deterioration of the metabolite disposal, acidosis and hyperosmolarity ensue [258], factors which can decisively worsen the deformability of erythrocytes. It is also probable that the flow properties of blood are impaired in the vicinity of inflamed ischemic ulcers or in the borderline zone of gangrene due to a local change in physicochemical parameters (osmolarity, pH value, etc.)

3.2.1.4 Circuli Vitiosi

Hemorheologic feedback mechanisms represent a special form of local deterioration of the microcirculation and are called circuli vitiosi ("vicious circles") [233]. For example, the deformability of the erythrocytes plays a special role in a poststenotic, vascular region which is characterized by low pressure gradients and low flow velocities [174]. If the erythrocytes are more difficult to deform, they form a relatively high resistance to the local driving forces which are already low, since their passage through the smallest capillaries (7.2 μm erythrocyte diameter, approxomately 5 μm capillary diameter) is impaired, while with normal pressure gradients they are easily pressed through the capillaries [259]. If the pressure gradient is poststenotically reduced (see Fig. 2), the rigid erythrocytes will flow particularly slowly at this point and, as a result of hypoxia [271], will increase an existing acidosis. Thus, a vicious circle begins which can possibly lead to a rheologically induced local stasis.

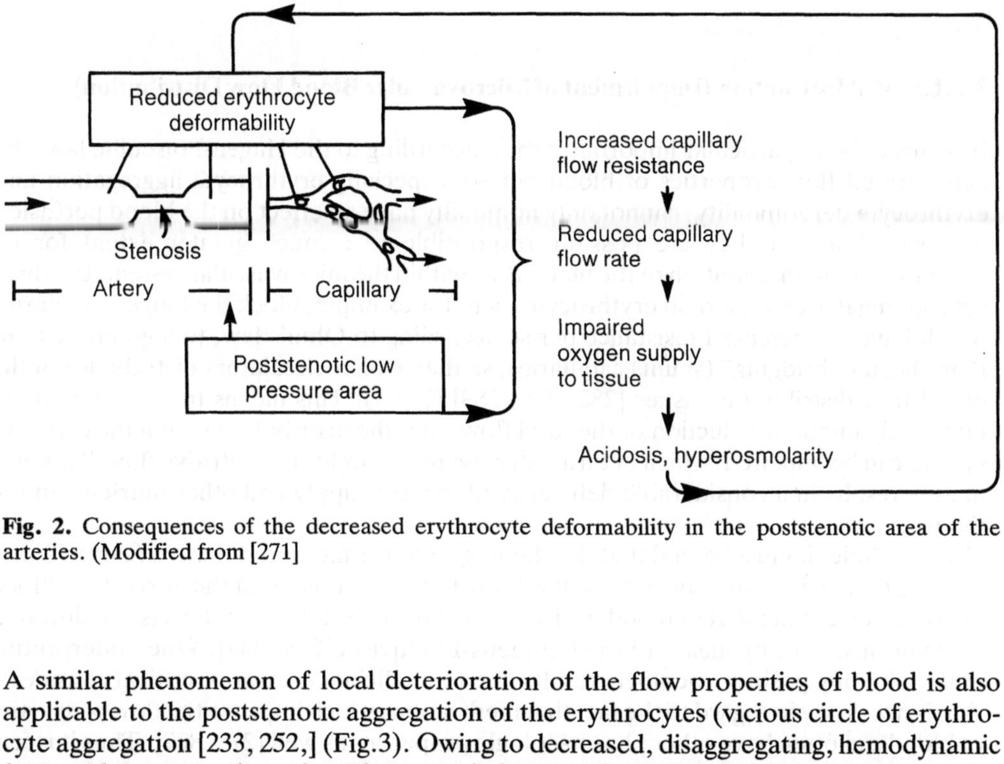

Fig. 2. Consequences of the decreased erythrocyte deformability in the poststenotic area of the arteries. (Modified from [271]

A similar phenomenon of local deterioration of the flow properties of blood is also applicable to the poststenotic aggregation of the erythrocytes (vicious circle of erythrocyte aggregation [233, 252,] (Fig.3). Owing to decreased, disaggregating, hemodynamic forces, the aggregation of erythrocytes is increased, leading in turn to a consecutive increase in the structural viscosity. These aspects are extremely important for rheologicotherapeutic considerations [252, 268, 273].

Comparative measurements of the circulatory rate and the tissue supply in ischemic extremities have shown that the tissue oxygen pressure may be decreased even though the flow values appear to be still adequate. This phenomenon, known as maldistribution or disturbance of microcirculatory blood flow distribution, can, among other things, be caused by impaired flow properties of blood (see Sect. 3.2.1.5). Thus, a seemingly small local hemorheologic disturbance can be succeeded by a relatively large disturbance in the tissue supply.

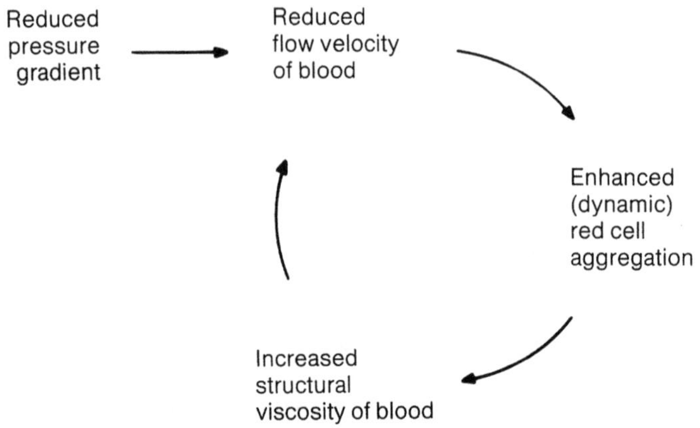

Fig. 3. Effect of erythrocyte aggregation in vascular low-flow areas with decreased pressure gradients in the sense of a vicious circle (from [233])

3.2.1.5 Maldistribution (Impoirment of Microvascular Blood Flow Distribution)

It seems to be of particular importance that, according to the Hagen-Poiseuille law, the deteriorated flow properties of blood per se, especially erythrocyte aggregation and erythrocyte deformability, cannot only nominally have an effect on the blood perfusion in tissue, but that they are possibly responsible to a much greater extent for an inhomogeneous distribution of the blood flow within the microvascular system. Erythrocyte aggregates or even rigid erythrocytes can, for example, block the longer capillaries which have an increased resistance per se according to Ohm's law, to a greater extent than shorter "bridging" (shunt) capillaries, so that a microcirculatory disturbance of the blood flow distribution ensues [282, 286, 288] (Fig.4). This means that, even with an effectively minimal reduction of the total flow rate, the distribution within the capillary system can be impaired, which in turn is disadvantageous for the nutritive flow. This ultimately results in a considerable deficiency of oxygen supply and other nutrients in the tissue.

On the whole, it must be said that the rheologic data as measured in vitro (extra vivum) must also be seen relatively because it is known, for example, that the viscosity of blood in vivo is lower than that of blood in vitro. This can be explained on the basis of different mechanisms, e.g., by means of the Fahraeus-Lindqvist effect [344]. When interpreting and evaluating positive rheologic results due to medicinal measures, critical consideration is necessary in regard to the fact that medications which improve the flow properties of blood can often have other pharmacologic properties as well [275, 827]. Thus it is feasible that these other pharmacologic properties contribute positively to the improvement in the flow properties of blood. But it is also possible that a drug which, for example, has a positive effect on the flow properties of blood is also able to reduce the blood pressure. Thus, in the end effect, it can compensate an improvement in the flow properties of blood measured in vitro, so that ultimately there is no improvement in the nutritive blood flow [297], i.e., in the tissue supply.

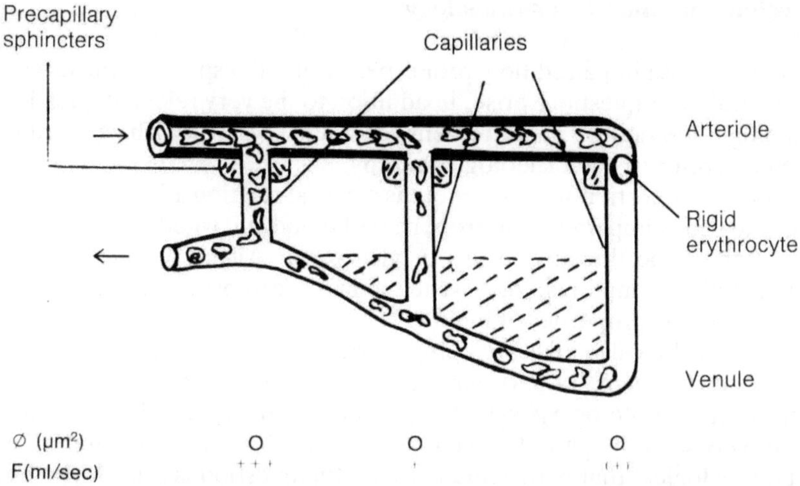

Fig. 4. Microcirculatory disturbance of the blood distribution (maldistribution) due to rigid erythrocytes in capillaries of different lengths and of the same diameter. The capillary impediment (e.g., rigid erythrocyte) is most pronounced in the longest, so-called nutritive capillary (greatest resistance). (from [282, 288])

3.2.1.6 Combined Systemic and Local Deterioration of the Flow Properties of Blood

The large group of patients with hemorheologic changes such as vascular diseases of peripheral, cerebral, and cardiac localization are not, as a rule, only afflicted with combinations of several, impaired, rheologic parameters such as blood viscosity and erythrocyte aggregation. In fact, it may be assumed that these parameters are also locally deteriorated in the systemic circulatory system as well as in certain organs and tissues. Moreover, the above-mentioned circuli vitiosi and maldistribution play a role in ischemic diseases [282, 288]. Thus, patients with diabetic microangiopathy and gangrene have pathologically altered values of whole blood viscosity, plasma viscosity, erythrocyte aggregation, and deformability of the erythrocytes in the circulating systemic blood [218, 887, 50, 73]. Furthermore, it may be assumed that in the vicinity of ulcers the flow properties of blood are also altered by inflammatory processes and by acidosis (see Sect. 3.2.1.3). Since these patients often do not show any substantial clinical limitation in the general circulation of the feet, a microcirculatory maldistribution is likely.

3.2.1.7 Compensation of Systemic Changes in the Flow Properties of Blood

In the large group of patients with paraproteinemia and especially Waldenström's disease, the plasma viscosity is decidedly increased in most cases, while the whole blood viscosity, as a rule, is usually decreased. This is due to a decrease in hematocrit values, which is typical for these types of diseases. The clinical symptoms, which only in some cases appear in the expected manner [915] (see Sect. 4.1.9) can be brought into direct relationship with the change in the flow properties of blood: raised plasma viscosity values do not necessarily lead to clinical symptoms (see below). The increase in plasma viscosity can obviously to a great extent be compensated for hemorheologically by the simultaneously existing anemia.

3.2.1.8 Open Questions on Clinical Hemorheology

When examining patients with impaired flow properties of blood, especially those with vascular diseases, a number of questions arise, in addition to the very relevant question as to the hemorheologic method of measurement (see Sect. 3.3.4) which, as a rule, should already be posed prior to hemorheologic therapy.
To date it remains an open question whether there is a causal relationship between the disease and the pathologically impaired flow properties of blood and whether the disease leads to a deterioration of the flow properties or vice versa. Also, the question still remains whether the clinical symptoms, e.g., ischemic pains, can be explained directly via a deterioration of the flow properties of blood.
Further experimental and clinical studies are being carried out concerning the question of whether there is a correlation or not between the course of the prognosis and the disease and the change in hemorheologic parameters, and whether the severity of the disease correlates positively with the pathologic changes in the blood viscosity and other parameters. Lastly, it is logical that with some diseases the question is raised whether deviations from the norm, as found in laboratory tests concerning the blood viscosity and other parameters of the flow properties of blood, have any pathophysiologic or pathogenetic importance at all or whether these are only epiphenomena that coincide by sheer chance. With some diseases the last question has been cleared up beyond any doubt. For example, in sickle cell disease, the so-called sickle cell crisis with its rigid and deformed erythrocytes is certainly not an epiphenomenon.
In recent times, studies have been carried out to test whether the abnormal rheologic data can be of use diagnostically, especially in the early diagnosis of diseases. In this connection, some authors are discussing the question of whether or not hemorheologic abnormalities are of significance as risk factors for cardiovascular diseases [37].
In principle, the questions remains open as to whether the hemorheologic results measured in vitro can be transferred quantitatively to the in vivo situation [916].

3.2.2 Validity of Statistical Evaluation of Changed Parameters in the Flow Properties of Blood in Patients

In the pertinent literature great emphasis is often put on the observation that with the most diverse diseases the flow properties of blood, usually assessed with the whole blood viscosity, are from the statistical point of view significantly deteriorated as compared to those found in healthy subjects. The relevance of such findings will be considered briefly in the following.
Using again the example of chronic arterial occlusive disease, various authors have shown that in this disease the flow properties of blood are often significantly reduced; however, these findings have not remained unchallenged. Seen absolutely, the degree of deterioration in the flow properties of blood is, nevertheless, on the average moderate [590, 504]. Closer analysis of these studies easily shows that in a small number of patients the flow properties of blood are unchanged, although no doubt remains regarding the diagnosis of severe chronic occlusive arterial disease. This implies that, in individual cases, the deterioration of the flow properties of blood is neither a necessity nor a prerequisite for the clinical picture [311].
Furthermore, we know that there are other diseases in which the flow properties of blood have deteriorated to a much greater degree without the appearance of symptoms in the sense of vascular disease [291] (see also Sect. 3.3.7). This situation begs the ques-

tion as to whether the changes measured in the flow properties of blood are the result rather than the source of the vascular disease (see also Sect. 3.2.1.8).

Particularly in the commonly described increases in the viscosity values of whole blood, it must be assumed that secondary influences such as accompanying local infections, a second disease, or general factors such as smoking also play a role.

As far as therapeutic hemorheology is concerned, these questions have been relatively unimportant up to now. The theoretical consideration of improving deteriorated flow properties and thereby improving the peripheral perfusion remains as is. It should also be emphasized that blood with "normal" flow properties can be rheologically optimized, e.g., by lowering the relatively high normal fibrinogen concentration of blood of healthy subjects [251].

For the statistical analysis of hemorheologic results from a therapy with hemorheologic drugs and measures (see Chap. 4), it is important to see that the mathematical compilation of data and the evidence of "statistically significant" changes cannot be compensated for by dubious measurement techniques or inadequate study design. The relevance and clinical validity of such results will be discussed in Sect. 3.3.7.

3.2.3 The So-called Hyperviscosity Syndrome

From the clinician's point of view, the commonly used terms "hyperviscosity syndrome" [343, 776, 808, 524, 522, 809, 180, 872], "blood high viscosity syndrome" [172, 175], "syndromes of hyperviscosity" [912] are open to criticism. A syndrome is defined as a clinical picture consisting of different symptoms. For example, if in vascular disease or plasmacytoma a hyperviscosity syndrome is being discussed, this only means that particular clinical symptoms exist and that, in addition, an abnormal laboratory result (e.g., raised blood viscosity values) has been found. *Fahey* [343] speaks of a serum hyperviscosity syndrome. Most patients with blood or serum hyperviscosity do not necessarily show clinical symptoms [915]. It is furthermore of interest that, with respect to comparable whole blood values as seen in polycythemia or polyglobulism on the one hand and paraproteinemia on the other, the clinical symptoms are much more common and stronger in paraproteinemia than in polycythenemia. From the clinical point of view, it does not seem reasonable to speak of a "hyperviscosity syndrome" if abnormal hemorheologic results have been found which probably have no causal correlation to the disease. From the same standpoint it would be just as correct to speak of a hypertransaminasis syndrome in hepatitis. The term "hyperviscosity", on the other hand, is correct (for a definition see Sect. 3.2.4). *Dintenfass* [183] has recently introduced the term "hyperviscosemia". *Case* [117] talks about so-called "symptomatic hyperviscosity" if patients have an increased viscosity with corresponding clinical symptoms with paraproteinemia (e.g., neuropathies, bleeding, etc.). I would like to support this wording and suggest for my part the formulation "asymptomatic hyperviscosity" for patients with increased blood or plasma viscosity but without corresponding clinical symptoms.

Guerrini [408] tried to make the phrase "hyperviscosity" more precise by simultaneously including the origin of the disease (so-called polycythemic hyperviscosity). *Stoltz* [829] and *Di Perri* (188) also attempted a pathogenetically related subdivision but are still using the incorrect term "hyperviscosity syndrome". *Ernst* [327] also uses the rather inconcrete term "hemorheologic deficit". *Stoltz* [830] and *Larcan* and *Stoltz* [550, 551] have, in addition to *Dintenfass* [179, 180], presented a copious documentation and classification of the so-called hyperviscosity syndrome. A good definition of the terms "hyperviscosity" and "hyperviscosity syndrome" can be found in *Baer* and co-workers [39].

3.2.4 Glossary of Important Rheologic Terms

It has already been pointed out that it cannot be the goal of this book to present theoretic and physiologic bases for rheology and especially for hemorheology (see Sect. 2.2). A certain basic knowledge of rheology is required; in some cases, the textbooks and reviews listed in the introduction can be consulted. It seems nonetheless reasonable to present the main rheologic terms in alphabetic order and to include a short definition of each.

Apparent viscosity: flow resistance of blood due to friction, i.e., the resultants of flow conditions and flow properties.

Defibrinogenation (defibrination): therapeutic measure for the controlled reduction of the plasma fibrinogen concentration by means of snake poison enzymes.

Endogenous hemodilution (the body's own "inner" hemodilution): increased entry of interstitial fluid into the intravascular space, e.g., after venesection or during vasodilation.

Endogenous hemoconcentration (the body's own "inner" hemoconcentration): increased flow of intravasal fluid from the intravascular space into the extravascular space, e.g., after stress with a rise in hematocrit values and plasma protein concentration.

Erythrapheresis: isolated removal of erythrocytes from the blood. The blood components, especially the plasma protein concentration, remain unchanged.

Erythrocyte aggregation: side to side attachment of erythrocytes as rouleaux-type structures, whereby secondary three-dimensional net-like structures can be formed. Clump aggregation is understood by some authors as a disorganized aggregation of erythrocytes, e.g., in diabetes mellitus.

Erythrocyte filtrability: measure for the filtrability of blood (mostly with an artificially determined hematocrit of 10%) using special filters (Nucleopore, Millipore) with pore diameters of about 5–8 μm. The deformability of the erythrocytes is indirectly determined from the amount of blood that is filtered or from the time it takes to flow through the filter.

Erythrocyte deformability (flexibility, fluidity): ability of the erythrocytes to passively adapt to the blood flow and to change their shape in the process. This enables them to pass through capillaries whose diameters are smaller than the diameter at rest of the erythrocytes.

Fahraeus-Lindqvist effect: reduction of the blood viscosity in narrow capillaries with a rapid flow rate. This effect is only important in capillary diameters less than 0.3 mm.

Fibrinolysis (fibrinogenolysis): enzymatic splitting of fibrin or fibrinogen molecules by means of streptokinase, urokinase, or other enzymes.

Flow conditions (flow prerequisites): hemodynamic and angio-architectonic parameters which determine blood flow, or make it possible, e.g., blood pressure level, pressure gradient, vessel diameter, branching, etc. The flow conditions and the flow properties together determine the flow behavior of blood.

Flow properties of blood: general term for the different, rheologic properties of blood:
1. Whole blood viscosity (hematocrit)
2. Plasma viscosity
3. Erythrocyte aggregation
4. Erythrocyte deformability

Fluidity (inverse viscosity): flow capability of a fluid; sometimes also used in the sense of erythrocyte deformability.

Hemodilution (blood dilution): dilution of the corpuscular and plasma components of blood as a result of infusion with cell-free (mostly colloidal) solutions causing a decrease in hematocrit in accordance with a reduction in the plasma protein concentration (hypervolemic hemodilution/isovolemic hemodilution (see p. 44 ff.).

Hemoconcentration: increase in the cell/plasma relationship (hematocrit), e.g., through water loss (exsiccosis, i.e., dehydration).

Hemorheologicum: drug whose mode of action and efficacy very likely improve the flow properties of blood (seldom used).

Hyperviscosity: increase in blood or plasma viscosity (mainly in diseases). In a wider sense, it can also mean an increased erythrocyte aggregation or a decreased erythrocyte deformability

Hyperviscosity syndrome: different kinds of clinical symptoms which can appear in some patients with extreme hyperviscosity (see Sect. 3.2.3).

Newtonian fluids: fluids in which the viscosity behaves according to Newton's law, i.e., the quotient of shear stress and shear rate is a constant.

Plasmapheresis (plasma exchange): removal of the usually increased concentration of pathologic proteins from the plasma followed by volume substitution.

Plasma viscosity: viscosity of cell-free blood plasma. The plasma viscosity level is mainly determined by the protein concentration in plasma (e.g., fibrinogen).

Relative blood and plasma viscosity: viscosity of blood or plasma relative to water (H_2O = 1) under the same experimental conditions. Viscosity of water at 20 °C = 1 cP at 37° = O.693 cP [915a].

Rigidity: = inverse of deformability

Shear rate: (velocity gradient, shear velocity, shear gradient, shear rate, dimension s^{-1}): unit for the shear of a fluid between two layers. It is the quotient of the differentials of velocity v and distance x (dv/dx).

Shear stress: (Shear force, dimension N/m^2). Tangential force which produces a mutual displacement of the fluid layers.

Simultaneous hemorheologic therapy: simultaneous application of hemorheologico-therapeutic measures with different modes of action.

Structural viscosity: this means an increase in blood viscosity while the shear rate is decreasing, caused by the aggregation of erythrocytes into rouleaux- and net-like three-dimensional structures.

Visco(si)meter: instrument for measuring the viscosity (internal friction) of fluids.

Viscosity (internal friction, toughness): physical property of fluids the flow behavior of which is mainly determined by it. Unit of viscosity: poise (P), or centipoise cP (1/100 P). In the SI system pascal · s (Pa · s), or millipascal · (mPa · s). 1 cP = 1 mPa · s.

Yield stress: if the blood flow stagnates or stops, the erythrocytes form three-dimensional aggregates which are rheologically very rigid. To surmount this stagnation of the blood flow, a force (yield stress) is needed which is much higher than the force which is normally required in order to keep the blood flowing.

3.3 Therapeutic Hemorheology

3.3.1 Short Historical Review of Therapeutic Hemorheology

In 1904 *Müller* and *Inada* [663] described a reduction of the blood viscosity in normals and in patients following the administration of potassium iodide (Fig. 5). Even though later studies by other groups could not confirm these results, this study is nevertheless important. It is the first publication in which an improvement of the flow properties of blood due to medication was reported and in which the resulting consequences for the blood flow are mentioned. In his introduction to the above-mentioned publication by *Müller* and *Inada, Romberg* (Tübingen) writes, "The use of iodine allows the blood to flow more easily. The impaired blood flow due to vascular disease can be compensated for to a certain extent or even fully by facilitating the blood to flow." Even disregarding the use of potassium iodide, this statement has become the scientific guideline for therapeutic homorhcology today.

In the early days of mankind, venesection was already known as the simplest way of reducing the hematocrit levels. It reached the climax of its application in the Middle Ages. *Kottmann* [526] was the first to find clinical evidence that venesection also reduced blood viscosity (Table 1). In appropriate cases, venesection is still a good and simple therapeutic measure for improving the flow properties of blood [96, 461].

A vital therapeutic advance was introduced by *Gelin* and co-workers in 1961 with the administration of plasma substitutes to improve the flow properties of blood [367, 368, 370]. Since then plasma substitutes have given great impetus to clinical hemorheology

Table 1. Chronologic order of the first description of major therapeutic principles (improvement of the flow properties of blood) based on rheologic findings

Therapy	Reference
Hematocrit reduction through venesection	*Kottmann* 1907 [526] *Burch* and *De Pasquale* 1963 [96]
Reduction of plasma protein concentration through plasmapheresis	*Schwab* and *Fahey* 1960 [779]
Hematocrit reduction (hypervolemic hemodilution) through plasma substitutes (e.g. LMW dextran)	*Gelin* and *Ingelmann* 1961 [367]
Hematocrit reduction (venesection) and simultaneous volume substitute with dextran solutions (isovolemic hemodilution)	*Zederfeldt* 1963 [939]
Reduction of the plasma fibrinogen concentration through	
1. Fibrinolysis (streptokinase)	*Ehrly* and *Lange* 1969 [235]
2. Defibrinogenation (ancrod)	*Ehrly* 1971 [239]
Drug-induced improvement of erythrocyte deformability (pentoxifylline)	*Ehrly* 1975 [266])

No. 48. Donnerstag, den 24. November 1904. 30. Jahrgang.

Zur Kenntnis der Jodwirkung bei Arteriosclerose.

Von Dr. **Otfried Müller** und Dr. **Ryokichi Inada.**

Mit einem Vorwort von Prof. **Romberg** in Tübingen.

Vorwort.

Eine stets zunehmende Anzahl von Aerzten sieht in den Jodpräparaten nützliche Mittel für manche Folgen der Arteriosclerose. Aber die Indikation für die Anwendung der Jodverbindungen konnte nicht scharf umgrenzt werden, solange die Art ihrer Einwirkung unbekannt war. Die Annahme Huchards, daß die Jodsalze die kleinen Gefäße erweitern, die Durchblutung der Organe erleichtern und vor allem das mit der erhöhten Gefäßspannung eng verknüpfte Fortschreiten der Arteriosclerose hemmen, ließ sich mit den experimentell festgestellten Tatsachen nicht vereinigen. Boehm und Berg[1]) und vor kurzem Stockman und Charteris[2]) haben gezeigt, daß das Jodkalium keine gefäßerweiternde Wirkung hat. Die Grundlage der ganzen Anschauungsweise wurde hinfällig, als zuerst anatomisch durch Hasenfeld[3]) und C. Hirsch[4]) und dann klinisch speziell durch Sawada[5]) gezeigt wurde, daß bei der großen Mehrzahl der Arteriosclerosen von einer dauernden Hypertension im arteriellen Gebiete, von einem dauernd erhöhten Arteriendrucke keine Rede ist, wenn nicht Nierenerkrankungen in dem Krankheitsbilde eine Rolle spielen. Andere Erklärungsversuche, die Annahme eines resorbierenden Einflusses auf die Produkte der Arteriosclerose, die von Rumpf[6]) vermutete Alkaliwirkung, die indirekte Einwirkung durch Beeinflussung der Schilddrüsen, konnten nur hypothetischen Wert beanspruchen. Von einer spezifischen Wirkung kann nur in einem Bruchteil der Fälle die Rede sein.

Wie ich auf dem Leipziger Kongreß[7]) mitteilte, war mir eine Wirkung der Jodsalze auf das Blut wahrscheinlicher als ein Einfluß auf die Gefäße. Die Untersuchungen von O. Müller und Inada in der Marburger Poliklinik haben gezeigt, daß in der Tat das Blut beeinflußt wird. Die Viscosität des Blutes nimmt durch Jodgebrauch allmählich ab. Die folgenden Mitteilungen der beiden Herren bringen die zahlenmäßigen Belege zunächst für normale Personen. Das Blut wird durch Jodgebrauch leichter flüssig. Die Erschwerung des Blutstromes durch die Gefäßerkrankung kann durch das leichtere Fließen des Blutes bis zu einem gewissen Grade oder auch völlig ausgeglichen werden. In gleichem Verhältnis zur Abnahme der Viscosität steigt nach der Poiseuilleschen Formel die Stromgeschwindigkeit.

Fig. 5. First description of therapy through the reduction of the blood viscosity. (from [663])

both scientifically and therapeutically. In 1969 the reduction of fibrinogen, caused by fibrinolysis with streptokinase [235], was introduced into therapy, followed in 1971 by defibrinogenation with snake poison enzymes [237]. Later other substances which are able to improve the rheologic properties of blood through a reduction in fibrinogen also became known [201, 595, 307].

In the mid-1970s it could be shown that there were drugs which could improve the pathologically impaired deformability of the erythrocytes. This led to therapeutic consequences like the use of pentoxifylline for the treatment of chronic arterial occlusive disease [266].

Recent studies indicate that the decrease in plasma viscosity obviously merits considerable attention [444, 776, 779, 534]. This decrease already occurs to a smaller degree in venesection, and in a narrower sense, also with fibrinogen-lowering therapy. Nowadays there are also plasma substitutes which decrease the viscosity of the suspension medium (plasma viscosity) [776, 534]. Therapeutic measures like plasmapheresis [779] are aiming in the same direction. With this method, the aggregation of the erythrocytes is also decreased considerably, just as with the defibrinogenation therapy with snake poison enzymes.

Recently various substances have been discovered which are supposed to improve the flow properties of blood. They cannot be classified into a standard group of substances and they may have different modes of action.

The question of whether the drugs used work directly like other hemorheologically effective measures, whether they act as metabolites, or whether such a therapy only indirectly leads to an improvement of the flow properties of blood often remains open. Moreover, the results of studies by *Dintenfass* [176, 181] pose the question as to what degree blood-group-specific influences on the flow properties of blood can modify their ability to be influenced by drugs.

Thus there are a number of therapeutic possibilities which have a positive influence on the flow properties of blood. They are based on a reduction of whole blood viscosity, a decrease in plasma viscosity, an inhibition of erythrocyte aggregation, and an improvement in the deformability (flexibility, fluidity) of the erythrocytes. The therapeutic techniques derived from these possibilities were already under discussion many years ago [343, 723, 236]: hemodilution with hematocrit reduction, disaggregation of the erythrocyte aggregates through defibrinogenation or plasmapheresis, and improvement of erythrocyte deformability.

In 1963 *Zederfeldt* [930] presented the first scientifically accurate rheologic measurements in simultaneous hemorheologic therapy (see above). This was the combination of venesection on the one hand and a dextran solution infusion on the other hand. Both therapeutic measures lead to "hemodilution". In 1977 *Ehrly* and *Saeger-Lorenz* [274] suggested the combination of two different hemorheologic principles: isovolemic hemodilution with simultaneous defibrinogenation. The simultaneous use of different hemorheologically effective principles of therapy or medications opens the door to further therapeutic possibilities (see Sect. 4.2).

3.3.2 Theoretically Plausible Modes of Action for Hemorheologic Therapy

The concrete possibilities and modes of action of hemorheologic therapy depend on the knowledge of those factors which can completely or partially impair the flow properties of blood. It cannot be ruled out that there exist mechanisms of influencing hemorheologic parameters other than those already known, but at the present time they are of neither practical nor theoretical consequence.

The easiest way to improve hemorheologic parameters is to lead the pathologically deteriorated hemorheologic factors back into the normal range by means of different measures. In fact, most of the therapies known today are based on this idea. Nevertheless, it is possible, and also practicable in therapy, to further optimize so-called normal values. This is, for example, the case for the reduction of hematocrit values from the normal 45% in the healthy male and 42% in the healthy woman to values of around 40%. With certain diseases, as, for example, with chronic arterial occlusive disease, the normal values of healthy individuals cannot be equated with a favorable hematocrit value. It is true that a therapeutically induced reduction of the number of erythrocytes in the circulating blood (hemodilution) reduces the number of oxygen carriers and thereby reduces the total oxygen-binding capacity, but, owing to the decrease in the blood viscosity, it also leads to a reduction in the peripheral resistance and accordingly to an improvement in the flow rate.

Therefore, the question of whether the oxygen transport capacity is improved by adjusting the hematocrit value is crucial. This should, if possible, be studied by directly measuring the tissue oxygen pressure values in the so-called ischemic tissue. The data available from literature up to now supports the idea that a moderate reduction of the so-called normal hematocrit really can result in an improvement of tissue oxygen supply [540, 541, 316].

In recent times it has been found that during generalized or local vasodilation there is an inflow of extravascular fluid into the intravascular system, which may be called "endogenous hemodilution" [306, 309]. In general, this mode of action may be assumed with vasodilating and vasoactive drugs (see Sect. 4.1.4.3 ff.).

At present little concrete information is available regarding a therapeutically induced decrease in the number of leukocytes and thrombocytes under the known normal values. However, it does seem feasible that in a few diseases such a therapy could bring an improvement in the clinical symptoms [330, 214, 756]. The question of whether a reduction of the so-called physiologic erythrocyte aggregation is of therapeutic use has not yet been clarified. However, without any doubt, the erythrocyte aggregation values can be decreased by reducing, for example, the usual fibrinogen values in plasma. In addition, little is known about the theoretically conceivable possibility of improving the normal deformability of human erythrocytes. This poses the question of whether the healthy human erythrocyte is not already optimally deformable to such an extent that further improvements of the mechanical properties are no longer possible.

With regard to the importance of the plasma viscosity for the perfusion in the microcirculation, it must be pointed out that a decrease in this parameter is quite possible through the reduction of, for example, the normal fibrinogen concentration in the blood. Such a therapeutic possibility ensues because the normal concentration of plasma proteins, especially that of fibrinogen, is relatively high. This was meaningful for evolutionary reasons because, with the frequent injuries, fibrinogen had to be present as a very important substance in the coagulation cascade. Thus, today, the fibrinogen concentration in healthy humans lies between 200 and 400 mg%; however, to activate the coagulation of the blood, 100 mg% are quite sufficient. Theoretically, a reduction of the complete protein concentration is feasible to a limited extent in order to improve the flow properties of the blood, as long as essential tasks of this system such as immune resistance or the oncotic activity of plasma are not altered.

The problem of the theoretically plausible mechanisms for improving the flow properties of the blood is easier to tackle when – as is the case with many diseases – the initial values of the flow properties are pathologically impaired. In the case of polyglobulism with hematocrit values of 60% and more, the reduction of the hematocrit by means of venesection and hemodilution is the obvious solution. With massive erythrocyte aggregation due to a very high concentration of fibrinogen or α_2-macroglobulin, an easy-to-understand therapeutic starting point presents itself here as well. If the erythrocyte deformability is impaired, a causal therapy must aim at improving or possibly even normalizing this very important function of the individual erythrocytes. The reduction of pathologic plasma viscosity values especially stands to reason with plasmacytoma and similar diseases in which the microcirculation is impaired due to the high total protein concentration.

If one summarizes the cited theoretic possibilities for improving the flow properties of blood, then these, on the whole, coincide with the methods of hemorheologic therapy which are already known today and also put into clinical practice (see Sect. 3.3.1). Some of the possibilities mentioned above have been studied only rather fleetingly up to now. This is particularly true in the case of the combination of several hemorheologic therapeutic principles.

3.3.3 The Relevance of In Vitro Experiments, Animal Experiments, and Studies in Healthy Individuals Using Rheologically Active Substances

As explained in Sect. 2.2 those publications were purposely left unconsidered which deal with the action of rheologically active substances and thus of potential rheologico-therapeutic agents on the flow properties of blood in classical in vitro experiments (admixture of agents to anticoagulated blood in vitro), in animal experiments, and in studies on healthy subjects.

A large number of scientific publications exists in which the influence of medications on the flow properties of blood were studied in vitro. Results of this kind from either in vitro experiments, animal experiments, or studies with healthy subjects cannot, of course, be transferred automatically to the situation with patients. In addition, it has to be clearly stated that the results of some in vitro studies have even led to false interpretations or wrong conclusions when applied to the situation of the patient. Thus, for example, it was found that in in vitro experiments with one substance a major improvement of erythrocyte deformability could be gained if "ischemic" blood was used (e.g., [158]). The so-called ischemic blood used, however, was not blood taken from ischemic tissue or from the collecting veins of an ischemic extremity but blood gained from the antecubital vein of healthy subjects after a tourniquet block of the upper arm following exhausting exercise of the lower arm muscles.

The situation is quite different with the so-called stress tests which are nowadays used in many forms and which were first described in 1974 [258]. In these in vitro tests, blood at 37 °C can be subjected to the same strain as that which can be shown to exist in in vivo conditions. Thus, for example, hyperosmolarity or slight acidosis is simulated, and as a rule the blood or erythrocyte suspensions are incubated. It would, however, be pointless to lower the pH of blood under in vitro conditions to 5 and then interpret the drug action in a therapeutic sense; a pH of 5 is known to be lethal. Moreover, in many in vitro experiments, the drug dose required is so high, in order to achieve a favorable hemorheologic effect, that such a dose could not be administered under in vivo conditions and in particular not with humans.

Nevertheless, in vitro experiments, and experiments with animals and with healthy subjects have their relevance in the development of substances which influence the flow

properties of blood. A series of substances which have meanwhile also found application with patients were initially discovered by way of in vitro experiments, and their development continued with animal experiments and with studies on healthy subjects. In vitro experiments are also well suited for clearing up questions on the mode of action and solving simple questions on the dose-response relationship.
As yet animal experiments have not been able to establish a position regarding the question of the efficacy of rheologically active substances. This is mainly due to the fact that the existing animal models do not present comparable conditions to the flow properties of blood in vivo in humans or do so only to a limited extent. The significance of studies carried out on healthy subjects with rheologically active substances is, of course, only limited. At any rate, an improvement in the flow properties of blood of healthy subjects after "therapeutic" measures is an indication that possibly the flow properties of blood can also be improved in patients. Whether this is connected with an improvement in the clinical symptoms is another important question which will be considered in a subsequent section (see Sect. 3.3.7.1).

3.3.4 Remarks on the Most Common Methods of Measuring a Hemorheologicotherapeutic Effect

The great number of different methods and different instruments used by the individual investigators is characteristic for the measurement of the flow properties of blood in patients. It is rare that the work of different authors or teams can be compared, because the methodologic techniques extra vivum are, as a rule, not identical. Thus, for the measurement of whole blood viscosity, sphere gravity viscosimeters have been utilized in addition to capillary viscosimeters and rotational viscosimeters according to the cone-plate or cone-cone systems. The capillary viscosimeter allows a very exact reproducibility of measurements; however, the shear stress is usually so high that the structural viscosity of blood, due to aggregation, cannot be measured.
Viscosimeters which permit measurements at very low shear rates are the cone-plate viscosimeters and the cone-cone viscosimeters. However, especially when assessing the blood of patients with pathologic erythrocyte sedimentation, there is a high risk that the results measured are seriously distorted due to phase separation (heterophase effect). On the other hand with the large surface covered by such viscosimeters the surface tension plays a role, and only in very few cases has a device been installed to minimize these effects (guardring). Therefore, viscosity values measured at low shear rates have to be interpreted with extreme care.
Blood viscosity measurements with falling ball viscosimeters are seldom carried out today owing to the many technical problems and difficult interpretation of the results.
Plasma viscosity is often measured with the same type of viscosimeter that is used for blood viscosity measurements. However, nowadays capillary viscosimeters are usually preferred since their sensitivity and precision are very high. In many cases, measurements with the cone-plate viscosimeters present fundamental technical problems owing to the low viscosity of the plasma itself.
Up to now, only few scientists have measured the aggregation behavior of the erythrocytes directly by means of light transmission [932, 87, 513, 756, 277, 492] in accordance with the technique of *Zijlstra* [932]. These methods are relatively new, but they do allow an assessment of the aggregation behavior of the erythrocytes (Fig. 6), especially in relation to the formation velocity of the aggregates after complete dispersion and the shear

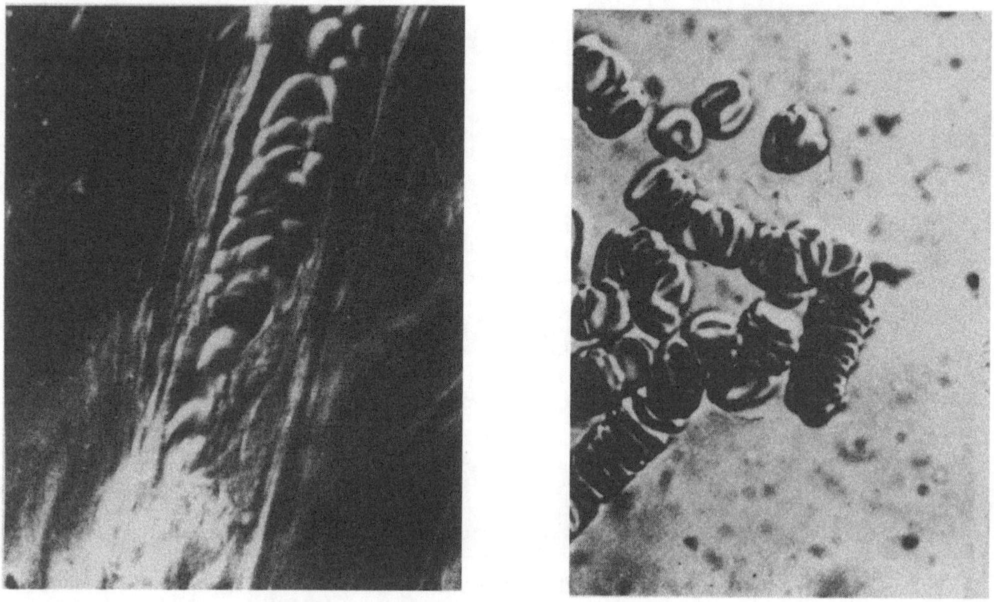

Fig. 6. Erythrocyte aggregation (left, in vivo; right, in vitro)

resistance of the aggregates. Nevertheless, even with this method other factors like plasma viscosity are bound to interfere to a certain extent.
Some authors have drawn their conclusions as to the behavior of the erythrocyte aggregates on the basis of the changes in the viscosity values at low shear rates. Keeping the above mentioned difficulties in mind, these conclusions must be interpreted with the utmost caution when measuring patients' blood with rotational viscosimeters. Other authors have taken the behavior of the erythrocyte sedimentation rate as an indirect measure for erythrocyte aggregation. In some cases, the changes were evaluated without hematocrit correction. This is impossible because of the extreme dependency of the erythrocyte sedimentation rate on the hematocrit (e.g., [241]). Of greater importance, on the other hand, is the evaluation of the blood cell sedimentation rate with corrected hematocrit (e.g., adjusting the hematocrit to a value of 40), especially, if these results are in agreement with the studies on the photometric aggregation measurements of the erythrocytes. More detailed data on erythrocyte aggregation in hemorheology can be found in the literature [832, 136]; a comparison of different aggregation methods has been presented by *Linderkamp* et al. [577].
Erythrocyte deformability (flexibility) (see Fig. 5) is nowadays usually measured by means of erythrocyte filtration through filters with minute pores of between 5 and 8 μm. Erythrocyte filtration is practically a flow method with which the number of erythrocytes or the amount of blood flowing through the filters per unit of time is measured or, on the other hand, with which the time needed for the blood to pass through the filters is determined. For this reason, the methods should more precisely be called filtration methods (microflow methods), and the result should be characterized as filtrability of the blood or of the erythrocytes. This presupposes that a reduced deformability of the erythrocytes impairs the flow through filters with a pore size around the diameter of the erythrocytes and thus reduces the flow rate.
Either Nucleopore filters with straight channels and a pore diameter of mostly 5 μm or cellulose ester filters (e.g., Millipore) with a sponge like network of slits and "capil-

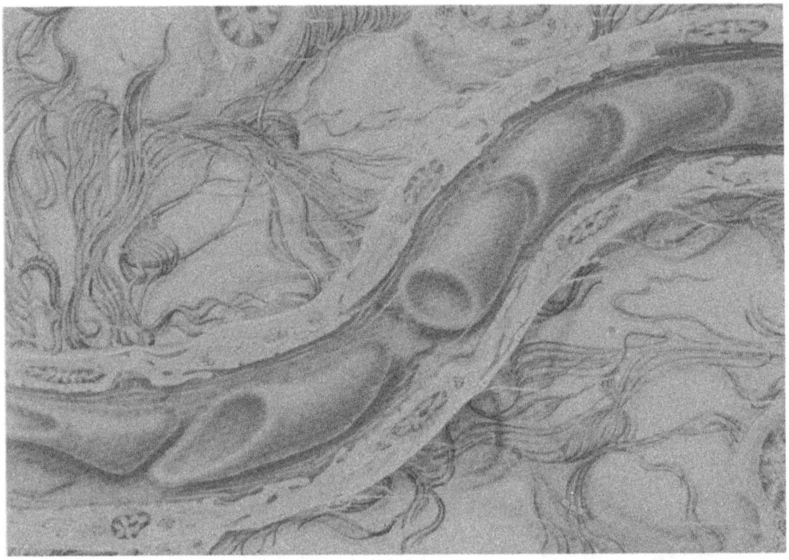

Fig. 7. Projectile-like deformation of fast-flowing erythrocytes in narrow capillaries (half-schematic presentation from photographs taken with high-speed film). (from [265])

laries" and an average pore diameter of 8 μm are used. In most cases, the blood is adjusted to a low hematocrit (e.g., hematocrit of 10%) with previous removal of leukocytes and thrombocytes. The results from whole blood filtration methods like those of *Reid* and co-workers [721] should therefore be evaluated critically. Despite the fact that it is known that the deformability of erythrocytes plays the biggest role in erythrocyte filtration tests, it must be remembered that other factors like plasma viscosity, erythrocyte aggregation and leukocytes can also influence the flow rate [467, 10, 846, 211, 607, 625, 626, 847]. Recently, devices have been developed to measure the deformability of individual erythrocytes [491]. Pipette methods (micropipettes) on the other hand, are rarely used in clinical hemorheology. Apart from the variety of methods used for measuring the flow properties of blood, it is also very difficult to compare the individual studies because, especially in the older publications, the complete spectrum of rheologic measurement methods (rheologic status) was not considered.

If only the whole blood viscosity or the plasma viscosity are measured, the effect may be, on the one hand, that rheologically and possibly also clinically important individual properties of medications (e.g., improvement of erythrocyte deformability) are overlooked. On the other hand, the different parameters of the flow properties of blood can also inhibit each other's effect or even make it redundant. An example of this – as proven in in vitro experiments [242] – would be that certain substances can reduce the erythrocyte aggregation to a considerable extent, but that in so doing the erythrocyte deformability is decreased. Which of these factors, in the end effect, plays the dominant role in the microcirculation cannot be said at the moment. However, it should not be expected that such competitive improvements or detriments of rheologic subparameters will lead to a global improvement in the tissue supply situation. On the other hand, it is possible that a simultaneous influence of different rheologic factors that have a favorable effect (e.g., hematocrit and erythrocyte aggregation) can also be therapeutically relevant (see Sect. 4.2.2).

In the studies on hemorheologically effective treatment measures referred to here, different temperatures of measurement were often chosen, different anticoagulants were used, or the time between blood removal and measurement was not the same. With many authors these details were often missing, so that the results can be utilized only to a limited extent and comparisons of the different studies become difficult. Many studies do not mention at what time of day the patients were examined and treated. Only in very few cases was the circadian rhythm of the flow properties of blood [785, 250, 312], which has become known in the meantime, taken into consideration. Thus, already under physiologic conditions due to biorhythmic divergencies, the rheologic parameters can partially undergo considerable changes depending on the time of day.

Independently of the measurement methods, the initial studies of a rheologically active medication are often quite different and difficult to compare because the groups of patients are different and the dosage of the rheologically effective medication was administered differently. In spite of all these difficulties, a deliberate attempt was made to make assessments as to the effect of measures to therapeutically influence the flow properties of the blood.

In the last few years, more and more modifications of measuring instruments and also new developments have appeared on the market making it possible to study partial aspects of the flow properties of blood more thoroughly. This brings up the question of whether, with the help of such very specific methods, new aspects of knowledge can be found for the clinical effect of hemorheologic drugs. It must be noted here that the most important hemorheologicotherapeutic principles were already found years ago with relatively simple methods (see Table 1). However, with modern methods it will be possible to study and examine subaspects better, e.g., in the therapy with substances to improve the erythrocyte deformability.

To complete the picture, it should also be mentioned in this connection that the practitioner can, in a high percentage of cases predict with relatively simple nonrheologic means whether a patient has rheologic abnormalities. This must always be assumed when the hematocrit values lie above 50% (with untrained persons who have not adapted to high altitudes), when the total protein concentration exceeds the normal values to a considerable extent, and especially when the fibrinogen concentration clearly lies above the norm. If the erythrocyte sedimentation rate has clearly increased while at the same time hematocrit values remain normal, then a deterioration of the flow properties of blood, especially an increased erythrocyte aggregation, must also be assumed. Only the decreased erythrocyte deformability cannot be determined using the usual routine laboratory methods.

3.3.5 Dependence of Hemorheologic Parameters on Blood Sampling

The flow properties of blood are usually studied with venous blood obtained from the anticubital vein. In the majority of studies cited in this book, changes in the flow properties of blood resulting from pathophysiologic reasons and also from therapeutic measures were measured from this mixed venous blood. This is done with the assumption that the flow properties of blood are identical or at least comparable throughout the whole circulatory system. There are, however, indications that the flow properties of blood are not the same in every area of the body, as, for example, when the body is in an upright standing position [443, 284]. Under relaxed resting conditions, the viscosity values tend toward lower levels. For comparative studies, like those which are important for the evaluation of therapeuticorheologic measures, it is therefore absolutely impera-

tive to keep to certain comparable prerequisite conditions. It may be assumed that, for example, in patients with intermittent claudication the flow properties of blood found locally in the working ischemic muscle are quite different from the mixed venous blood obtained from the anticubital vein. This may be explained by the fact that the local metabolic situation, particularly of the pH and/or the osmolarity, influences the flow properties of blood and especially erythrocyte deformability. This decrease in the deformability is reversible and can then no longer be measured when the above-mentioned physiochemical parameters in the mixed venous blood are normalized again (see also Sect. 3.2.1.3).
An additional factor which can distort the results of hemorheologic studies in connection with therapeutic measures is the fact that the flow properties of blood are subject to deviations depending on the time of day. In studies with healthy subjects, it was found that under resting conditions and without therapeutic or other influences the blood viscosity values showed maxima and minima. Thus there is, for example, an increase in blood viscosity during the course of the morning while the blood and plasma viscosity values reach their minimum at three a.m. (see Sect. 3.3.4). Since most studies are carried out at different times after medication, spontaneous changes in rheologic results must also be taken into account. If the different publications are checked for this aspect, it will be seen that spontaneous changes in blood viscosity are only taken into consideration in very few studies.

3.3.6 Pathophysiologic Validity of Various Subparameters of the Flow Properties of Blood

The four most important parameters of the flow properties of blood, whole blood viscosity, plasma viscosity, erythrocyte aggregation, and erythrocyte deformability, are evaluated differently in accordance with their importance in the framework of clinical hemorheology. There remains no doubt that the increased whole blood viscosity, which in general is the result of an increase in hematocrit, may be numerically, clinically, and therapeutically decreased when the hematocrit level is decreased by various methods.
In earlier times, whole blood viscosity was considered to have maximal priority based on the total viscosity as a factor of Hagen-Poiseuille's law. For this reason, in most studies on the therapeutic improvement of the flow properties of blood, the behavior of whole blood vicosity was measured with a defined shear rate or using various shear rates. The pathophysiologic validity of an increased whole blood viscosity is rarely disputed in polyglobulism or polycythemia since with venesection (hemodilution) the corresponding clinical symptoms generally improve rapidly. On the other hand, there are no doubt healthy subjects (dwellers of the Andes, marathon runners) as well as patients with increased hematocrit levels without corresponding clinical symptoms (see Sect. 3.2.2).
The pathophysiologic and pathogenetic importance of the plasma viscosity has, in the author's opinion, been underestimated up until now. For example, as early as 1960 *Bayliss* [53]) and later *Merrill* [636, 637] and others showed the importance of plasma viscosity within the framework of the apparent whole blood viscosity, while *Begg* and *Hearns* [57] believe that the concentration of plasma protein has no essential influence on the viscosity. In 1966, in their studies with plasma substitutes from a clinical point of view, *Groth* and *Thorsén* [395] and *Schülke* and *Härtel* [776] pointed out that plasma viscosity plays an important role because the hematocrit is lower in the capillaries than in the larger blood vessels, and that the blood viscosity in these microvessels is mainly determined by plasma viscosity. *Barras* [46], *Lutz* and *Barras* [600] *Ehrly* and *Landgraf*

[292], *Ernst* and *Matrai* [325], *Landgraf* and *Ehrly* [535] and *Bartolo* [51] have also placed great value on the clinical importance of plasma viscosity. *Rieger* [730], on the other, hand expressed his doubts as to the therapeutic value of a reduction in plasma viscosity. In animal experiments, *Gustafsson* and co-workers [410] were able to show that the apparent blood viscosity in the skeletal muscle essentially depends on plasma viscosity. *Browning* and co-workers [90] found a correlation between increased plasma viscosity and sensoneural deafness, but hematocrit values did not correlate. The therapeutic effects of a reduction in plasma viscosity by means of plasmapheresis, e.g. with paraproteinemia, or by means of defibrinogenation should be taken as an incentive to show more interest in this subparameter of the flow properties of blood. This is especially true because in Hagen-Poiseuille's law the plasma viscosity factor comes into its own in the terminal vessels (Fahraeus-Lindqvist effect). The serum viscosity, which was often measured in earlier studies in connection with therapeutic measures, seems, in the author's opinion, to be only of importance with paraproteinemia: the plasma viscosity level is of greater relevance since it corresponds to the situation in vivo.

The question as to which of the individual subparameters of the flow properties of blood is pathophysiologically the most important one remains open. It must also be taken into consideration that, in different areas of the circulation, the individual hemorheologic factors are of varying significance. Thus, for example, an increased rigidity of the erythrocytes in the macrovessels is insignificant, while it may play a major role in the microcirculation (e.g., sickle cell crisis). On the other hand, the pathologically increased erythrocyte aggregation is theoretically of greater importance in the postcapillary venules and veins as well as in the poststenotic region of arteries than an increased rigidity of erythrocytes.

For pathophysiology as well as for therapeutic hemorheology of diseases (e.g., shock, peripheral arterial occlusive diseases) the erythrocyte aggregation is important for two reasons. First of all, it is known that erythrocyte aggregation causes the so-called structural viscosity, i.e., an increase in the apparent whole blood viscosity at low shear rates. This type of aggregation of the erythrocytes may be presumed particularly if the shear stress is clearly decreased in poststenotic regions, for example, as a result of stenoses [273, 304]. Moreover, it is presently being discussed whether erythrocyte aggregates are of significance as flow hindrances especially in the smallest veins. In addition, erythrocyte aggregates can be the cause of maldistribution in low-flow areas (see below).

In the last 15 years, the deformability of erythrocytes has attained special recognition in the pathophysiology of various ischemic diseases [270, 722, 291]. It is assumed that a deteriorated erythrocyte deformability in the microvessels can slow down or even stop the microflow. Thus, a therapeutic improvement in the impaired erythrocyte deformability correspondingly decreases the flow resistance in the periphery. Rigid erythrocytes, moreover, could influence the microvascular blood distribution negatively, so that this also results in a disturbance in the microcirculatory blood distribution (maldistribution) (see Sect. 3.2.1.5).

It must also be remembered that an improvement in a single parameter of the flow properties of blood could also be disadvantageous for another one. Thus, for example, an increased erythrocyte aggregation can be therapeutically managed by making the erythrocytes more rigid. Which factors have priority cannot be decided upon with rheologic methods alone (see Sect. 3.5).

Owing to the various rheologic components, different therapeutic measures have also been suggested for improving the flow properties of blood by means of improving the individual parameters of the flow properties (e.g., erythrocyte aggregation or erythrocyte deformability) (see Sect. 3.3.2).

3.3.7 Evaluation of Hemorheologic Data Obtained from Patients During Treatment

3.3.7.1 General Remarks

If one considers the conclusions which some authors have drawn from the measurements of hemorheologic parameters, one finds, in addition to the descriptive analyses of therapeutically induced changes in the flow properties of blood ("therapeutic hemorheology"), publications in which an obviously drug-induced clinical effect is concluded from an improvement in the hemorheologic parameters either directly or in analogy to them. In the latter case, as far as the situation in the microcirculation or even the clinical symptoms are concerned, a simple numerical or even percentage-wise deduction based on Hagen-Poiseuille's law is no doubt no more valid than a general rejection of the importance of the flow properties of blood in the pathophysiology of different diseases – an opinion occasionally voiced in former times.

Hemorheologic results gathered in the course of therapy are after all "extra vivum data", since the measurements of rheologic parameters are almost without exception carried out in vitro. However, the situation in vivo must be differentiated much more subtly because there are a number of variables which influence the actual viscosity on the spot. In vivo the flow properties of blood must not be considered in isolation from other hemodynamic parameters (flow prerequisites, see Sect. 3.2.1.2) and from the angio-architectonic conditions. It is feasible that within the framework of rheologic therapy the expected influence on pathophysiologic rules like maldistribution (see Sect. 3.2.1.5) is possibly more important than a numerically minimal decrease in the systemic whole blood viscosity.

Therefore, there is no lack of critical voices who warn against overestimating the clinical relevance of positive hemorheologic effects of different drugs in diseases [275, 210, 291, 353, 294, 730, 321, 749, 840]. While there is nowadays no doubt that a purely rheologic therapy is effective with certain diseases (polycythemia, polyglobulism, paraproteinemia with corresponding symptoms) (see Sect. 3.2.1.1), the clinical efficacy of other types of therapy with hemorheologic modes of action has only been partially proven up to now. The reason for this is, among other things, that clinical studies of adequate design are difficult to carry out and are usually time consuming so that, with the relatively new rheologic principles of therapy, it will still take some time to clarify the situation. But there are also considerable methodologic and ethical problems in proving the clinical efficacy of hemorheologic substances.

In this connection it must be emphasized that with a study that is wrongly set up, nothing can be said regarding the clinical relevance of a hemorheologic therapy (see also Sect. 3.3.7.1). However, it also does not seem very helpful to correlate the efficacy of hemorheologic therapy with hemodynamic parameters like Doppler pressure values in the ankle region or with plethysmographic measurements of the flow rate. The fact that a reduction in blood viscosity is not automatically followed by an increase in the flow rate [252, 263] does not allow the conclusion that there is no clinical effect, since the conditions at the microcirculatory level cannot be determined by means of such global measurements (see Sect. 3.3.7.3.2).

3.3.7.2 Examples of the Discrepancy of Hemorheologicotherapeutic Measurements on the One Hand and the Clinical Effect or Efficacy on the Other Hand

Especially in the area of comparative measurements within the framework of hemorrheologic therapy, it can be shown that the improvement of hemorheologic parameters does not necessarily lead to an improvement in the clinical symptoms. Thus, it is possible that within the framework of therapy in which the flow properties of blood are modestly deteriorated – quasi as a side effect – the clinical situation is improved (sympathicomimetics, erythrocyte transfusion).
Studies with etilefrin, an α-mimetic substance showed [314], that in patients suffering from intermittent claudication and at the same time from systemic hypotension, it was possible to achieve an improvement in the pain-free walking distance. The infusion with etileferin leads to a modest but statistically significant increase in hematocrit levels and blood viscosity, which can be explained in the sense of the so-called rheoregulation (see Sect. 3.4). In addition to the increase in the pain-free walking distance these patients may show an increase in the tissue oxygen supply in the ischemic calf muscles [314] because the systemic blood pressure may increase and with it the poststenotic pressure.
In patients with severe anemia intermittent claudication symptoms may ensue because the hematocrit level is reduced and with it the number of oxygen carrying erythrocytes. The clinical symptoms of pain in the calf after physical exercise can be alleviated immediately with a transfusion of washed erythrocytes. The author was able to observe this in an eighty year old patient with chronic myeloid leukemia. As expected the whole blood viscosity increased quantitatively owing to an increase in the hematocrit level [313].

3.3.7.3 Exceptional Features of Hemodilution

3.3.7.3.1 Optimal Hematocrit

As already mentioned in Section 3.3.2 each hemodilution is followed by a decrease in hematocrit and with it in the number of oxygen-carrying cells per unit of volume. Through the reduction of the blood viscosity there ensues an increase in the flow rate which ensures the net transport capacity for oxygen. There is a level of hematocrit values which is especially suited for an optimal oxygen transport capacity. It lies between the extremes of a very high hematocrit with its accompanying high blood viscosity and a very low hematocrit with not only low blood viscosity but also low oxygen-binding capacity. This hematocrit level is called the optimal hematocrit. From the animal experiments of different authors it is known that in healthy animals the hematocrit can be lowered to well under 30 without deterioration in the oxygen transport capacity [153, 851, 360]. However, this is only true if the myocardial reserve allows an increased cardiac output and if the vascular system is intact, i.e., there are no occlusions or stenoses. Patients with chronic occlusive arterial diseases often have a deteriorated myocardial function; but, above all, the stenoses or occlusions limit the possibility of a compensating increase in the blood flow in the ischemic extremity to a considerable extent. As early as 1968 *Bollinger* and co-workers [81] found that hemodilution with plasma substitutes in patients with chronic occlusive arterial diseases resulted in a reduction of the viscosity with an accompanying increase in the circulatory rate in the extremities. The increase in the circulatory rate was, however, considerably higher in the healthy extremity than in the afflicted one, which was to be expected for the above-mentioned reasons. Clinical

experience shows that the optimal hematocrit level in patients lies way above that found in animal experiments.
Only recently have concrete data been available as to what hematocrit level to strive for in ischemic diseases, i.e., ever since it has been possible within the framework of hemodilution to measure the hemorheologic parameters as well as the tissue oxygen pressure in the ischemic muscle tissue [539, 540, 316]. From the data available up to now, it can be concluded that this level should lie slightly above 40% for muscle tissue. These results are in agreement with the studies carried out by *Linderkamp* et al. [576] who theoretically assumed an optimal hematocrit level around 40%. The information on discrepant clinical successes with hemodilution in patients with intermittent claudication must also be seen in the light of the above statements.
The question of the optimal hematocrit in the course of hemodilution therapy of ischemic diseases depends not only on the essential oxygen supply, but also on an improved wash-out of metabolites, which can then result in an improvement in symptoms. The question of whether hemodilution therapy is carried out hypervolemically or isovolemically is also of special significance. Especially with hypervolemic hemodilution, where infusion of plasma substitutes or plasma is carried out but no erythrocytes are removed, a raised cardiac output going beyond that incurred with an isovolemic hemodilution. Thus, perfusion is also improved in the poststenotic and postocclusive areas. Furthermore, it should be noted that the question of the optimal hematocrit cannot, in general, be answered for the whole body, because the conditions in organs with high oxygen consumption such as the cardiac muscle or the peripheral muscular apparatus are quite different from those in bradytrophic tissue such as the skin, for example. Thus the therapy with isvolemic hemodilution for coronary insufficiency must be relativized [730] since the uncritical use of this therapy can provoke angina pectoris attacks and myocardial infarction. However, if the hematocrit levels in adults lie above 50%, then a marked clinical improvement can be achieved in most cases by means of a hemodilution therapy that reduces the hematocrit level.

3.3.7.3.2 Maldistribution due to Hemodilution Using Human Albumin Infusions

Regardless of the question of the optimal hematocrit, it has been shown [301] that, after an infusion of 500 ml 5% human albumin solution, patients with intermittent claudication whose hematocrit reduction did not exceed 40% showed lowered tissue oxygen pressure values in the ischemic calf muscles. On the other hand, the hemorheologic parameters, i.e., whole blood viscosity, plasma viscosity, erythrocyte aggregation, and erythrocyte filtrability, were improved to a considerable extent.
However, if a 10% hydroxyethyl starch solution is used, the tissue oxygen pressure in the calf muscles is raised even though the plasma viscosity is not lowered by this type of plasma substitute. We suspect that traces of serotonin and kinins in the human albumin solution, which influence the regulation of the microvascular tone, are responsible for the fact that the microcirculation is altered in the sense of maldistribution (see Sect. 3.2.1.5). As a result the blood flow distribution becomes inhomogeneous in the capillary region leading to a deterioration of tissue supply.

3.3.7.4 Conclusions

The above-mentioned examples clearly show that the results of extra vivum measuremnts can only be transferred to the in vivo situation to a limited extent and even less to the pathophysiologic situation [275]. In this context, it should be pointed out again that the flow properties of blood depend to a considerable degree on the flow prerequisites (flow conditions) (see Sect. 3.2.1.2). Owing to different anatomic and functional hemodynamic conditions in the various vascular regions, hemorheologic conditions exist which presently cannot be measured in vitro. These new scientific findings corroborate in principle with what *Burton-Opitz* [103] already pointed out in 1911: "While no doubt viscosity may play a very important role under certain conditions, it must also be said that clinicians have attached altogether too great an importance to it. To be sure the viscosity can develop into a mighty circulatory factor, but only when other dynamic conditions favour its development." This means that hemorheologic factors, as a rule, only have pathophysiologic relevance if other, e.g., hemodynamic parameters such as the blood pressure gradient, are also impaired distally by arterial stenoses.
This shows how complex the assessment of the significance of the flow properties of blood is in diseases and especially in therapeutic measures. Proof of an improvement in the clinical symptoms cannot therefore be produced with solely rheologic methods. In addition to clinical studies, the direct measurement of an improvement in the tissue supply, e.g., by determining the tissue oxygen pressure, is called for [281, 295, 315]. This is not only true for success in treatment by means of plasmapheresis in cases of the most severe paraproteinemia or by means of venesection within the framework of polycythemia and/or polyglobulism. However, for therapy using medications the hemorheologic chain of evidence as to the mode of action does not, in my opinion, suffice. Evidence for an improvement in the tissue supply must also be offered with adequate methods, and, finally, its clinical efficacy must be evident (see Sect. 3.3.9, Fig.8).

3.3.8 Indications for Therapy with Rheologically Active Measures

A therapy with rheologically effective medications will, in general, be undertaken when it is likely that a deterioration in the flow properties of blood or one of their parameters can either completely or partially be held responsible for the pathogenesis of the disease. Based on rheologic studies, modified or deteriorated flow properties of blood have been shown to exist in many diseases and groups of diseases (e.g., [343, 523, 200, 251, 722]). Many authors therefore recommend a therapy with rheologically active substances even though perhaps only one parameter of the blood is normalized with this method and thus the symptoms of the disease cannot be influenced. Nonetheless, in recent years a series of diseases or groups of diseases have been found that today can be treated partially or primarily with rheologically effective measures. Vascular diseases form the first and largest group. These include the general circulatory disorders resulting from various types of shock (e.g., burn shock, immunologic shock, hemorrhagic shock) in which especially the flow properties in the microcirculatory regions are deteriorated (e.g., increased erythrocyte aggregation, decreased erythrocyte deformability). As a rule, shock is accompanied by hypovolemia which, according to current opinion, is very important for the genesis and maintenance of this disease. Rheologically effective medications which, on the one hand, increase the intravascular blood volume (e.g., by plasma substitutes) [367, 368, 370] and which, at the same time, improve the flow properties of blood are therefore indicated.

Table 2. Indications for a hemorheologic therapy (examples)

Polycythemia		
Polyglobulism	1.	Inborn valvular heart disease in children, respiratory distress disease
	2.	Heart and lung diseases (cor pulmonale) in adults
Plasmocytoma, myeloma, and others		
Vascular diseases		
	1.	Chronic occlusive arterial disease
		Cerebrovascular disease
		Coronary heart disease
	2.	Venous disorders
	3.	Disturbances in the mircrocirculation (e.g., Raynaud's disease).
Circulatory shock		
Pre-, intra-, and postoperative conditions		
Diabetes mellitus		
Hematologic diseases with deterioration of the microcirculation (e.g., sickle cell crisis)		

In addition to circulatory shock with general deterioration of the blood flow situation, vascular diseases, in particular, are treated hemorheologically today. Here chronic arterial occlusive disease, i.e., obliterative arteriosclerosis, plays the most important role. In a wider sense this also includes occlusive diseases of the extremities as well as the corresponding ischemic states of the brain and the heart. Pathophysiologically the flow properties of blood should also play a role in venous diseases [231, 238, 240, 199, 594].

The available hemorheologic literature on groups of diseases shows that "circulatory impairment" (in the widest sense, including circulatory shock) plays the greatest role. Several years ago clinical hemorheology was therefore considered to be part of angiology, for very good reasons [232]. It is, however, impossible to overlook the fact that other diseases, and especially also hematologic diseases, such as for example, polycythemia or paraproteinemia, are pathologically closely connected with a deterioration of the flow properties of blood and can, therefore, be approached therapeutically. The same applies to sickle cell anemia, which is a special form of hematologic disease in which the deteriorated erythrocyte deformability is impaired and which becomes clinically evident through microcirculatory disturbances in the form of sickle cell crisis. Apart from these groups of diseases (Table 2), additional individual clinical pictures have been described with which a hemorheologically based therapy can be of benefit. These diseases are explained in detail in the individual sections of the specific part.

3.3.9 Aim of Therapeutic Hemorheology and Verification of the Clinical Results

As with every therapeutic measure. the goal of therapeutic hemorheology is also primarily aimed at the improvement of the clinical symptoms. A considerable therapeutic advance may be assumed especially when in the same groups of diseases other therapeutic measures have shown little success. An example of a concrete aim in hemorheologic therapy is the elimination of pains at rest in severe arterial occlusive disease. If the pain disappears coincidently with rheologic therapy and if this is reproducable in therapy-free phases, without knowing or having measured any direct analgetic components of the drug, then a logical connection between the therapeutic goal and the therapeutic effect seems reasonable. Difficulties arise, as in many areas of therapy, when trying to objectivize clinical observations made in individual cases during hemorheologic therapy. On the one hand prospective clinical trials may have been planned in which the most objective clinical parameters are to be tested under valid statistic precautions. Such a result then logically says very little about the question of whether the mode of action (in our case the improvement of the flow properties of blood) actually shows the rational cause for the success of therapy or whether possibly the rheologic results only show an incidental epiphenomenon. After all, it would then be possible, allowing for some reservations, to close the chain of evidence whenever the rheologic parameters have probably resulted in an improvement in the flow situation.

On the other hand, the problem is quite different with the simpler and more commonly practiced procedure of statistically ensuring the changes in the flow properties of blood and then bringing them into a more or less conscious correlation to favorable clinical results observed at the same time. This is no doubt a risky enterprise which cannot convince the critical clinicians. It has already been suggested with the example of hemodilution (see Sect. 3.3.7.3) that, owing to the reduced number of oxygen-carrying erythrocytes, a statistically significant improvement of the flow properties of blood, e.g., of the subparameter whole blood viscosity, need not necessarily be followed by an improved tissue supply. During such a therapeutic measure, great care must be taken not to let the hematocrit fall below a particular level. Moreover, in the case of hemodilution with some colloidal plasma substitutes, an increase in the plasma viscosity must also be taken into consideration; its possible detrimental effects on the microcirculation are still under discussion.

The difficulty is often that at an early stage in the onset of drug treatment prospective studies on the "efficacy" (clinical effectiveness) of rheologically effective medications are not yet available, while parameters concerning the "effects" of therapy (e.g., on the flow properties of blood) can be obtained relatively quickly and it is relatively easy to test them statistically. Nonetheless, it must not be overlooked that changes in any hemorheologic parameter (e.g., viscosity of blood from the antecubital vein, measured in vitro) reflect only to a very limited extent the situation in the microcirculation of an ischemic extremity. Even if it were possible to sample blood directly from the ischemic area of, for example, the lower leg and to analyze it rheologically, this would still not prove that the supply to tissue was really improved as a result of rheologic therapy.

In Table 3 the worthwhile types of rheologic improvement as well as their possible effects and consequences for the macrocirculation and the microcirculation are summarized.

It may be deduced from Table 3 that the decisive methodologic criterion of any hemorheologic therapy is an improvement in tissue supply. If a hemorheologically effective therapy is to have any value for the tissue supply, then it must lead to a measurable increase in the tissue oxygen pressure. For this reason, the author's team has, for a number of years now, chosen a modus vivendi which lies in between the rheologic in vitro results on the one hand (which in themselves have relatively little predictive therapeutic value) and the not easily available, long-term, prospective clinical studies on the other hand. Our primary consideration was that an improvement of the blood flow in the capillaries, especially of the nutritive flow, must be indicated in a parameter which can be measured in vivo, i.e., directly in the patient. This is the case with the measurement of the tissue oxygen pressure [261] (see Sect. 3.3.7.4).

Since the correlation between tissue oxygen pressure and function (performance) is good [285], it is possible to measure directly in the tissue and thus perhaps settle the

Table 3. Improvement of rheologic parameters, their possible effects and consequences for the micro- and macrocirculation

Reduction of (apparent) whole blood viscosity Reduction of plasma viscosity Decrease in erythrocyte aggregation Improvement in erythrocyte deformability
↓
Increase in micro- and macrocirculatory flow rate Reduction of peripheral flow resistance More homogeneous (nutritive) capillary blood flow distribution = Management of the microcirculatory blood maldistribution
↓
Improvement in the tissue supply with oxygen and nutritives (supply) Improvement of metabolite removal (wash-out)
↓
Improved tissue function Prevention of cell death (e.g., necrosis)

question of whether a rheologically effective therapy does actually lead to an improvement in the tissue supply. The improvement of the tissue supply with oxygen and inevitably also with nutritives logically has a much closer connection to the expected clinical improvements than rheologic in vitro findings (see below).
The effects of rheologic therapy must therefore be studied under all circumstances and confirmed with a direct in vivo method like the measurement of the oxygen tissue pressure [305, 310, 278]. If there is no improvement in the tissue supply, the rheologic therapy must in fact be classified as rheologically "effective", in the sense of "rheologic

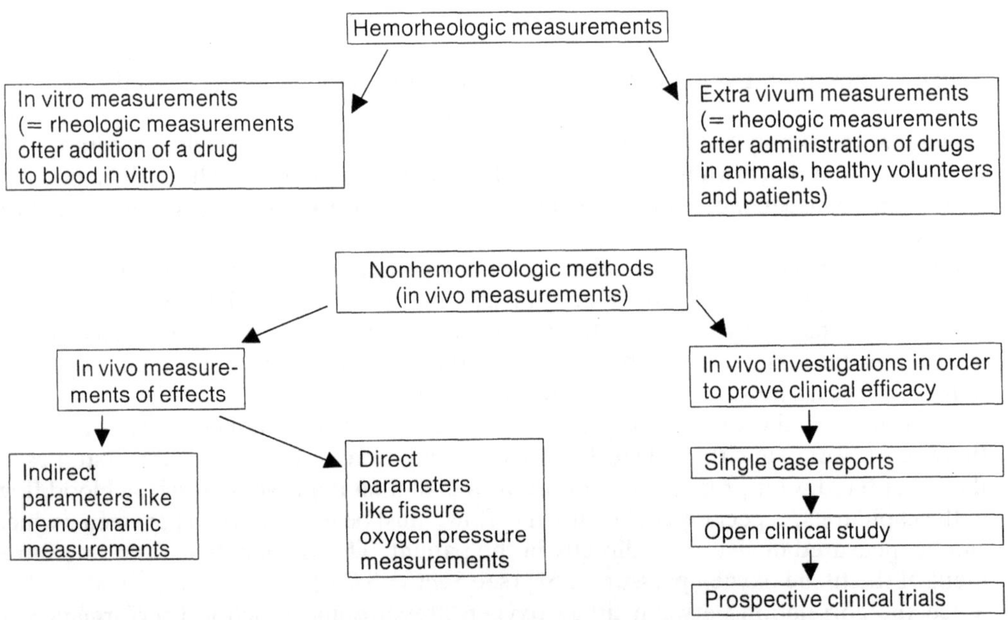

Fig. 8. Rational sequence of experimental and clinical studies of a rheologically active substance

cosmetics", but then the clinical value would nonetheless be dubious. The sequence of the experimental and clinical studies of a rheologically effective substance would then be as shown in Fig. 8.

It mus be noted in this connection that even statistically proven, prospective studies on the clinical efficacy of rheologic measures are practically worthless if, for example, the clinical evaluation criteria are incorrectly chosen, or if the dosage or the time of therapy are inadequate, or if a hemorheologically valid medication is in fact applied but the clinical indications were chosen incorrectly. Such examples are known from the literature (e.g., an 8-day control study of a defibrinogenation therapy which as a long-term study should usually last four weeks, or a control study with too low a pentoxifylline dosage). In these cases, adequate in vivo parameters that were obtained directly (see Table 3) are often more valuable than inadequately planned prospective clinical trials.

3.3.10 Deterioration of the Flow Properties of Blood as a Result of Medication

In addition to the great number of medications for which an improvement of the flow properties of blood and a therapeutic efficacy in different diseases is claimed or with which this effect could be objectively measured, the literature shows that there are medications and clinically applied substances which lead to a deterioration of the flow properties of blood.

The deterioration of the flow properties of blood resulting from such therapeutic and diagnostic measures is then also discussed as the cause for clinical symptoms. In this connection, contraceptives must be mentioned in particular. The statistically more frequent deep vein thromboses and their complications are attributed by many authors, among other causes, to a deterioration of the flow properties of blood. The situation is similar with a diuretic therapy that is not indicated or is falsely dosed possibly leading to dehydration of the patient and which may even be followed by a hemoconcentration with deterioration of the flow properties of blood [247].

Further examples in this group are the ionic contrast media. The application in patients can, at least temporarily, lead to an impairment of erythrocyte deformability owing to the high hyperosmolarity of those contrast media. Substances and medications which have a negative effect on the flow properties of blood are not mentioned separately in the specific part but are rather listed according to the standard chronologic order.

3.3.11 Future Development and Limitations of Therapeutic Hemorheology

A series of measures have been presented in the last 20 years which have as their goal the improvement of the flow properties of blood and thus to successfully treat patients. Although the most important, general principles for the improvement of the flow properties of blood have been known for some time now, e.g., decrease in hematocrit, decrease in fibrinogen concentration, improvement of erythrocyte deformability (see Sect. 4.1), it seems that these principals have not as yet been maximally utilized. This is especially true for hemodilution with regard to the question of what hematocrit level the blood of patients should be lowered to in order to reach an optimal correlation between the reduction of the flow properties of blood on the one hand and the oxygen transport capacity on the other hand (see Sect. 3.3.7.3.1). With therapeutic defibrinogenation, the high costs of the medications up to now have limited a wider clinical use. In the development and use of medications for the improvement of erythrocyte deformability, it also seems that not all possibilities have as yet been exhausted.

It cannot be ruled out that, in addition to the general possibilities known to improve the flow properties of blood, new ones are also on their way. Independently of these, it seems to be of particular importance to treat patients with various diseases and special hemorheologic situations in a differential manner, using the different hemorheologic therapeutics available. This will no doubt be a major task for the future. In the past, the possibility of combining several types of hemorheologically active substances and their influence on the flow properties of blood, as measured with tissue supply measurements, was rarely dealt with. Although some clinical attempts have already been made in this connection (see Sect. 4.2.2), these possibilities have not been fully realized as yet. For the future, the question will be whether it is possible to recognize the patients who respond well to hemorheologic therapy (responders) before the onset of therapy in order to improve the therapeutic result.

All in all, the situation ought to develop in such a way that the original importance of the improvement of the flow properties for the flow in larger vessels, will be gradually surpassed by the significance of the improvement of the individual rheologic subparameters for the microcirculation. Of critical clinical importance will be the task of making it possible to quantify the therapeutic value of hemorheologically effective substances in vivo, i.e., directly in the patients. For some time, now promising initial steps have been undertaken in this direction (see Sect. 3.3.9).

Further findings on the pathophysiology of the diseases responsible will most likely be the basis for the application of today's and tomorrow's drugs for the improvement of the flow properties of blood in patients. In the next few years and decades, hemorheologic therapy will have to be seen in close connection with other therapeutic principles, in particular microcirculatory ones. This not only indicates the limitations of hemorheologic therapy, but also, at the same time, opens a wide field of new therapeutic possibilities.

3.4 Survey of the Most Important Hemorheologico-therapeutic Measures According to Their Modes of Action

In the course of the last 80 years, many possibilities for the improvement of the flow properties of blood in different diseases have become known while others have been forgotten again. The reason for this was, among other things, that, on rechecking the seemingly favorable rheologic results, a positive effect was no longer reproducible. It is, therefore, important to present a summary of those substances and groups of substances, as well as other therapeutic possibilities for improving the flow properties of blood, which are generally agreed upon with regard to their hemorheologic effects (Table 4).

In this connection, it should be mentioned that the listed drugs or other measures for improving the flow properties of blood can have this effect either directly or indirectly. The group of the directly effective measures includes, for example, fibrinogenolysis with streptokinase, with which the rheologically important fibrinogen molecule is split enzymatically, thus improving the blood flow properties multifactorially (see Sect. 3.3.2). While with homodilution there is also a direct decrease in blood viscosity after the infusion with plasma or plasma substitutes, there is an influx of interstitial fluid into the intravasal space when certain vasodilating medications such as isosorbide-dinitrate are used, the effect of which is secondary to an endogenous hemodilution (see below and Sect. 4.1.43). Strictly speaking, drugs must also be considered to be acting indirectly, when not the substance itself is effective but its metabolites. The question of a direct or indirect mode of action is only discussed peripherally in the following.

Venesection is followed by a decrease in hematocrit owing to a compensating influx of interstitial fluid into the intravasal space. As a result of this endogenous "inner" hemodilution, the plasma protein concentration is also decreased because the interstitial fluid is low in protein. Especially in newborns with polyglobulism resulting from inborn valvular heart disease, the clinical effect of such a venesection is quite obvious and undisputed. In addition, this form of treatment can also be surprisingly effective in adult patients with polyglobulism, e.g., in chronic cor pulmonale. As far as obliterative arteriosclerosis is concerned, the clinical value of venesection therapy has not been conclusively proven for all types and stages. However, without a doubt, a reduction of the whole blood viscosity occurs, and usually also a reduction of plasma viscosity and a decrease in the aggregation tendency of the erythrocytes. A special kind of hematocrit reduction is achieved with **erythrapheresis,** where the erythrocytes of the blood are mechanically separated and only the plamsma is reinfused.

It is nowadays quite common to reduce the hematocrit level by means of **hemodilution with the infusion of erythrocyte-free solutions,** especially of human plasma, plasma protein fractions such as albumin or the so-called plasma substitutes (see Sect. 4.1.8, 4.1.10, 4.1.18, 4.1.36). The indications of this ("hypervolemic") hemodilution correspond to those of venesection (see above). In addition, a further significant indication must be added – circulatory shock therapy, where, in addition to volume replacement an improvement in the flow properties of blood most likely plays a significant role as well. Depending on the type of infusion and especially on its viscosity, changes occur in

Table 4. Summary of some important rheologicotherapeutic measures, their indications, as well as their modes of action and proof of efficacy

Therapeutic measures	Indications	Rheologic parameters	Modes of action	Evaluation of efficacy
		Hemorheologic monotherapy		
Venesection	Polycythemia, polyglobulism, chronic arterial occlusive disease, etc.	Whole blood viscosity, plasma viscosity, erythrocyte aggregation	Hematocrit, plasma protein concentration	Clinical
Hemodilution with				
a) Plasma	Chronic arterial occlusive disease, etc.	Whole blood viscosity	Hematocrit, total protein concentration	Clinical
b) Albumin solutions	Chronic arterial occlusive disease, etc.	Whole blood viscosity, plasma viscosity, erythrocyte aggregation	Hematocrit, total protein concentration	Clinical
c) Dextran solutions	Chronic arterial occlusive disease, etc.	Whole blood viscosity, plasma viscosity,	Hematocrit, total protein concentration	Clinical
d) Hydroxyethyl starch solutions	Chronic arterial occlusive disease, etc.	Whole blood viscosity, plasma viscosity, erythrocyte aggregation	Hematocrit, total protein concentration, erythrocyte aggregation	Clinical, tissue pO_2 measurements
e) Endogenous hemodilution with vasodilators	Cardiovascular diseases	Whole blood viscosity, plasma viscosity	Hematocrit, total protein concentration	Open
Plasmapheresis	Macroglobulinemia, myeloma, Raynaud's syndrome, etc.	Whole blood viscosity, plasma viscosity, erythrocyte aggregation	Total protein concentration	Clinical
Fibrinogenolysis	Thrombosis, myocardial infarction, arterial and venous diseases	Whole blood viscosity, plasma viscosity, erythrocyte aggregation	Fibrinogen reduction (total protein concentration)	Clinical
Enzymatic defibrinogenation (e.g., ancrod)	Chronic arterial occlusive diseases, etc.	Whole blood viscosity, plasma viscosity, erythrocyte aggregation	Fibrinogen (total protein concentration)	Clinical, tissue pO_2 measurements
Improvement of erythrocyte deformability (e.g., pentoxifylline)	Chronic arterial occlusive diseases, etc.	Whole blood viscosity, erythrocyte deformability	Improvement of erythrocyte deformability	Clinical, tissue pO_2 measurements
		Simultaneous hemorheologic therapy		
Isovolemic hemodilution (venesection + infusion of e.g., colloidal plasma substitutes)	Chronic arterial occlusive diseases, etc.	Whole blood viscosity, plasma viscosity, erythrocyte aggregation	Hematocrit, fibrinogen, total protein concentration	Clinical
Isovolemic hemodilution + defibrinogenation	Arterial occlusive diseases	Whole blood viscosity, plasma viscosity, erythrocyte aggregation	Hematocrit, total protein concentration	Clinical

plasma viscosity, which may then result in a decrease as well as in an increase in the values.
The effect on the rheologic subparameters erythrocyte aggregation and erythrocyte filtrability is – if studied at all – also variable according to the type of plasma substitute used. Opinions still diverge on the clinical merit of therapeutic hemodilution for individual indications; for some types of diseases there are still no prospective studies and investigations with objective parameters available.
While only a modest hemodilution effect is to be expected from the easily applied infusion therapy with cell-free solutions (plasma, plasma protein fractions, plasma substitutes, electrolytes), the effects on the hematocrit value and vice versa also on the whole blood viscosity values are much more pronounced with a combination of venesection followed by reinfusion of plasma and/or plasma substitutes. Contrary to "hypervolemic" hemodilution (see above), we talk about *"isovolemic" hemodilution* which is the combination of blood removal and refilling with cell-free plasma or plasma substitutes. A reduction of the hematocrit may be achieved in this case without any major change in the circulating blood volume and, according to the technique used, not just for a short period of time but for periods of several weeks (chronic isovolemic hemodilution). Chronic arterial occlusive disease at stage IV has been mainly studied as an indicator for isovolemic hemodilution. In most of the publications on isovolemic hemodilution, dextran solutions were used as plasma substitutes (see Sect. 4.1.7). In addition to the clear-cut reduction of whole blood viscosity there are different behavior patterns for plasma viscosity and also for the tendency to erythrocyte aggregation depending on the type and dosage of the plasma substitutes used. In general, it may be said, that, from the rheologic standpoint, the unfavorable effect on plasma viscosity as well as the rise in the aggregation tendency of the erythrocytes are the more pronounced the higher the concentration and the molecular weight of the infusion fluids are. As far as the clinical efficacy is concerned, the same may be said for hypervolemic hemodilution. There are a great number of publications as well as positive individual observations; as yet valid parameters are rare, which is in part due to methodologic reasons. In addition, prospective statistically sound studies are rare or cannot be carried out for ethical reasons.
The so-called *endogenous hemodilution* may be seen as a special type of hemodilution therapy with drugs having a high vasodilative effect (especially α-blockers and nitrates). Similar to the effect after venesection (see above) the use of vasodilators results in a decrease in the hematocrit levels and the total protein concentration. This can be explained by a temporary shift of interstitial fluid into the intravasal space [350, 28, 107, 300, 306, 91]. This "vasodilative" hemodilution without an exogenous fluid supply leads to a reduction of blood and plasma viscosity caused by a "dilution" of the blood as a result of the influx of low-protein extravascular fluid.
Vasoconstrictive drugs such as, for example, the α-mimetic drug etilefrine lead, on the other hand, to a displacement of the intravascular fluid into the extravascular tissue and thus to an endogenous hemoconcentration followed by an increase in blood and plasma viscosities. The alternating change in the flow properties of the blood by vasodilators on the one hand and vasoconstrictive drugs on the other hand is considered to be an additional mechanism for controlling the peripheral resistance and is called rheoregulation [306, 309], (see Sect. 3.3.2). This endogenous hemodilution with vasodilators such as nitrates and other substances is possibly of significance for cardiovascular diseases and for primary Raynaud's disease.
With the second major group of measures for the improvement of the flow properties of blood, there is a global or selective reduction of the plasma protein concentration, whereas the number of oxygen-carrying erythrocytes (hematocrit) remains unaffected

(see Table 4). Historically, *plasmapheresis* was the first measure to be used – in particular with paraproteinemias of different kinds (see Sect. 4.1.9). However, this therapy is also carried out with particular forms of arterial occlusive disease (e.g., Raynaud's syndrome). By eliminating a part of the body's own plasma proteins and substituting low-viscosity plasma substitutes, a reduction of plasma viscosity, whole blood viscosity at low shear rates, as well as a decrease in the erythrocyte aggregation are achieved. The interest in plasmapheresis has been increasing in recent times. Clinical improvements have also been shown in certain ischemic diseases. As far as the scientific definition of favorable therapeutic effects is concerned, the same applies as for venesection and hemodilution therapy. It should, moreover, be noted that with paraproteinemias a number of authors combine plasmapheresis or plasma exchange with cytostatic therapy.

Owing to the special rheologic significance of the plasma protein fraction fibrinogen a third form of rheologic therapy has been developed: the *reduction of the fibrinogen concentration in plasma via fibrinolysis* with, for example, streptokinase (see Sect. 4.1.19). This is based mainly on the decrease in the aggregation supporting fibrinogen fraction in the plasma, while the hematocrit levels remain unchanged. Fibrinogen, which affects the flow properties of blood in a multitude of ways, is reduced to split products via proteolysis. The resulting reduction of the fibrinogen concentration in the blood leads to a reduction of the so-called structural viscosity at low shear rates. The decrease in the erythrocyte aggregation is particularly pronounced and can be easily measured technically. The knowledge that this favorable rheologic influence exists has paved the way for further therapeutic possibilities for, for example, shock, myocardial infarction, and various forms of arteriosclerotic diseases.

A much more pronounced reduction of the fibrinogen concentration, which, moreover, can be precisely controlled, may be achieved through so-called *defibrinogenating substances* such as ancrod and batroxobin (see Sect. 4.1.19). The mechanistic rationale, as well as the effects on the flow properties of blood, must be seen in the same light as those with fibrinolysis, i.e., there ensues a decrease in blood and plasma viscosities and, especially, a pronounced reduction of the erythrocyte aggregation.

Compared to streptokinase which also acts thrombolytically, the clinical improvements from the defibrinogenating substances can be attributed solely to the favorable rheologic effects. In addition to the many single-case observations, there is in fact a whole series of publications with favorable results. In the case of ancrod, the measurement of the tissue oxygen pressure in muscle tissue also showed a direct objective improvement in the microcirculation. On the other hand, it is also true of the fibrinogen-reducing substance ancrod that adequate prospective clinical studies still have to be carried out. To complete the picture, the fibrinogen-reducing substances clofibrate and stanozolol should also be mentioned. Both substances have a slight fibrinogen-reducing effect; correspondingly, the rheologic changes are also relatively small. In recent times the combined therapy with hemodilution and defibrinogenation has also become of interest; so far the clinical results with particular types of diseases seem to be promising (see Sect. 4.2).

The fourth group of rheologically active substances has a completely different mechanism of action which has neither an effect on the hematocrit nor on the total protein concentration. It concerns the *pharmacologic improvement of erythrocyte deformability*. The studies of various authors have shown that erythrocyte deformability, as judged by the filtration of blood through filter systems with 3–8 μm pore diameter in patients with arterial occlusive disease is statistically significantly decreased in comparison to healthy subjects. This implies that especially in the poststenotic area with lower pressure gradients the erythrocytes have more difficulty in passing through the 5 μm capillaries.

Most studies on the improvement of the erythrocyte filtrability (= deformability) were performed with the substance pentoxifylline (see Sect. 4.1.27). There have been prospective clinical trials on pentoxifylline as well as measurements of the tissue oxygen pressure which show that the blood flow in the microcirculation can be improved with this substance.

A further hemorheologicotherapeutic possibility is to *decrease the erythrocyte aggregation tendency*. According to recent opinion, the erythrocyte aggregation, i.e., the reversible attachment of erythrocytes into two- or three-dimensional rouleaux-like groups, plays a special role in the pathophysiology of ischemic diseases. Within the framework of fibrinogen reduction and a reduction of the plasma protein concentration through plasmapheresis, there ensues, in addition to the reduction of the plasma viscosity and the structural viscosity, in particular a decrease in the erythrocyte aggregation tendency. Recently a report has also come out on a plasma substitute (hydroxyethyl starch), which shows a specific disaggregating effect on the erythrocyte aggregates (see Sect. 4.1.35).

Some medications effect a reduction of the concentration of the plasma proteins, especially of fibrinogen, without having a direct fibrinolytic property. These are, for example, antiphlogistics and antirheumatics which suppress the acute phase reaction and thus indirectly improve the blood flow properties. Hormones like insulin, cortisone, and thyroxin also belong to this group (drugs with indirect hemorheologic effects).

In summary, it may be said that at the present time there are many different ways of improving the flow properties of blood in patients (see also Table 4), many of which have in the meantime been introduced into therapy to a great extent. These are:

1. Reduction of the hematocrit values via venesection or hemodilution
2. Reduction of the total protein concentration via plasmapheresis
3. Reduction of the fibrinogen concentration in plasma by fibrinolysis and in particular by defibrinogenation
4. Improvement in the deformability of erythrocytes by substances such as pentoxifylline
5. Possibly a specific disaggregation of erythrocyte aggregates
6. Combination of drugs with various hemorheologic modes of action

It may theoretically be assumed that venesection is the most favorable therapy for patients with high hematocrit levels, that fibrinogen-reducing therapy is most promising with high plasma fibrinogen concentrations in the blood, and that therapy for the improvement of erythrocyte deformability is best for patients with proven deterioration of the erythrocyte deformability. Since combinations of impaired rheologic factors are pathophysiologically common, simultaneous applications of therapeutics with various means of rheologic action are also possible.

4 Specific Part

Hemorheologicotherapeutic Measures and Medications

4.1 Monotherapy (Single Substances, Single Measures)

4.1.1 Potassium Iodide

4.1.1.1 Introduction

***Huchard, Erb** and **Erlenmeyer** (cited in [736]) reported favorable clinical experiences in the treatment of moderate cardiac insufficiency with dyspnea, cardiac asthma, light to moderate angina pectoris, and intermittent claudication with the application of potassium iodide. Gangrenous skin ulcers could, however, not be influenced by the application of potassium iodide. The question of whether the mechanism of action could primarily be seen in the dilation of vessels as **Huchard** et al. postulated, or rather in the reduction of the blood viscosity was first tested by **Müller** and **Inada** of the **Romberg** group (see below). This publication on the improvement in the flow properties of blood through potassium iodide was a stimulus for several successive investigators to publish studies of their own on the effect of potassium compounds on hemorheologic parameters and represents the first attempt at drug-induced therapeutic hemorheology. Potassium iodide is still being applied from time to time in diseases such as arteriosclerosis and senile emphysema.*

4.1.1.2 Publications

The first studies on the influence of potassium iodide on the flow properties of the blood in two patients with cerebral and peripheral arteriosclerosis were presented by **Müller** and **Inada** [663] from the Romberg team in the year 1904. The patients received 0.3–0.5 g potassium iodide three times per day for 10–14 days per os. The blood and serum viscosity were measured with a Hirsch and Beck capillary viscosimeter (H_2O = 1) at a temperature of 38 °C. During treatment the authors found a reduction of the blood viscosity while the serum viscosity remained unchanged.

Note: Further rheologic parameters were not measured.

Clinical findings: The patients reported an improvement in their symptoms (dizz, headaches, forgetfulness and attacks of anginal pain).

In 1907 **Kottmann** [526] treated two patients with Friedrich's disease and gastric carcinoma using an oral application of 14 g potassium iodide over 11 days. The author also noted a reduction of the blood viscosity (Ostwald capillary viscosimeter, H_2O = 1; 38 °C). Hirudin was used as an anticoagulant.

Note: Further rheologic parameters were not measured.

On the other hand, in 1908 **Determann** [168] reported nonuniform rheologic results after the oral application of potassium iodide. Twelve patients with various diseases (bronchitis, anemia) received a daily dose of 1.0–1.5 g potassium iodide per os. Blood viscosity was measured with a Hirsch and Beck capillary viscosimeter.

Note: No data on the temperature of the viscosity measurements; no measurement of other rheologic parameters.

In the same year **Lindmann** [578] confirmed **Determann's** results. Twelve patients with arteriosclerotic vascular changes and partly also with other diseases received 18 g potassium iodide within a treatment period of 12 days. The blood viscosity was measured with a newly constructed capillary viscosimeter (H_2O = 1) according to Determann. The oral administration of potassium iodide did not lead to any uniform change in the blood viscosity.

Note: No temperature data; no measurements of further rheologic parameters.

In 1909 a publication by **Adam** [2] appeared on the same topic. Patients with various diseases received a daily dose of 1.5–2 g potassium iodide. The blood and plasma viscosities of the blood samples anticoagulated with hirudin were measured with a Hirsch and Beck capillary viscosimeter (H_2O = 1). **Adam** could also not find a viscosity-reducing effect of the oral application of potassium iodide. The temperature for the viscosity measurements was 38 °C.

Note: No measurements of further rheologic parameters.

In 1913 **Koga** [519] treated patients with spontaneous gangrene of the upper and lower extremities using potassium iodide. **Koga** was able to measure a reduction of the blood viscosity with potassium iodide.

Note: The type of viscosimeter used was not mentioned; no measurements of further rheologic parameters.

Clinical findings: No favorable clinical results were achieved with potassium iodide.

In a publication by **Drossbach** [217] in 1923, a note appeared on the oral application of 0.1–0.2 g potassium iodide and sodium iodide 5 times daily and of 0.7–1.0 g iodipine 3 times daily. The blood viscosity was determined with a Hess capillary viscosimeter (H_2O = 1). During therapy there was no decrease in blood viscosity.

Note: The temperature of the viscosity measurements was not given. It was not mentioned whether the subjects treated were healthy or sick; no measurement of further rheologic parameters.

In 1939 **Alwall** [12] treated 21 patients with hypertension and arteriosclerosis with an oral administration of 0.5–4 g iodine. The blood and plasma viscosities were determined with a capillary viscosimeter (H_2O = 1) at 18 °C from blood samples that were in part anticoagulated with heparin and novirudin. The author did not find a significant reduction of the blood and plasma viscosities after oral administration of iodine.

Note: The viscosity measurements were carried out below body temperature; no measurements of further rheologic parameters.

In 1985 **Moser** and co-workers [657] studied the influence of diet, physical therapy, and the administration of iodine on the flow properties of the blood of patients with poorly stabilized diabetis mellitus. An initial group of patients was subjected to a physiotherapeutic program with gymnastics, ergometric training, and swimming. A second group received a daily dose of 200 ml iodine saline concentrate equal to about 9 mg iodine per patient and day. A third group was submitted to the physiotherpeutic program as well as the iodine saline therapy. The blood viscosity was measured with a Contraves low-shear viscosimeter at a shear rate of 27.7 s^{-1}; the hematocrit was determined using a microhematocrit method. A reduction of the whole blood viscosity, as well as a decrease in the plasma viscosity, was found in all three groups. The changes were of minimal importance percentage wise, but they do have statistical significance. In addition, there was a statistically modest decrease in the hematocrit level. The authors pointed out that the values cannot be considered as iodine specific without reservation, since the effect of the other therapeutic measures must also be considered within the framework of therapy.

Note: No measurement of other hemorheologic parameters.

4.1.1.3 Discussion

Contrary to *Müller* and *Inada* and to *Kottmann,* the subsequent investigators found no changes in the blood viscosity following oral administration of potassium iodide. The most recent studies in this field show a certain effect of potassium iodide which, however, cannot be isolated from other therapeutic measures carried out at the same time. Therefore, it is improbable that a reduction of the blood viscosity is a direct causal effect of the administration of potassium iodide.

4.1.2 Venesection, Erythrapheresis

4.1.2.1 Introduction

Venesection is one of the oldest methods of medical treatment. The first reports on venesection come from ***Hippocrates*** *(approximately 460-322 BC). He recommended venesection for all sorts of diseases (acute infections, pneumonia, headaches).* ***Celsus*** *(20–30 A.D.) also endeavoured to promote the spread of this method of treatment.* ***Galen*** *(160–210 AD) put forward the idea of a topographic correlation between the place of affliction and the location of the venesection. The golden age of venesection was in the Middle Ages.* ***Paracelsus*** *(approximately 1500) and* ***Marcello Malpighi*** *(1628–1694) supported the idea of venesection. In 1907 a first publication on the hemorheologic effect of venesection on the flow properties of blood was published.*

The classical indications for venesection were polycythemia and polygobulism. Since these diseases are accompanied by a marked increase in the blood viscosity values, circulatory disturbances such as dyspnea, cyanosis, ischemic cerebral attacks, etc. are to be expected. As a result, the blood flow slows down, which in turn can lead to deep vein thrombosis. Therefore, theoretically, the improvement in the flow properties of blood by means of a reduction of the hematocrit is of major clinical importance.

In recent times a reduction of the hematocrit has also been achieved by extracting erythrocytes from the blood with the help of cell separators. In the so-called erythrocytapheresis (erythrapheresis) the erythrocytes are separated from the blood, and endogenous plasma and, if necessary, plasma substitutes are reinfused (see Sect. 4.1.9 for a definition of the term "erythrocytapheresis").

4.1.2.2 Publications

The first reports about the hemorheologic influence of venesection on the flow properties of blood were made by **Kottmann** in 1907 [526]. In a female patient with cirrhosis of the kidney, a venesection of 320 cm^3 blood was carried out. The blood and plasma viscosities were measured with a Hirsch and Beck capillary viscosimeter ($H_2O = 1$) at 38 °C. The author reported a reduction of the blood and plasma viscosities which was measurable for more than 21 days.

Note: No measurement of further rheologic parameters.

In 1911 **Welsh** [914] removed 300–400 ml blood from patients with various diseases (cardiac insufficiency, eclampsia). The relative blood viscosity ($H_2O = 1$) of the blood samples anticoagulated with hirudin was measured with a Determann capillary viscosimeter. A reduction of the blood viscosity was accompanied by a simultaneous reduction of the hematocrit.

Note: The temperature of measurement was not mentioned; no measurement of further rheologic parameters.

In 1936 **Markson** [613] treated patients with inborn valvular heart disease by means of phlebotomy. The amount of blood removed was not mentioned. The relative blood viscosity ($H_2O = 1$) was determined with a Hess capillary viscosimeter at temperatures of between 17 °C and 20 °C. A reduction of the blood viscosity was found to be in "correlation" with a decrease in the hematocrit.

Note: The viscosity measurements were performed at below body temperature; no further rheologic measurements.

Clinical findings: The clinical symptoms of inborn valvular heart disease, manifested in cyanotic conditions and in dyspnea, was decisively improved by venesection.

In 1937 **Albers** [8] treated patients with decompensated cardiac insufficiency and increased viscosity values. The blood viscosity of blood samples anticoagulated with novirudin was measured with a Hess capillary viscosimeter. The author observed a simultaneous reduction of the blood viscosity and the hematocrit value.

Note: The temperature of the viscosity measurements and the amount of blood removed by venesection were not mentioned; no further rheologic measurements.

Clinical findings: The existing cyanosis and the respiratory disturbances disappeared.

Another publication on the improvement of the flow properties of blood appeared in 1963 by **Burch** and **De Pasquale** [96]. Between 200 and 425 ml of blood were removed several times from patients with angina pectoris as well as with acute myocardial infarction in order to thereby lower the hematocrit levels from 50% to 45%–40%. The blood viscosity was measured with a rotational viscosimeter with four shear rates of 23, 46, 115, and 230 s^{-1}. A reduction of the blood viscosity was found at all shear rates, but it was especially strong at low shear rates. A further report on similar results in patients with coronary heart disease was published by **Burch** and co-workers in 1965 [97].

Note: The temperature of the viscosity measurements and the type of viscosimeter used were not mentioned; no measurement of further rheologic parameters.

Clinical findings: A reduction of the frequency and intensity of anginal attacks of pain ensued, and in the patients with acute cardiac infarction the retrosternal pains disappeared.

In 1956 **Rakita** and co-workers [717] treated patients with secondary polycythemia and chronic pulmonary emphysema using repeated venesections of 300–1000 ml blood. The blood viscosity was measured with an Ostwald capillary viscosimeter. A reduction of the blood viscosity was found simultaneously with a decrease in hematocrit.

Note: The temperature of the viscosity measurements was not mentioned; no measurements of further rheologic parameters.

In 1966 **Segal** and **Bishop** [786] studied patients with chronic bronchitis, polycythemia, and pulmonary hypertension. 500 ml blood was removed several times within 10 days. For the blood viscosity measurements a Pirofsky capillary was inserted into the antecubital vein. A significant reduction of the blood viscosity was found simultaneous to the decrease in hematocrit.

Note: No measurements of further rheologic parameters.

In 1970 **Kontras** and co-workers [522] treated children with increased blood viscosity and inborn valvular heart disease as well as increased hematocrit levels using repeated venesections. The blood viscosity was measured at 37 °C with a Brookfield micro-cone-plate viscosimeter at 11.5 s^{-1}. Simultaneous to the reduction of the hematocrit values, there was a significant decrease in the blood viscosity.

In 1972 **Kontras** [523] reported the results of venesection in newborns with polycythemia and increased blood viscosity. Within the framework of this study there was a simultaneous reduction of the blood viscosity and the hematocrit level.

Note: No further measurements of rheologic parameters.

Clinical findings: The authors observed an improvement in the plethoric symptoms which were manifested in a decrease in respiratory difficulties, lethargy, cramps, and cyanosis.

In 1977 **Thomas** and co-workers [857, 858] reported on patients with manifested polycythemia, all of whom suffered from decreased cerebral circulation. The patients suffered cerebral ischemic attacks, mild strokes, dementia, and headaches. Two patients were also suffering from severe intermittent claudication with pains at rest. The blood and plasma viscosities of the blood samples anticoagulated with heparin were measured with a coaxial cylinder viscosimeter (Contraves Low Shear 2) at four shear rates, 0.2, 0.67, 26.7, and 91 s^{-1}. Following repeated venesections of 200–250 ml blood at intervals of 2 days to 5 weeks, there was a significant reduction of the blood viscosity parallel to a decrease in hematocrit. In this case, the reduction of the blood viscosity was the highest at low shear rates. The decrease in plasma viscosity was not significant.

Note: The temperature of the viscosity measurements was not given.

Clinical findings: There was an improvement in the clinical symptoms. Both of the patients with pains at rest in their legs reported that these symptoms disappeared.

In 1979 **Stangel** and co-workers [815] reported on the influence of erythrapheresis on the blood viscosity of patients with polycythemia vera. Fourteen patients were studied. The measurements were performed with a Contraves Low Shear viscosimeter with a very low shear rate (0.288^{-1}). The erythrapheresis itself was performed in such a way that, following the removal of 440 ml venous blood, 150 ml isotonic saline solution was first infused along with 60 ml ACD solution, followed by the patient's own plasma in order to ensure normal blood volumes. In all patients there was a decrease in the hematocrit values and a statistically significant reduction of the blood viscosity (abstract, no further details).

In 1980 **van den Berg** and co-workers [871] as well as **Sgries** and co-workers [792] published studies on the influence of erythrapheresis on the flow properties of the blood of patients with polycythemia vera and pathologically raised whole blood viscosity. A total of 440 ml blood was removed from the patients and, following centrifugation of the blood, the endogeous plasma was reinfused together with an additional 150 ml physiologic saline solution. The viscosity of the blood anticoagulated with EDTA was measured with a Contraves Low Shear 1000 viscosimeter, at shear rates of 0.0452 and 0.288 s^{-1} (37 °C). Simultaneous to the hematocrit reduction, the authors found a significant decrease in the blood viscosity.

Note: No measurements of further rheologic parameters.

Clinical findings: The authors reported a reduction of the complaints that had been caused by pathologically raised blood viscosity (headaches, numbness, nausea, impaired vision, paresthesias, claudication, and Raynaud symptoms, as well as angina pectoris.).

In 1980 **Humphrey** and co-workers [460] reported on the treatment of patients with cerebral ischemic attacks, slight strokes and polycythemia. Approximately 200 ml blood was removed at weekly intervals from the patients until the hematocrit decreased to values of less than 45%. The blood and plasma viscosities of the heparinized blood samples were measured at 37 °C with a coaxial cylindrical viscosimeter (Contraves Low Shear) at two shear rates of 0.67 and 91 s^{-1}. Simultaneous to the decrease in the hematocrit levels, there was a significant reduction of the blood viscosity which was the highest at low shear rates. The plasma viscosity did not change significantly.
In the same year a further report appeared from the **Humphrey** team [461] with similar results.

Note: No measurements of further rheologic parameters.

Clinical findings: The patients reported a decrease in the cerebral ischemic attacks.

In 1981 **Challoner** and co-workers [120] studied the effect of several minor venesections on two groups of patients with increased hematocrit values. One of the groups had increased blood pressure at the same time. When the hematocrit was decreased by about 40% from the initial values, the decrease in the viscosity was of the same order of magnitude at shear rates of 0.675 s^{-1} and 48.18 s^{-1}. There was no difference between both groups in the change in viscosity; however, the diastolic blood pressure decreased statistically significantly in the group of hypertensive patients (abstract, no further details).

Pollok and co-workers [704] studied the question of whether the deformability of the erythrocytes was changed due to the expected iron deficiency during repeated removal of whole blood. Studies were performed in patients who were simultaneously suffering from increased hematocrit values and vascular diseases. A modification of the filtration method according to Reid was applied. Although there was no significant change in the average cell volume, there was a minor, statistically significant decrease in the deformability index (abstract, no details).

Humphrey and co-workers [462, 463] found a statistically significant reduction of the blood and plasma viscosities following venesection therapy carried out in patients with increased hematocrit values and vascular diseases. In the majority of the patients, hypovolemia could not be found (abstract, no details).

In 1982 **Milligan** and co-workers [649, 650] reported on 14 patients with primary or secondary polycythemia from whom 500 ml blood was removed in several sessions. The blood viscosity was measured with a Contraves Low Shear viscosimeter and a Deer rheometer with a range of between 0.1 and 220 s^{-1}. These measurements, as well as the simultaneous plethysmographic measurements of the blood flow, were carried out, at the earliest, two weeks following the last venesection. The authors found a statistically significant decrease in blood viscosity with a simultaneous increase in the peak flow during reactive hyperemia, while, on the other hand, the blood flow under resting conditions showed no statistically significant increase. These viscosity measurements were performed at 37 °C.

Note: No measurement of further rheologic parameters.

In 1983 **Wade** [490] studied 20 patients with primary polycythemia and found a reduction of the whole blood viscosity at shear rates of 91 s^{-1} as measured by a Contraves Low Shear viscosimeter at 37 °C. At the same time the cerebral blood flow was measured via ^{133} Xenon clearance.

In 1984 **Moscatelli** and co-workers [636] studied whole blood viscosity in polycythemic patients using erythrapheresis and found a reduction of this parameter. The viscosity measurements were carried out with a Brookfield viscosimeter at 225 s^{-1} (abstract, no details).

Cavestri and co-workers [119] treated 18 patients with stroke using erythrapheresis and studied plasma viscosity, eryhtrocyte aggregation and erythrocyte filtrability (abstract, no details).

In 1986 **Lanzer** and **Tilz** [547] described a reduction of the whole blood viscosity in polycythemic patients following therapy with a Haemonetics 30 cell separator.

Note: No details on the methodology.

In 1986 **Challoner** and co-workers [121] reported on the effect of venesection on the flow properties of blood. Patients with vascular diseases and hematocrit values between 46% and 55 % were treated with weekly venesections in order to achieve a hematocrit reduction to 40%. The hematocrit was determined with a microhematocrit centrifuge; the whole blood viscosity was measured with a Contraves Low Shear 2 viscosimeter at low and high shear rates. The removal of 250 ml blood resulted in a short-term moderate decrease in the hematocrit values and in a statistically significant decrease in the whole blood viscosity. If, within the framework of long-term treatment, the venesections are repeated without subsequent applications of colloidal or crystalline solutions, then there is also a decrease in the hematocrit values as well as a significant decrease in the whole blood viscosity. Marked changes in the plasma protein concentration were not observed. The authors conclude from their results that venesection alone, without substitution, is a reliable method of improving the flow properties of the blood.

In 1986 **Wallis** and co-workers [899] studied the effect of erythrapheresis in patients with secondary polycythemia using a cell separator. They studied the whole blood viscosity with a Wells-Brookfield cone-plate viscosimeter at 37 °C. The whole blood viscosity was clearly reduced both at 23 and at 230 s^{-1}.

4.1.2.3 Discussion

The first viscosimetric measurements by *Kottmann,* in 1907, showed a reduction of the blood and plasma viscosities which could be observed for more than 3 weeks. *Welsh* and later *Markson, Albers,* and others attributed the reduction of the blood viscosity to the decrease in the hematocrit.
The clinical improvement of patients with cardiac insufficiency and other diseases is certainly not only due to a

reduction of blood viscosity but also to a volume reduction. We now know that the results of *Kottmann* in 1907 may be explained by the fact that with venesection not only erythrocytes are removed but plasma protein is as well. This results in a decrease in the plasma viscosity because following venesection the body tries to normalize the reduced volume of circulating blood by an influx of extravascular low-protein fluid into the vascular system (inner hemodilution). That the effect of therapy ended after 21 days, as reported by *Kottmann,* could be explained by the fact that after that time the major proportion of the erythrocytes lost via venesection is replaced again by fresh red blood cells.

Burch and *De Pasquale* were the first to treat patients with coronary heart disease by reducing the hematocrit from 50% to 45% and then to 40%.

In subsequent years venesection therapy was mainly used for the management of primary and secondary polycythemia. *Burch* and *De Pasquale* and later *Thomas* and co-workers pointed out that the reduction of blood viscosity is most pronounced at low shear rates. *Thomas* and co-workers found that there was no significant plasma viscosity reduction. The findings which were especially discrepant in regard to capillary viscosimeter studies of earlier times are probably due to the difficulties in the management of the complex low-shear viscosimeters which allow only rough measurements of plasma viscosity.

In fact, there was probably just a slight reduction in plasma viscosity, since the amount of blood removed, for example, by *Humphrey* and co-workers, was only 200 ml in the acute phase. Despite various measurement techniques and the lack of temperature data in some cases, all the authors agreed on a decrease in the hematocrit and blood viscosity values. Changes in the blood and plasma viscosities (hematocrit reduction and reduction of the total protein concentration) may be explained in the sense of an autohemodilution in which the influx of extravascular low-protein fluid probably plays the major role.

The viscosity studies on venesection (without simultaneous or subsequent infusion of fluid) show the close correlation between hematocrit and whole blood viscosity. The studies of *Pollok* and co-workers could be an indication for the fact that repeated venesections over a longer period could lead to an undesirable rheologic effect, i.e., to a decrease in the erythrocyte deformability.

The results of *Challoner* and co-workers also indicate that in the course of long-term venesection therapy microcytosis is incurred which could lead to a decrease in erythrocyte deformability. The studies of *Milligan* and co-workers as well as those of *Wade* show the same results in regard to a decrease in viscosity. However, in the discussion of both papers the problematic nature of these studies is obvious, i.e., a decrease in blood viscosity does not necessarily lead to an increase in flow rate or to an improvement in the tissue oxygen supply.

Milligan and *Davies* contest that, despite a marked decrease in blood viscosity, there is an improvement in oxygen supply to the tissue in patients with polycythemia, while *Wade,* using other means of calculation, found an increase in the oxygen transport to the brain following venesection. However, he warns against an overestimation of the role of the hematocrit in the flow properties of blood, in particular in the area of cerebral blood flow.

Recently hematocrit reduction in patients suffering from diseases resulting from high hematocrit values has increasingly been carried out with cell separators with which the erythrocytes are selectively separated and the remaining plasma is reinfused. The rheologic results of *Wedzicha, Lanzer,* and *Wallis* correspond with the data of conventional venesection therapy. Most of the authors speak of an improvement in the clinical symptoms.

At this stage it should be noted that many studies on the influence of venesection on the flow properties of blood were carried out with a simultaneous administration of fluids (saline solutions, plasma substitutes). These publications are listed under the respective fluid substitute or in the section on combined (simultaneous) rheologic measures (see Sect. 4.2).

4.1.3 Saline Solutions, Crystalloid Solutions

4.1.3.1 Introduction

For a long time isotonic saline solutions have been used for the therapy of fluid loss. Moreover, they constitute the carrier solutions for plasma substitutes (e.g., dextran) and amino acids. In 1907, **Kottmann** *was the first to study to what extent the flow properties of blood were also changed by saline solutions. Ever since then isotonic saline solutions of various types have been used for the therapy of dehydration, blood volume deficit, and shock.*

4.1.3.2 Publications

The first reports on the influence of saline infusions in patients with various diseases (arteriosclerotic cirrhosis of the kidney, nephritis, arteriosclerosis, pernicious anemia) came from **Kottmann** [526] in 1907. The patients were infused intravenously with 267–800 ml isotonic saline solution. The plasma and serum viscosity (H_2O = 1) was measured with a Hirsch and Beck capillary viscosimeter at 37 °C. After the infusion of saline the author found a decrease in the blood and plasma viscosities.

Note: Measurements of further rheologic parameters were not carried out.

Mayesima [630] studied the flow properties of blood of a patient with rheumatoid arthritis in 1912. The blood viscosity was determined with a capillary viscosimeter at 25 °C–30 °C. Following 3–21 infusions of an unknown amount of isotonic saline solution, there was a decrease in the blood viscosity and in the hematocrit value.

Note: The viscosity measurements were performed below body temperature; no measurements of further rheologic parameters.

In 1913 **Koga** [619] introduced the infusion of isotonic saline solutions and Ringer solutions into the therapy of spontaneous gangrene of the upper and lower extremities. Blood viscosity values were reduced to normal values by repeated infusions.

Note: The type of viscosimeter used, as well as the number and amount of saline infusion used, were not mentioned; no measurements of further rheologic parameters.

Clinical findings: The patients reported that the pains at rest and the intermittent claudication stopped and that toe and finger necroses healed.

In 1966 **Schülke** and **Hartel** [776] postoperatively studied patients having undergone cholecystectomy and infused 1000 ml physiologic saline solution as well as other plasma substitutes (low- and high-molecular weight dextran, gelatine). Plasma viscosity was measured with a capillary viscosimeter by Coulter Electronics Ltd. at 37 °C. The authors found a clear decrease in the plasma viscosity after the infusion of saline solutions, while the infusion of the usual colloidal plasma substitutes could not produce this reduction in viscosity.

Note: No measurements of further rheologic parameters (e.g., blood viscosity).

In 1966 **Boyan** and co-workers [85] intravenously infused patients having undergone cancer surgery with 1000 ml isotonic saline solution for 30 min. The blood viscosity was measured with a Brookfield micro-cone-plate-viscosimeter at shear rates of 46, 115, and 230 s^{-1} at 37 °C. The authors did not observe a significant change in the blood viscosity, although the hematocrit fell significantly at the end of infusion.

Note: No measurements of further rheologic parameters.

In 1967 **Shabanov** and co-workers [793] treated patients with arteritis obliterans with infusions of 5% NaCl solution (50–300 ml/per day). The blood viscosity was measured with a capillary viscosimeter. During the 3-week treatment period, the raised blood viscosity values normalized.

Note: No measurements of further rheologic parameters; the temperature of the blood viscosity measurements was not given.

In 1971 **Dormandy** [198] reported on the hemorheologic effect of an infusion of 500 ml Hartmann's solution as well as 500 ml 10% dextran in patients who were due to have an abdominal operation. The initial viscosity measurements of the blood samples anticoagulated with heparin were performed after the premedication. The blood viscosity measurements were carried out with a Brookfield micro-cone-plate-viscosimeter at a temperature of 37 °C and a defined shear rate of 230 s^{-1}. At the end of the 10–20 minute infusion, the investigators found a significantly stronger and longer-lasting decrease in blood viscosity after the dextran solution than after the Hartmann's solution. Since blood viscosity changes are correlated with the changes in hematocrit values, the author sees the blood viscosity-reducing effect of dextran and saline solutions solely as a result of hemodilution.

Note: No measurements of further rheologic parameters.

In 1971 **Warstat** [900] reported on an improvement of the flow properties of blood after an intravenous infusion of 500 ml saline solution in patients with a marked increase in the erythrocyte sedimentation rate. The blood and plasma viscosities of the heparinized blood samples were measured with an Ostwald capillary viscosimeter at 23 °C; erythrocyte deformability was determined by the Ehrly and Rossbach filtration method (Millipore filter made of cellulose ester, pore diameter eight μm, driving force 12.9 mm H_2O, 23 °C, hematocrit 10%). At the end of the infusion a minimal change was found in the blood and plasma viscosity, as well as in the hematocrit value. However, the erythrocyte filtration rate was increased.

Note: No measurement of further rheologic parameters.

In 1975 **Dawidson** and co-workers [157] reported an improvement in the flow properties of blood after a postoperative intravenous administration of 1000 ml isotonic saline solution in patients. The blood and plasma viscosities were determined with a Brookfield micro-cone-plate viscosimeter, model LTV, at 37 °C and at a shear rate of 11.5 s^{-1}. The authors found a simultaneous decrease in the blood and plasma viscosities as well as in the hematocrit value.

Note: No hematocrit correction for the blood viscosity measurements; no measurements of further rheologic parameters.

In 1978 **Störmer** and co-workers [820] also reported changes in the flow properties of blood following administration of electrolyte solutions in patients with aorto-femoral bypass operations. The whole blood viscosity was measured with a Wells-Brookfield cone-plate-viscosimeter at high shear rates and with a Contraves rheometer at low shear rates. In 15 patients there was a reduction of the blood, plasma and serum viscosities during the operation which was more marked after the infusion of an electrolyte solution than without it.

Note: No hematocrit correction; no measurement of the erythrocyte aggregation and erythrocyte deformability. The relevance of this study is limited by the fact that, on the one hand, the patients were anesthesized and that, on the other hand, the blood loss certainly played a role. It is therefore difficult to quantify the degree of reduction of blood and plasma viscosities due to electrolyte solutions.

In 1978 **Heidrich** and **Wachta** [427] published a report on the treatment of patients with arterial and venous diseases. The subjects received a single infusion of 500 ml 0.94% NaCl solution within a period of 2–6 h. Whole blood viscosity was determined with a Brookfield micro-cone-plate viscosimeter (model LVT) at four shear rates of 23, 46, 115, and 230 s^{-1} and at a temperature of 37 °C. Compared to the initial value, physiologic saline solution only led to a significant viscosity reduction at the end of the 4–6 h infusion measured at a shear rate of 230 s^{-1}. A significant reduction of hematocrit was registered.

Note: No hematocrit correction of the viscosity measurements; no further measurements of rheologic parameters.

In 1980 **Harke** and co-workers [417] also studied the influence of various infusion solutions on the flow properties of blood in patients undergoing hysterectomy. In this case an isotonic electrolytic solution was infused as a control. The blood viscosity was determined with a capillary viscosimeter at 37 °C. During the infusion and 1 h later, there was a decrease in blood viscosity. However, compared to the infusion of colloidal plasma substitutes this reduction was relatively small.

Note: No hematocrit correction for the viscosity values; other rheologic parameters were not measured. The effect of electrolyte solutions on the blood viscosity cannot be selectively gathered from these studies, since the anesthesia could have had an effect during the operation and the blood loss could have lead to an “inner” hemodilution.

In 1981 **Litwin** and co-workers [581] reported that during surgery in 39 patients an infusion of saline solution (no quantities given) effected a decrease in the blood viscosity and in the hematocrit (abstract, no further details).

In 1982 **Perego** and co-workers [692] reported on results from patients with peripheral vascular disorders after saline infusions (500 ml). Blood and plasma viscosities were determined with a rotational viscosimeter of an unknown type. At the end of the infusion there was a significant decrease in blood viscosity (450, 90, 11.25 s^{-1}) and in hematocrit. The following parameters did not change significantly: blood viscosity (hematocrit corrected at 450 s^{-1}), plasma viscosity, erythrocyte deformability (no detailed data on the method used), hematocrit-corrected erythrocyte sedimentation rate.

Note: No measurements of the erythrocyte aggregation; no temperature data.

4.1.3.3 Discussion

The measurements of *Kottmann* in 1907, performed at body temperature, showed a reduction of blood and plasma viscosities at least for a short time, an effect which seems to be quite plausible as a result of hemodilution. *Kottmann* and later *Koga* were the first to also use isotonic saline solutions for the therapy of chronic arterial occlusive disease of the extremities and they reported a clinical improvement with a simultaneous reduction of the viscosity values.

Schülke and *Hartel* found a decrease in plasma viscosity after infusions with saline solutions, but not after the infusion of colloidal plasma substitutes, which may be attributed to the clearly different viscosities of the solutions themselves.

Boyan and co-workers did not find a significant decrease in the blood viscosity, even though the hematocrit was reduced. This discrepancy is difficult to explain and may be due to technical problems of measurement.

In 1971 *Dormandy* found that, following the infusion of 500 ml Hartmann's solution, there was a decrease in the hematocrit and the blood viscosity in patients having undergone an abdominal operation. This effect, as well as the even stronger effect of the low molecular plasma substitutes, was explained in the sense of a hemodilution.

In 1974 *Warstat* was the first to study not only the blood and plasma viscosities but also the erythrocyte filtrability (deformability) following the infusion of 500 ml saline solution in patients. In addition to a reduction of the blood and plasma viscosities and the hematocrit there was an improvement in the filtrability of the blood through 8 μm pore filters. The decrease in the blood and plasma viscosities was explained in the sense of a hemodilution, the improved filtration rate on the basis of the low viscosity values of the plasma of the suspension medium.

In later years *Dawidson* and co-workers, as well as *Heidrich* and co-workers and *Perego* and co-workers, were also able to measure reductions of the hematocrit and the blood viscosity after the infusion of isotonic saline solutions. Moreover, *Störmer* and co-workers, *Harke* and co-workers, and *Litwin* and co-workers also found reductions of the blood viscosity following the infusion of electrolyte solutions during operations. This reduction of the viscosity values could, however, also be explained by the influence of the anesthesia or the blood loss during the operation.

In summary, it may be said that most studies showed a decrease in the blood, plasma, and serum viscosities after saline infusions in patients with various diseases. The cause of the decrease in the viscosity values can be explained as hemodilution. Of great interest seems to be the study of *Koga* in 1913 in which he infused patients with stage IV (according to Fontaine) chronic arterial occlusive disease with several saline or Ringer solutions and reported surprising success in gangrene of the extremities. On the whole, it may be said that a decrease in normal as well as raised blood viscosity values may be presumed to be due to low-viscosity saline solutions, and this is true both for whole blood viscosity as well as for plasma viscosity. In this case we are dealing with "hypervolemic hemodilution". However, since there is a rapid fluid flow from the intravasal into the interstitial and intracellular compartments, the duration of the reduction in blood viscosity is severely limited.

4.1.4 Calcium Gluconate

4.1.4.1 Introduction

Calcium gluconate is administered to patients with metabolic calcium disturbances as well as with calcium deficiencies in pregnancy, bone fractures, and tetanus. Moreover, it is used in allergic diseases such as hay fever, urticaria, Quincke's edema, as well as in drug and serum exanthema.

4.1.4.2 Publications

In 1940 **Rogen** [735] studied the effect of calcium gluconate on patients with cardiac disease or insufficiency. He used a Hess capillary viscosimeter; further data were not given. A total of 10 cm^3 10% solution of calcium gluconate was injected intravenously within 2 min. In patients with cardiac insufficiency but without the presence of edema, there was a reduction of the blood viscosity values following calcium gluconate.

Note: No exact data on the measurements; no measurement of further rheologic parameters.

4.1.4.3 Discussion

Based on this publication no statement can be made on whether calcium gluconate changes the flow properties of blood or not.

4.1.5 Heparin and Heparinoids

4.1.5.1 Introduction

In 1916 ***McLean*** *isolated a substance from different tissues which inhibited blood coagulation. Since the presence of this substance was first found in the liver, it was called heparin. The mast cells of the reticuloendothelial system were considered to be the most important location of heparin production. Heparin is not a chemically homogeneous substance, but rather a mixture of various combinations of different chain lengths and different molecular weights which have a saccharide structure. The average molecular weight is between 10000 and 20000. The smallest units are glucosamine, glucuronic acid, and iduronic acid, which are estered with sulphuric acid (mucopolysaccharide polysulphuric acid ester).*

Heparin is used in the prophylactic treatment of thromboembolic diseases of the arterial and venous system. If thrombi do already exist, heparin can only hinder further growth. Heparin is, moreover, used for the thrombotic prophylactic treatment of acute myocardial infarction until the simultaneously started coumarin prophylactic treatment becomes fully effective. Without heparin an extracorporal circulation is not conceivable. The so-called heparinoids are semi-synthetically produced sulphatized polysaccharides in different chemical modifications. The most well-known is pentosanpolysulphate. According to the manufacturers, these substances have lipolytic, fibrinolytic and anticoagulative properties and are applied in different vascular diseases.

4.1.5.2 Publications

4.1.5.2.1 Heparin

In 1950 **Nalefski** and co-workers [667] reported on a possible hemorheologic effect of heparin on the flow properties of blood in in-patients with coronary heart disease. The authors found that heparin has a blood viscosity-reducing effect. They suspected that the clinically favorable influence on the symptoms of patients with coronary thrombosis found by some investigators is to be attributed to a change in the vessel tone.

Note: No data on the type and means of viscosimetric studies; no reference on the dosage of the anticoagulant.

In 1959 **Swank** [852] studied the effect of subcutaneous heparin injections on the blood viscosity in ten patients with cerebrovascular diseases. The dosage level of heparin was chosen so that the blood coagulation time was doubled or tripled. The period of treatment lasted 2–6 months. The viscosity of the native blood was measured with a Swank and Roth capillary viscosimeter. An attempt was made to consider the hematocrit changes mathematically. The authors found that heparin had a significant viscosity-reducing effect on the blood viscosity.

Note: The temperature of measurement was not given; no measurement of further rheologic parameters.

In 1957 **Watson** [902] found no change in the blood viscosity following intravenous administration of 5000 units heparin in reconvalescents. A Hess capillary viscosimeter was used for the viscosity measurements at room temperatures of between 14 °C and 22 °C. After heparin medication, no blood viscosity changes could be measured either after fasting or after the intake of high-fat cream.

Note: The blood viscosity measurements were not performed at body temperature, and further measurements of rheologic parameters were not carried out. No correction for hematocrit.

In 1959 **Gousios** and **Shearn** [389] studied the effect of intravenous heparin infusions (80 mg) on the flow properties of the blood in 15 fasting subjects without hyperlipidemia or blood coagulation disorders. The blood viscosity was measured in vivo using a glass capillary inserted into an arm vein. With the help of a rubber cuff the venous pressure was maintained at values of between 270 and 300 mm water. In their study the authors described only a minimal reduction of the blood viscosity, to which neither a clinical nor a physiologic importance was attributed.

Note: Further parameters of the flow properties of blood were not studied. The hematocrit was not corrected.

In 1965 **Pavlovski** [688] carried out an intramuscular and intravenous anticoagulation therapy in 65 operated patients. The patients received 5000 units of heparin every 4–6 h in the first 2–3 postoperational days. This was followed by dicumarol (0.1 g three times daily) or henindione (0.03–0.05 g three times daily). A capillary viscosimeter ($H_2O = 1$) was used for the blood viscosity measurements. Following the administration of heparin, the author observed a reduction in blood viscosity (no evaluation of statistical significance). With dicumarol basically the same, although weaker, viscosity-reducing effect was registered.

Note: No temperature data for the viscosity measurements; measurements of further rheologic parameters were not carried out; no correction of the hematocrit.

In 1967 **Dintenfass** and **Rozenberg** [173] published a study in which 5000 IU heparin were intravenously administered to two patients with myocardial infarction. A Dintenfass cone-in-cone rotational viscosimeter with shear rates of between 0.01 and 150 s^{-1} at a temperature of 37 °C was used for the blood viscosity measurements of the native blood samples. In one patient there was an increase in the blood viscosity, in one patient a decrease.

Note: Measurements of further rheologic parameters (e.g., plasma viscosity) were not carried out; no correction of the hematocrit.

In 1967 **Gelin** [372] published a report on the rheologic effects of intravenous heparin injections (100 mg) in patients with hypercholesterinemia and increased blood and plasma viscosities values compared to healthy controls. The blood viscosity of the heparin-anticoagulated blood samples was measured with a Brookfield micro-cone-plate viscosimeter at 37 °C and shear rates of 23–230 s^{-1}. Following the intravenous administration of heparin, there was a significant decrease in the blood viscosity of the patients and of the control group, especially at 23 s^{-1}. Moreover, the authors found a decrease in the triglyceride concentration and an increase in the free, nonesterified fatty acids. The cholesterol and phospholipid levels did not change. The plasma viscosity decreased minimally.

Note: There was no correction for hematocrit; no measurements of further rheologic parameters.

In 1968 **Moulopoulos** and co-workers [659] treated patients with acute myocardial infarction and also some with arterial, peripheral, and cerebral thromboses using oral and intravenous doses of heparin (dosage not stated). The viscosity of uncoagulated blood was measured with a cone-plate viscosimeter at shear rates of 11.7–175.5 rotations per minute and at a temperature of 36.6 °C. In the patients with increased blood viscosity, there was a decrease (no data on statistical significance) in the blood viscosity due to heparin.

Note: Measurements of further rheologic parameters were not carried out.

In 1969 **Chazan** [122] reported on a 13-year-old female patient with diabetic acidosis. The blood viscosity was measured with a capillary viscosimeter. After the injection of an unknown amount of heparin, there was a slight increase in the blood viscosity.

Note: Measurements of further rheologic parameters were not carried out; no correction for hematocrit.

In 1976 **Erdi** and co-workers [318] reported on the effect of subcutaneous injections of heparin on the flow properties of blood. Sixteen patients received a dose of 5000 units heparin subcutaneously before surgery and at 8-h-intervals, followed by 10 days of treatment with a daily dose of 5000 units heparin. A control group of healthy volunteers received the above-mentioned dose in one single application. The blood viscosity of the heparinized blood samples with a mathematically corrected hematocrit value of 45% was measured with a Contraves Low Shear viscosimeter at shear rates of 0.77 and 2.62 s^{-1}. A few hours after the heparin injection, there was a significant decrease in the blood viscosity in patients at all shear rates measured, but especially at the lowest shear rate of 0.77 s^{-1}. In a comparable group of patients who did not undergo any anticoagulation therapy, no viscosity reductions could be determined. Changes in the fibrinogen and triglyceride levels were not found.

Note: Correction for hematocrit was carried out mathematically, and the blood viscosity measurements were only made at low shear rates. Temperature data of the viscosity measurement are not available. Measurements of further rheologic parameters were not carried out in this study.

In 1976 **Girolami** and **Cella** [376] studied the blood viscosity in patients having undergone heart valve operations and in patients with thromboembolic diseases using a Brookfield micro-cone-plate viscosimeter (37 °C; 46, 115, 230 s^{-1}) following subcutaneous injections of 0.1 ml heparin per 20 kg body weight twice a day. The duration of the treatment period was not mentioned. Significant changes in the blood and plasma viscosity could not be verified rheologically. The slight decrease in the blood viscosity was considered to be due to a corresponding decrease in the hematocrit.

Note: No correction for hematocrit; the anticoagulant used was not mentioned. Measurements of further rheologic parameters were not made.

In 1978 **Girolami** and co-workers [377] again reported on the lack of improvement in the flow properties of blood with heparin in patients with insufficiency of the heart valves. The patients received 0.1 ml calcium heparin per 20 kg body weight twice a day subcutaneously for 10 days. The blood and plasma viscosities were measured in EDTA-anticoagulated blood samples with a Brookfield micro-cone-plate viscosimeter at shear rates of 23, 46, 115, and 230 s^{-1} at a temperature of 37 °C. The investigators found no changes in the blood and plasma viscosities nor in the relative blood viscosity (blood viscosity divided by plasma viscosity), but they did find a slight decrease in the hematocrit values.

Note: No measurements of further rheologic parameters.

Motta and **Ratto** [658] studied the blood viscosity in patients with peripheral arterial occlusive disease of the lower extremities after an intravenous heparin infusion of 5 mg per kilogram body weight in 100 ml physiologic saline solution. The blood viscosity was measured at a corrected hematocrit value of 45% using a capillary viscosimeter at a temperature of 37 °C. Changes in the blood and plasma viscosities were not found.

Note: The blood viscosity measurements were carried out at high shear rates; no further rheologic measurements.

Ruggiero and co-workers [745] gave heparin intravenously to 32 patients with acute myocardial infarction or severe angina pectoris and later on subcutaneously in the follow-up period [dosage: 4 · 5000 units per day intravenously during the first 2 weeks, followed by 2 ml heparin (10 000 units) subcutaneously in 12 h periods during the next 2 weeks]. The blood plasma and serum viscosities were measured with Cannon, Fenske, and routine viscosimeters. Moreover, parallel to this, the hematocrit, the total protein concentration, the fibrinogen, the platelet count, and other factors were determined. Based on the high values for the blood, plasma, and serum viscosities, there was a statistically highly significant decrease in all these values following heparin. The results became more marked with increasing duration of treatment. At the same time there was a decrease in the hematocrit, the fibrinogen concentration, the platelet count, and the total protein concentration. Only some of these results could be interpreted.

In 1981 **Fontányi** and co-workers [351] reported on the influence of long-term treatment with heparin on the relative blood viscosity measured with an Ostwald capillary viscosimeter. In 13 patients with intermittent claudication, there was a decrease in the blood viscosity in the majority of the cases. The authors also found an increase in the pain-free walking distance following heparin (abstract, no details).

In 1985 **Pernigotti** and co-workers [698] reported on the effect of an intravenous application of 100 units heparin per kilogram body weight in patients with myocardial infarction and high fibrinopeptide-A levels. There was a decrease in the blood viscosity, measured with a Wells-Brookfield viscosimeter at 450 s^{-1}, and an improvement in the filtrability of the blood also ensued (Nucleopore filters). On the other hand, healthy subjects and patients with myocardial infarction, without raised fibrinopeptide-A level did not show any hemorheologic changes.

4.1.5.2.2 Heparinoids

In addition to heparin the so-called heparinoid sodium pentosan-polysulphate has also been studied as to its effect on the flow properties of blood.

Fendler and **Horvath** [348] applied 100 mg SP54 orally for 100 days to 16 patients with primary hyperlipoproteinemia. There was a reduction of the blood viscosity as well as a decrease in the lipid values (abstract, no details). As early as, 1981 the same group [623] had reported on a reduction of the blood viscosity with a combination therapy of sodium pentosan-polysulphate and nicotinic acid for 30 days and had found a reduction of the blood and plasma viscosities.

In 1985 **Boisseau** and co-workers [80] described the influence of pentosan-polysulphate on the filtrability of the blood (method according to Reid and co-workers) in patients with vascular diseases. In older patients the authors found a tendency toward a normalization of hemorheologic abnormalities (abstract, no details).

In 1985 **Buffet** and co-workers [94] studied the influence of pentosan-polysulphate in 26 patients during the course of a 7-day-therapy with intramuscular injections of 50 mg twice a day. The blood viscosity was measured with a Couette viscosimeter and the erythrocyte deformability with an Ektacytometer. The authors could not find any changes in these hemorheologic parameters in the course of therapy with pentosan-polysulphate (abstract, no details).

In 1984 and 1985 **Montefusco** and co-workers [653, 654] described the effect of a mucopolysaccharide mixture on the flow properties of blood in patients with hyperlipidemia using a Brookfield LTV viscosimeter at different shear rates and at 37 °C. In addition, the erythrocyte filtration was measured according to the method of Reid et al. The authors reported a reduction of the blood viscosity at a shear rate of 11.25 s^{-1}, while the plasma viscosity remained unchanged. The erythrocyte filtrability could be improved.

Calabro and co-workers [111, 112] studied the effect of sulodexide (glucoronylglucosamineglycane sulphate) in a double-blind study in patients with peripheral occlusive diseases. Whole blood viscosity was studied with a Wells-Brookfield viscosimeter at 37 °C and a shear rate of 11.25 s^{-1}. Plasma viscosity was measured with the same viscosimeter at 405 s^{-1}. The filtrability of the blood was measured according to a method by Reid and co-workers. There was a reduction of the whole blood viscosity and of the plasma viscosity and an improvement in the erythrocyte filtrability. The triglyceride levels decreased at the same time.

In 1985 *Ciuffetti* and co-workers [133] described the effect of sulodexide on the flow properties of blood in patients with chronic arterial occlusive disease. Thirty patients received 900 lipase units of this special type of mucopolysaccharide per day orally during the course of 90 days, or 1200 units per day intramuscularly during the course of 15 days. The whole blood viscosity and the plasma viscosity were measured with a Wells-Brookfield viscosimeter; the erythrocyte filtrability was measured according to the whole blood method of Reid and co-workers. The authors found a slight but statistically significant decrease in the plasma viscosity and in the whole blood viscosity.

Note: The decrease in the plasma viscosity of about 3% to 5% could lie within range of the measurement accuracy of the method used.

In 1985 **Pernigotti** and co-workers [698] studied the effect of a single dose of heparin-mesoglycan (1 mg per kilogram body weight) on the whole blood viscosity (Wells-Brookfield viscosimeter at 450 s^{-1}) and the erythrocyte filtrability (Nucleopore filters) in patients with myocardial infarction and found an improvement in the blood filtrability and a decrease in the blood viscosity, while healthy subjects showed no corresponding changes.

4.1.5.3 Discussion

The results available concerning the influence of heparin on the flow properties of blood are somewhat discrepant. While most investigators have not reported any major reductions, others have found a significant reduction of the blood viscosity following intravenous heparin administration. A correction of the hematocrit (not mathematically) has only been performed by one team [658]; the temperature of the viscosity measurements was not always reported; or the viscosity measurements were carried out below body temperature. With the exception of *Fontanyi* and *Ruggiero*, the investigators who also determined the plasma viscosity values did not find any changes in this subparameter. Additional findings of *Fontanyi* and *Ruggiero,* especially the synchronous decrease in the hematocrit, the total protein, the fibrinogen concentration, and the platelet count indicate a nonspecific effect of heparin, somewhat like an endogenous hemodilution. In addition, these patients were initially in a stress situation (acute myocardial infarction or severe angina pectoris) which may have been accompanied by a relative hemoconcentration. Since a comparative group was not studied, the results of this work must be relativized.
In this connection, the observations of *Pernigotti* and co-workers are interesting. They only found a reduction of the blood viscosity due to heparin in those patients who had a high fibrinopeptide-A level. These patients also showed an improvement in the filtrability of the blood. None of the groups carried out measurements on erythrocyte aggregation.
According to the studies carried out, it is, on the whole, improbable that there is a decrease in blood and plasma viscosities due to an intravenous administration of heparin; there is possibly an

influence in cases with a high fibrinopeptide-A level. The single mention by *Pernigotti* of a change in the filtrability of the blood does not allow an evaluation, since no concrete data are given as to the method.
The results with regard to the influence on the flow properties of blood through so-called heparinoids are not uniform. While *Fendler* and *Horvath, Montefusco* and co-workers, *Calabro* and co-workers, as well as *Ciuffetti* could show a reduction of the blood viscosity, *Buffet* and co-workers did not find any changes in the flow properties of blood. *Pernigotti* and co-workers only found an influence on the blood viscosity of patients with myocardial infarction when there was a high fibrinopeptide-A level at the same time. The statements made by *Montefusco* and co-workers with regard to the filtrability of the blood must be relativized because of the method used. On the whole, it can only be said that the question as to the effects of heparinoids on the flow properties of blood must still remain open to discussion.

4.1.6 Coumarins

4.1.6.1 Introduction

Different chemical modifications of coumarin (phenprocoumon, warfarin) are used as oral anticoagulants (vitamin K antagonists).
Coumarins are used for the prophylaxis and therapy of thromboembolic diseases and for the long-term prophylaxis of myocardial infarction.

4.1.6.2 Publications

Initial hemorheologic reports on dihydroxycoumarin were published by **Nalefski** and co-workers [667] in 1950. The authors were not able to prove a blood viscosity-reducing effect of dihydroxycoumarin in in-patients with coronary heart disease.

Note: Statements as to the type and manner of viscosimetric studies were not made; no measurement of further rheologic parameters; no data on dosage.

In 1965 **Pavlovskii** [688] reported on anticoagulant therapy with dihydroxycoumarin in 65 patients having undergone surgery. The patients received 0.1 g dihydroxycoumarin intravenously and intramuscularly. This dose was administered three times a day. The blood viscosity measurements were carried out with a capillary viscosimeter (H_2O = 1). The investigators were not able to find any blood viscosity-reducing effect of the anticoagulant.

Note: The temperature of the viscosity measurements was not given; no measurements of further rheologic parameters.

In 1965 **Rosenblatt** and co-workers [739] studied the effect of oral warfarin on the whole blood viscosity of patients following cardiac infarctions. The studies were performed with EDTA-anticoagulated blood at 37 °C; a Brookfield cone-plate viscosimeter, type LVT, was used at a shear rate of 230 s^{-1}. No significant differences in the whole blood viscosity values were found with warfarin.

In 1967 **Kallio** and co-workers [484] studied the effect of an oral anticoagulant therapy on four patients (no exact data on the disease) and found no changes in the blood viscosity during a 5-day-period of treatment using a Wells-Brookfield microcone-plate-viscosimeter at various shear rates and at a temperature of 37 °C.

Note: No figures on the dosage; no measurements of other rheologic parameters.

In 1976 **Mayer** [629] reported on the improvement in the flow properties of blood in 31 patients after myocardial infarction and with anginal pains. Dihydroxycoumarin was given orally at different dosages. A capillary viscosimeter was used for the blood and plasma viscosity measurements at 38 °C. Both in the patients and in a control group a significant decrease was found in the blood and plasma viscosities as well as in the hematocrit values.

Note: No correction for hematocrit; no measurements of further rheologic parameters.

Clinical findings: The anginal pains disappeared.

In 1982 **Marshall** and **Reiser** [615] compared patients treated with phenprocoumon with a control group and found lower viscosity values in the patients than in the control group. The measurements were performed with an automatic capillary viscosimeter at 42 s^{-1} and 85 s^{-1}.

Note: No temperature data on the viscosity measurements; no measurements of further hemorheologic parameters.
Similar results were presented by these authors elsewhere [614].

4.1.6.3 Discussion

The above-mentioned studies on the effect of coumarin on the flow properties of blood are not uniform. While *Nalefski* and co-workers, *Pavlovskii, Rosenblatt* and co-workers, and *Kallio* and co-workers were not able to find any decrease in the blood viscosity following oral anticoagulant administration, *Mayer* reported on a reduction of the blood and plasma viscosities. Since *Mayer* found at the same time a decrease in the hematocrit value, it may not be excluded that the hematocrit reduction in his studies lead to the blood viscosity reduction. The studies of *Marshall* and *Weiser,* which were performed without any hematocrit comparisons, also give no indication that coumarin has a specific effect on the flow properties of blood. From the publications available there are therefore no definite findings for a blood viscosity-reducing effect of oral anticoagulants.

4.1.7 Dextran Solutions

4.1.7.1 Introduction

Modern therapeutic hemorheology got a definite boost from the introduction of plasma substitutes (plasma expanders) for the therapy of microcirculatory disturbances. Up until that time, plasma substitutes had only been used to replace inadequate blood volume. Initially low molecular weight (LMWD) and mid-molecular weight dextran (MMWD) with average molecular weights of 40 000 or 60 000 or 75 000 were used for some decades for

most rheologicophysiologic and rheologicotherapeutic studies. First, only single studies were carried out, but recently studies have been increasingly performed on the rheologic effect of other types of plasma substitutes. This is particularly true for hydroxyethyl starch solutions.
From a historical point of view, the extensive research on the rheologic effects of dextran solutions may be considered to be a major factor in all of the research in the field of clinical and therapeutic hemorheology of the last decades. After ***Gelin*** *and co-workers had reported in 1961 on special rheologic properties of a dextran solution with a molecular weight of 40 000, there followed a wave of publications by various teams throughout the world who were pursuing research of their own in this field.*
It is to ***Gelin's*** *credit that he pointed out the clear necessity of improving the deteriorated flow properties of blood in states of shock. That is also how his definition of LMWD as as "flow improver" is to be understood. There is no doubt that the widespread application of LMWD for the therapy of hypovolemic shock conditions exemplified a major advance, since up to then one had attempted to compensate fluid losses with blood conserves. That this can lead to a further deterioration of the already existing unfavorable flow properties of blood was already recognized by* ***Gelin's*** *Swedish team in the early 1960s. A plasma substitute based on LMWD was not the first therapeutic measure in the sense of "therapeutic hemorheology"; it was, however, the first commercially available substance which came to be known as being therapeutically particularly useful for the improvement in the flow properties of blood.*
The dextrans belong to a group of bacteriologically produced polysaccharides ($C_6H_{10}O_5$) of the d-glucopyranose type with a dominant α-1-6 glycoside bond. Among the strains of bacteria which are used for the industrial production of dextrans, those of the Leuconostoc mesenteroides *have to be mentioned.*
The dextran solutions commonly used today contain either LMWD with an average molecular weight of 40 000 (e.g., Rheomacrodex or MMWD with an average molecular weight of 60 000 to 75 000). The dextran solutions are marketed as 6% or 10% solutions in isotonic saline as well as in glucose solution.

4.1.7.2 Publications

In 1961 **Gelin** and **Ingelman** [367] for the first time reported on the utilisation of LMWD (average molecular weight 40 000) and presented their rheologic results. The blood and plasma viscosities were measured with a Brookfield micro-cone-plate viscosimeter at shear rates of 6–60 s^{-1} at a temperature of 37 °C. In one patient with biliary peritonitis there was a decrease in the blood and plasma viscosities following the infusion of 500 ml 15% LMWD solution. In another patient with burns and a high hematocrit value, a simultaneous reduction of the hematocrit value and the blood viscosity, which was particularly marked at low shear rates, could be observed after an infusion of 1000 ml 15% LMWD.

Note: Further rheologic parameters were not measured.

In 1961 **Gelin** and **Thorén** [368] studied the blood viscosity of patients with various types of diseases and highly increased erythrocyte sedimentation rates using a Brookfield micro-cone-plate viscosimeter at shear rates of 6–60 s^{-1} and at a temperature of 37 °C. The blood samples for the viscosity measurements were anticoagulated with heparin. The patients received an intravenous infusion of 15% MMWD with an average molecular weight of 75 000 at a dosage of 1 g per kilogram body weight. Both dextran solutions reduced the hematocrit. With MMWD the blood viscosity values were slightly increased, whereas with LMWD a significant decrease was found in the blood viscosity.

Note: Further rheologic parameters were not measured.

In 1962 two studies by **Gelin** [369, 370] appeared in which results similar to those in 1961 were reported.

In 1963 **Bergentz** [66] from **Gelin's** team reported on a patient with extreme trauma in whom there was a decrease in the blood viscosity due to an infusion of LMWD (no quantitative data, no percentual data). The measurements were made with a Brookfield micro-cone-plate viscosimeter at a temperature of 37 °C and shear rates of 6–60 s^{-1}.

Note: Measurements of further rheologic parameters were not carried out.

Clinical findings: The neurologic symptoms of the patients disappeared.

In 1965 **Gelin** [371] published an article in which the blood viscosity-reducing and hematocrit-reducing effect of LMWD was described in a patient with trauma after a serious road accident (Brookfield micro-cone-plate-viscosimeter, 37 °C, shear rate range 6–60 s^{-1}). Reports with similar results were published by **Gelin** in 1969 [373] and 1971 [374].

In 1965 **Groth** and **Thorsén** [394] studied the effect of a 10% and a 15% LMWD solution (molecular weight 40 000) and a 6% MMWD solution (molecular weight 75 000) in patients with an increased erythrocyte sedimentation rate and compared this to a 5% glucose solution with regard to the flow properties of blood. The patients were infused intravenously with 500 ml of the corresponding dextran or glucose solution in the course of 25–30 min. The plasma viscosity of the heparinized blood samples was measured with an Ostwald capillary viscosimeter at 37 °C and presented as the relative viscosity compared to distilled water. The infusion of LMWD increased the viscosity of the plasma; with high initial values of the plasma viscosity, a decrease was found. MMWD, on the other hand, increased the plasma viscosity in all patients independent of the initial viscosity (see Sect. 4.1.15 for the effect of glucose solutions).

Note: Measurements of further rheologic parameters and of blood viscosity were not carried out.

In 1966 **Groth** [396] studied the effect of 10% LMWD (molecular weight 40 000) and 10% albumin solution on the flow properties of the blood in patients with increased viscosity values as a result of various diseases. The patients were given 10 ml plasma substitute per kilogram body weight intravenously for 1 h.
The blood and plasma viscosities of the heparinized blood samples were measured with a Brookfield micro-cone-plate viscosimeter (model LVT) at 37 °C and at four shear rates of 23, 46, 115, and 230 s^{-1}. After the infusion of the LMWD solution, there was a reduction of the hematocrit and a significant reduction of the blood viscosity at all shear rates, which was most prominent at low shear rates. Contrary to the blood viscosity values, the author found an increase in plasma viscosity. When the initial values for the plasma viscosity were high, then there was no change in that parameter (see Sect. 4.1.16 for the hemorheologic effect of albumin solutions). A study with similar results was published by **Groth** [397] in 1968.

Note: Quantitative erythrocyte aggregation and filtration measurements were not made.

In 1966 **Ehrly** [230] studied the blood viscosity of three patients with a marked sludge phenomenon and a highly increased erythrocyte sedimentation rate. The test was carried out at 37° using an Ostwald capillary viscosimeter Ostwald ($H_2O = 1$). The patients received 500 ml 10% LMWD infused intravenously for 1.5 h. Erythrocyte aggregation was determined according to the microscopic observations of the so-called blow-out smears. In addition to a decrease in the blood viscosity and the hematocrit values, there was an increase in the erythrocyte sedimentation rate (no hematocrit correction). While the erythrocyte aggregation in the conjunctival vessels, as seen by capillary microscopy, was decreased, the erythrocyte aggregation on the smear was not influenced by the infusion. Based on these results and on additional in vitro studies, the conclusion was reached that the decrease in the erythrocyte aggregation (sludge phenomenon) was due to the improved flow conditions and that there is no specific disaggregating effect of LMWD.

Note: Measurement of the plasma viscosity, the erythrocyte filtrability (erythrocyte deformability), and direct measurements of the erythrocyte aggregation were not carried out.

In 1966 **Yao** and **Shoemaker** [928] compared patients having undergone surgery with a healthy control group following LMWD infusions. They studied the blood and plasma viscosities with a Brookfield micro-cone-plate viscosimeter, model LVT, at various shear rates between 21 and 212 s^{-1} and at a temperature of 37 °C. The blood was heparinized and adapted to a standard hematocrit. Following the infusion of 500 ml (time not mentioned), the authors found a significant decrease in the blood viscosity at the same time as the decrease in the hematocrit. The reduction of the viscosity was, as a whole, more distinct at low shear rates. Even when the hematocrit was set to a standard value before and after administration of LMWD, the authors observed a decrease in the viscosity values due to LMWD, while this was not the case with healthy subjects. Therefore a specific viscosity-reducing effect of LMWD in shock situations is presumed. Plasma viscosity only decreased when the patients' initial plasma viscosity values were raised.

Note: Direct measurements of the erythrocyte aggregation and filtrability were not carried out.

In 1966 **Schülke** and **Hartel** [776] studied the effect of infusions of 1000 ml each of different plasma substitutes on the relative plasma viscosity compared to that of physiologic saline solution (Harkness capillary viscosimeter, $H_2O = 1$; 37 °C) in 20 cholecystectomized patients. They found a reduction of the plasma viscosity following saline infusions. On the other hand, following a 3.5% gelatin infusion the authors observed a slight increase in the plasma viscosity. The infusion of LMWD solution led to a moderate increase and that of MMWD to a strong increase in the plasma viscosity. LMWD (10%) at the dosage given led to a less distinct increase in the plasma viscosity than MMWD (4%, 10%), although the relative viscosity of the solutions was much higher with LMWD, owing to the higher concentration of the solution, than with LMWD. The authors placed great weight on the significance of the plasma viscosity because, when the hematocrit in the capillaries is reduced, the whole blood viscosity is, to a great extent, determined by the plasma viscosity. They therefore recommended using low-percentage solutions of low-molecular weight plasma substitutes.

Note: Measurements of the blood viscosity and the erythrocyte aggregation were not carried out.

In 1966 **Boyan** and co-workers [85] studied the effect of various plasma substitutes on the blood viscosity in patients at the end of a tumor operation. The heparinized blood samples were measured with a Brookfield micro-cone-plate viscosimeter at various shear rates between 46 and 230 s^{-1} and at 37 °C. Following the infusion of 500 ml 15% LMWD solution, following 500 ml 6% MMWD solution, and following 500 ml plasma protein solution, the authors found no decrease in the blood viscosity at the end of the 30 min infusion even though the hematocrit was significantly decreased. This was explained the fact that the viscosity of the infused solution per se was too high and this could compensate the effect of the hemodilution.

Note: Measurements of further rheologic parameters such as, for example, the plasma viscosity were not carried out.

In 1966 **Bernstein** and **Castaneda** [68] treated 25 patients, who were connected to a heart-lung machine during a heart operation, with solutions of 10% LMWD and 5% dextrose solutions. The temperature of the blood viscosity measurements was, in accordance with the degree of hypothermia, at 30°–31 °C (Brookfield micro-cone-plate viscosimeter, 1.3–130 s^{-1}). Simultaneously with a decrease in the hematocrit, the blood viscosity was reduced more strongly and longer by LMWD than by a 5% dextrose solution. The reduction of the blood viscosity was most distinct at low shear rates. The authors believed that the viscosity-reducing effect of the LMWD may be traced back to its colloido-osmotic effect and to an increased electrostatic charge of the erythrocytes which causes a disaggregation of erythrocyte aggregates.

Note: Measurements of further rheologic parameters such as, for example, plasma viscosity were not carried out.

In 1966, 1967 and 1968 [543–545] rheologic findings after infusions of LMWD in acute myocardial infarction were presented by **Langsjoen** for the first time. At the same time as a hematocrit decrease, a blood viscosity reduction, which was the most distinct at low shear rates (rotational viscosimeter of an unknown type; 2, 6, 11, 23, 46, 115, and 230 s^{-1}), was observed after 5% dextrose solutions and even more so after infusions of 10% LMWD. A reduction of the mortality after acute myocardial infarction was dicussed.

Note: There are no data on the temperature at measurement and the type of viscosimeter used.

In 1968 **Bollinger** and co-workers [81] studied the effect of 10% LMWD on the behavior of the flow rate distal to arterial occlusions of the limbs and simultaneously measured the blood and plasma viscosities at 37 °C with a Brookfield micro-cone-plate viscosimeter at shear rates between 11.5 and 230 s^{-1}. After the infusion the authors found a significant decrease in whole blood viscosity and an increase in plasma viscosity. Immediately after the administration of LMWD, the average increase in the blood flow at rest and reactive hyperemia (measured with the help of venous occlusion plethysmography) were less marked in patients with vascular disease than in patients without chronic occlusive arterial disease. The more severe the ischemic disease, the milder the reactive hyperemia following infusion with LMWD; in some cases a decrease in the blood flow was observed. It is of interest that the increase in the blood flow did not reach its maximum immediately after the end of the infusion.

Note: No measurements of further rheologic parameters.

In 1968 **Lim** and co-workers [575] carried out an intravenous infusion of 500 ml 10% LMWD and 6% MMWD in 15 patients with pathologically increased erythrocyte sedimentation rates. The blood and plasma viscosity measurements were made with a Brookfield micro-cone-plate viscosimeter at various shear rates between 23 and 230 s^{-1} and at a temperature of 37 °C. For the blood viscosity measurements, the hematocrit was set at different levels of standardized values (15%,

30%, 45%, 60%). In the LMWD group, the authors observed, especially at low shear rates, a reduction in the blood viscosity with corrected hematocrit and a slight increase in the plasma viscosity. Following MMWD infusions, an increase was found in the plasma and the blood viscosity. A specific disaggregating effect of the LMWD was postulated by the authors.

Note: Measurements of further rheologic parameters were not carried out.

In 1968 **Ditzel** and co-workers [189] presented a double-blind study on the influence of 10% LMWD on the blood viscosity in acute myocardial infarction (Brookfield micro-cone-plate viscosimeter, 46 s^{-1}). They were able to confirm earlier studies by **Langsjoen** (see above) according to which the hematocrit and the whole blood viscosity were decreased. The mortality rate, however, remained unchanged, so that a favorable therapeutic effect for acute myocardial infarction was contested.

Note: Measurements of further rheologic parameters were not made. No temperature data for the viscosity measurements.

In 1971 **Dormandy** [198] studied heparin-anticoagulated blood from premedicated patients before an abdominal operation using a Brookfield micro-cone-plate viscosimeter at a shear rate of 230 s^{-1} and at a temperature of 37 °C. Ten patients were infused for 10–20 min with 10% LMWD in saline solution and 10 further patients with Hartmann's solution. In addition, the blood flow in the lower calf was measured plethysmographically. During and following the infusion, the blood viscosity decreased in both groups. Ten minutes after the end of the infusion, the blood viscosity reduction was significantly greater in the dextran group than in the control group. The higher viscosity reduction with dextran was explained by a stronger hemodilution effect. The correlation between the decrease in viscosity and the increase in total blood flow was good.

Note: Measurements of further rheologic parameters were not carried out.

According to **Carey** [116] the administration of 500 ml 10% LMWD led to a decrease in the hematocrit and in the blood viscosity and to an increased cardiac output in three patients after esophagogastrectomy due to carcinoma (Brookfield micro-cone-plate viscosimeter, shear rate 23 s^{-1}).

Note: Measurements of further rheologic paramters were not carried out. No temperature data for the viscosity measurements.

In 1973 **Langsjoen** [546] published further results on the decrease in mortality from cardiac infarction due to the improvement of the flow properties of blood by using LMWD. These results correspond to earlier studies (see above).

In 1973 **Amtoft** [13] reported on 22 patients with intermittent claudication whose blood viscosity was measured with a Brookfield micro-cone-plate viscosimeter at various shear rates between 5.75 and 230 s^{-1}. Six patients received a single infusion 500 ml 10% LMWD solution for 1 hr. The same quantity of dextran was administered intravenously to the other patients daily over a treatment period of 10 days. The authors observed a significant decrease in the blood viscosity with the single dextran infusions. Moreover, an improvement was observed in the diffusion capacity (^{51}Cr-EDTA) after dextran infusion.

Note: There are no data on the temperature of the viscosity measurements. Further rheologic parameters were not mentioned.

Clinical findings: Despite the 10-day treatment with LMWD the authors found no improvement in the clinical symptoms as judged by the pain-free walking distance, the change in color of the cyanotic extremities, and the arterial pulses. The treatment showed success in only one patient.

In 1973 and 1974 studies were carried out in Hess' laboratory on the influence of 10% LMWD on the blood viscosity.

Jauch [475] infused 500 ml 10% LMWD for 2 h. The blood viscosity was measured with a Reiss and Müller falling ball viscosimeter (data in falling rate units, 37 °C). During and at the end of the infusion, the author found a decrease in the blood viscosity and in the hematocrit.

Note: Further rheologic parameters were not determined.

In a dissertation **Franke** [356] reported on similar results.

Also in 1973, **Ott** [676] reported on the effect of 10% LMWD on blood viscosity. Among the test persons there were healthy subjects and patients with vascular and other diseases. A Haake falling ball viscosimeter was used. The measurements were made at 37 °C. Five subjects were infused with 500 ml LMWD for 1 h. The viscosity measurements were performed 4–8 h after the end of the infusion. In the meantime, the blood samples were stored at 4 °C. At the end of the infusion the authors found a 15.2% blood viscosity reduction and a hematocrit reduction of 13%.

In 1974 and 1975 **Heidrich, Ott,** and **Schlichting** published major parts of this study [421, 423].

Note: Measurements of further rheologic parameters were not carried out.

In 1974 **Warstat** [900] from Ehrly's group studied the influence of an infusion of 10% LMWD on the flow properties of blood in patients with a moderate and a strong increase in the erythrocyte sedimentation rate. The patients received 500 ml LMWD intravenously for 30–45 min. The blood and plasma viscosities of the heparinized blood samples were measured with an Ostwald capillary viscosimeter at 23 °C; the erythrocyte deformability was determined with a filtration method (Millipore filters made of cellulose ester, pore diameter eight μm, hydrostatic pressure of 12.9 mm H_2O, 23 °C). At the end of the infusion, the author found a strong reduction of blood viscosity and parallel to this a reduction of hematocrit. On the other hand, the plasma viscosity rose slightly. Erythrocyte filtration measurements showed no changes, but rather an increased perfusion through the capillary filter system, which had not been expected owing to the increased plasma viscosity after the infusion with LMWD.

Note: Measurements of the erythrocyte aggregation were not carried out, and the temperature of the rheologic measurements was below 37 °C.

In 1975 **Dawidson** [157] treated ten patients postoperatively with 500 ml 10% LMWD solution. The blood viscosity was measured with a Brookfield micro-cone-plate viscosimeter at 37 °C and at a shear rate of 11.5 s^{-1}. In addition, the blood viscosity was also measured at a hematocrit of 98%–99% (1.15 s^{-1}). Compared to a control group and to a group of patients who received only an infusion of 1000 ml 0.9% NaCl, there was a significant reduction of the blood viscosity as a result of LMWD, which correlated with a drop in the hematocrit. A specific disaggregating property of LMWD corresponding to the results with a high hematocrit was discussed.

Note: Measurements of further rheologic paramters were not carried out.

In 1976 **Humphreys** and co-workers [464], within the framework of a by-pass operation, infused 500 ml LMWD (percentage not mentioned) intravenously for 10–15 min in 12 patients who were suffering from an occlusion of the superficial femoral artery. The viscosity measurements of the EDTA-anticoagulated blood samples were made with a Contraves Low Shear 2 coaxial cylindrical viscosimeter at shear rates of 0.675, 0.786, and 124 s^{-1}. After the infusion there was a significant decrease in the blood viscosity at all shear rates, which correlated with a drop in the hematocrit. If, however, the hematocrit was corrected mathematically to a value of 45%, then no blood viscosity changes could be observed at low shear rates. When the viscosity was measured at 124 s^{-1}, a significant viscosity increase could be determined after hematocrit correction. Just like Ehrly in 1966 [230] and Dormandy in 1971 [198], the authors consider the mode of action of LMWD to be the result of hemodilution.

Note: Temperature data were not given, and measurements of further rheologic parameters were not carried out.

In 1978 **Heidrich** and co-workers [426, 427] studied 54 patients with peripheral arterial and venous disorders. The behavior of the blood viscosity was tested (Brookfield micro-cone-plate viscosimeter at different shear rates between 23 and 230 s^{-1} and at a temperature of 37 °C, heparin as anticoagulant) during intravenous infusions of 2, 4, and 6 h with 500 ml 10% LMWD or 500 ml 0.9% NaCl solution (control group). After 2-, 4-, and 6-h long infusions, the dextran infusions led to a significant decrease in the blood viscosity, compared to the saline control group, which was most distinct at low shear rates. The comparison to dextran solutions showed that following 4- and 6-h infusion a significantly stronger reduction of the viscosity values was observed than after a 120-min infusion. The viscosity changes registered after 2, 4, and 6 h of saline infusion were significantly decreased compared to the initial values only at the end of the 4- and the 6-h infusion interval at a shear rate of 230 s^{-1}.

Note: No measurements of further rheologic parameters.

In 1978 **Störmer** and co-workers [820] studied the behavior of the blood, plasma, and serum viscosity in patients undergoing aortofemoral bypass surgery. The blood, plasma and serum viscosity was measured with a Wells-Brookfield cone-plate viscosimeter or a Contraves rheometer for low shear rates. The temperature at measurement was 37 °C. In a total of 15 patients with operations of varying length, the viscosity of whole blood, plasma, and serum was determined before and after the start of the operation, after the resection of the aorta, and after volume expansion (dextran 40). There was a drop in these values during the entire operation. However, after the infusion of 1000 ml dextran 40, the reduction of the whole blood viscosity was particularly marked, while the plasma viscosity did not change very much.

Note: The relevance of these experiments is limited because the patients were under anethesia. Furthermore, it is known that the loss of blood during the operation leads to a reduction of the blood viscosity values as the result of inner hemodilution.

In 1979 **Burke** and co-workers [98] reported on patients with increased blood and plasma viscosities who, because of an intercranial aneurysm, had undergone a craniotomy. Seven patients received 500 ml 10% LMWD solution in the first 6 h after the operation and in the next 48 h two further infusions corresponding to the above mentioned dosage and were compared to a control group. The viscosity measurements were performed with a modified, coaxial cylinder viscosimeter according to Gilinson and co-workers at a temperature of 37 °C and at shear rates of 0.5, 5.2, and 52 s^{-1}. On the first day after the operation, there was a significant decrease in the blood viscosity and in the hematocrit value in the dextran group.

After the operation the plasma viscosity did not decrease in all 13 patients. With a standardized hematocrit of 45%, a reduction of the blood viscosity was also observed in only the dextran group on the first day. This was particularly distinct at low shear rates. The relative blood viscosity (blood viscosity with hematocrit of 45% divided by plasma viscosity) decreased significantly in the first 24 h, however, only at a shear rate of 0.5 s^{-1}. The measurements of the erythrocyte deformability (erythrocyte flexibility) with polycarbonate filters (pore diameter 3 μm) at a pressure of 20 cm H_2O and at a hematocrit of 1.5% showed no changes. In the opinion of the authors, the blood viscosity-reducing effect of LMWD was due to hemodilution as well as to a specific, disaggregating mode of action.

Note: No quantitative measurement of the erythrocyte aggregation.

In 1980 **Harke** and co-workers [417] carried out an open study on the effect of an infusion of 1000 ml of three different solutions (electrolyte solution, 10% dextran 40 solution, and 10% hydroxyethyl starch solution) in 30 patients after hysterectomy. The duration of infusion was 2 h. Among other parameters the blood viscosity was determined with a capillary viscosimeter at 37 °C. Blood sampling to determine the blood viscosity and the hematocrit took place before, during, and after the infusion. Following the administration of electrolytes or dextran 40, there was a statistically insignificant reduction of the blood viscosity, while the hematocrit decreased statistically significantly. The authors attributed this to the fact that, in that particular series of experiments, a 12-h food and fluid abstinence was maintained leading to a relative dehydration of the patients so that the hyperoncotic dextran solution could not be sufficiently diluted with interstitial fluid. The hemodilution effect of the infusion lasted for several hours.

In 1980 *Motta* and co-workers [658] studied the effect of the infusion of 500 ml LMWD on blood and plasma viscosities (capillary viscosimeter, 37 °C) in 20 patients with peripheral arterial occlusive disease. After an infusion of 30 min, a decrease in blood viscosity and an increase in plasma viscosity could be observed. When the hematocrit was set to a value of 45%, there was an increase in the blood viscosity.

Note: The blood viscosity was measured at high shear rates; no determination of other rheologic parameters.

In 1980 **Humphrey** and co-workers [460] reported on the treatment of patients with ischemic cerebral attacks and polycythemia using LMWD. The patients were infused intravenously with 500 ml LMWD for 1 h. The blood and plasma viscosities were measured in heparin-anticoagulated blood samples with a Contraves Low Shear 2 rotational viscosimeter at shear rates between 0.67 and 92 s^{-1}. Simultaneous to the drop in hematocrit, the authors found a significant drop in blood viscosity, which was particularly distinct at low shear rates. The plasma viscosity measurements showed a slight increase in these values.

Note: A correction of the hematocrit was not carried out; no measurements of further rheologic parameters.

In 1981 **Litwin** and co-workers [581] studied the influence of an infusion of dextran 40 in postoperative patients and found a decrease in the whole blood viscosity with a simultaneous decrease in the hematocrit (abstract, no details).

In 1982 **Tsyganyi** [864] reported on the influence of plasma substitutes available in the USSR on the flow properties of blood of patients after cardiac surgery and found that rheopolyglucin (corresponding to the usual dextran solutions) improved the flow properties of blood (original in Russian).

In 1982 **Perego** and co-workers [692] reported on the therapy with saline and LMWD solutions in patients with peripheral chronic occlusive arterial disease (no classification of stages). The blood and plasma viscosities were measured with a rotational viscosimeter of an unknown type. After an infusion of 500 ml 10% LMWD solution, the authors found a significant decrease in the blood viscosity at high and low shear rates (450, 90, and 11.25 s^{-1}), in the hematocrit-corrected blood viscosity (450 s^{-1}), and in the hematocrit. The plasma viscosity showed a minimal increase. The erythrocyte deformability (no details given as to the method) was decreased significantly. The erythrocyte sedimentation rate (hematocrit corrected) did not change significantly. Contrary to the LMWD solution, there was no significant decrease in the blood viscosity (hematocrit corrected) after saline infusion (500 ml).

Note: No measurements of further rheologic parameters; no temperature data.

Karasawa and co-workers [485] studied the blood viscosity in pregnant women with a cone-plate viscosimeter at 150–375 s^{-1} and found that the whole blood viscosity decreased after the intravenous infusion of dextran.

Siekmann and co-workers [797] found that there was a reduction of hematocrit as well as a decrease in the erythrocyte aggregation after a single infusion of 500 ml dextran 40 solution in 40 patients (abstract, no details).

In 1983 **Balabanova** and co-workers [41] described the influence of an intravenous infusion of 400 ml LMWD solution on the flow properties of blood of 20 patients suffering from chronic systemic scleroderma. The authors found a reduction of the whole blood viscosity (original in Russian, English abstract, no details).

In 1983 **Landgraf** and **Ehrly** [535] carried out a cross-over study on the influence of an infusion of 500 ml LMWD (9% solution) compared to the corresponding quantity of hydroxyethyl starch with a molecular weight of 200 000 on the flow properties of blood in patients with intermittent claudication. After a 60-min infusion there was a statistically significant decrease in the whole blood viscosity which lasted throughout and beyond the entire 2-h test. The relative plasma viscosity remained high until the end of the testing period. The value of the erythrocyte aggregation was decreased significantly up to 1 h after the end of the infusion.

Heilmann [435] described the influence of hypervolemic hemodilution with 500 ml LMWD per hour in patients with pre-eclampsia and found a slight increase in plasma viscosity (capillary viscosimeter at 37 °C), as well as a decrease in erythrocyte aggregation (rheoaggregometer). An improvement in the clinical situation could be noted at the same time. These data were also presented elsewhere [433].

In 1983 **Witte** and co-workers [919] reported on the influence of a 10-day therapy with LMWD on the viscoelasticity of the whole blood and the viscosity of the plasma in patients with stroke. The rheologic studies were carried out with an oscillating capillary rheometer at a fixed hematocrit of 45%, at a shear rate of 10 s^{-1}, and a measurement temperature of 23 °C. The authors could not find any significant effects on the viscoelastic parameters (abstract, no further details).

Luther [599] studied the blood viscosity of patients with surgical diseases before and after the infusion of 500 ml 6% dextran 40 solution using a falling ball viscosimeter. The author found that the distinctly raised viscosity values could be decreased particularly well after the infusion, but that the viscosity reduction as a whole may best be described as minimal.

Note: No measurements of other rheologic parameters.

In 1984 **Krepp** and co-workers [528] described studies on the infusion of 500 ml 10% LMWD solution daily versus a 10% solution of mid-molecular weight hydroxyethyl starch solution (molecular weight 200 000) in patients with cerebrovascular disease. The rheologic tests were carried out on the 4th day after the beginning of hemodilution therapy and at the end of therapy. The plasma and the urine viscosity were determined with a Ubbelohde capillary viscosimeter. In this prospective randomized study there was an increase in the plasma viscosity in the dextran group while the plasma viscosity decreased following infusion with hydroxyethyl starch solution. Urine viscosity showed a statistically significant increase after the administration of dextran 40; in this case the urin viscosity did not rise as much with mid-molecular weight hydroxyethyl starch solution.

Note: Measurements of further hemorheologic parameters were not carried out.

Clinical findings: The authors reported a clinical improvement of the ischemic cerebral symptoms due to hypervolemic hemodilution.

In 1985 **Apsatarov** and co-workers [25] studied the influence of hemodilution with dextran in 110 patients with Leriche syndrome and found a decrease in the whole blood viscosity and an improvement in the tissue perfusion (original in Russian, abstract in English, no details).

Before a planned operation **Gottschalk** [84] infused children from 3 to 14 years of age with 6% dextran with a molecular weight of 40 000. The blood viscosity was measured at 38 °C with a Höppler rheoviscosimeter. There was a distinct decrease in the blood viscosity.

Note: No further rheologic tests.

In 1985 **Landgraf** and co-workers [539–541, 316] reported on a cross-over study with patients with intermittent claudition who had been infused hypervolemically (i.e., without previous venesection) with different plasma substitutes. The infusion lasted 30 min. Intraindividually 10% dextran 40, 6% hydroxyethyl starch (molecular weight 40 000), and 10% hydroxyethyl starch (molecular weight 200 000) were infused. Between the individual infusions there was a washout period of 8 days. An infusion of 500 ml physiologic saline solution served as a control. All infused solutions caused a reduction of the whole blood viscosity, as did the saline solution to a lesser extent. The plasma viscosity was only reduced with the saline solution and the low molecular weight hydroxyethyl starch solution. After the administration of hydroxyethyl starch with a molecular weight of 200 000, the plasma viscosity increased minimally, after the infusion of LMWD 40 distinctly. The erythrocyte aggregation was decreased immediately after the infusion of dextran 40 and hydroxyl starch 40; and more towards the end of the testing period with hydroxyethyl starch with a molecular weight of 200 000 and with saline solution.

In 1986 **Kiesewetter** and co-workers [501] studied the influence of LMWD solution compared to low-molecular weight hydroxyethyl starch (average molecular weight of each 40 000) in patients with chronic arterial occlusive disease during a 14-day hemodilution period in combination with physical therapy. The erythrocyte aggregation, the erythrocyte rigidity, and the yield stress were studied. With dextran there was a marked decrease in the rigidity of the erythrocytes and a minimal decrease in the erythrocyte aggregation, while plasma viscosity and yield stress remained unchanged by dextran.

Staedt and co-workers [812–814] treated 15 patients with acute cerebrovascular insufficiency within the framework of a controlled randomized cross-over study with 500 ml 10% hydroxyethyl starch (200/0.5) or dextran 40 for 6–8 h for a period of 2 days. The hemorheologic parameters were measured in EDTA-blood and -plasma at two different shear rates with a low shear rheometer. The erythrocyte aggregation was determined with an instrument by Myrenne and Co. There was a decrease in the whole blood viscosity and the plasma viscosity, but no influence on the erythrocyte aggregation.

Note: No measurement of the erythrocyte filtrability.

4.1.7.3 Discussion

From a modern point of view the therapeutic term for the so-called hemodilution is closely connected with the application of plasma substitutes, in particular of dextran solutions. If we disregard the so-called inner (endogenous) hemodilution, which occurs as a spontaneous regulatory mechanism, e.g., following blood loss or as a result of vasodilation (see Sect. 4.2.1.2), a hemodilution is in general obtained by infusing fluids, especially plasma or so-called colloidal plasma substitutes, into the vascular system. The first infusions of dextran solutions were made in the year 1961 as simple infusions into the circulating blood so that the blood volume was increased (hypervolemic hemodilution) and a viscosity-reducing effect was achieved.

Hypervolemic hemodilution, which is achieved by the infusion of plasma, plasma substitutes, or to a limited extent by electrolyte or glucose solutions, is the consequence of an infusion which is carried out without previous blood removal resulting in an increase in the amount of circulating blood.

In 1963 *Zederfeldt* from the *Gelin* team described the first isovolemic hemodilution. He removed 1000 ml venous blood from patients with burns, whose whole blood viscosity was much increased, and reinfused an equivalent volume of LMWD. This type of combined therapy of different rheologically active techniques will be discussed later (see Sect. 4.2.1). Despite different methods of measurement, all the authors reported a decrease in blood viscosity and in hematocrit values after an infusion of 9%–15% dextran solutions (mostly with a molecular weight of 40 000) in patients with various diseases; one group of authors did not find an influence on the viscoelastic parameters after dextran infusion [919]. The degree of reduction of the whole blood viscosity as well as of the hematocrit, i.e., the degree of hemodilution, of course depends on a series of factors which the authors treated differently, e.g., infusion quantity, infusion duration, concentration of the infusion solution, molecular weight, degree of hydration of the patient, retention time, etc. As regards the discussion on the most appropriate plasma substitute or the most appropriate solution for hemodilution, the given symptoms of the various diseases must also be taken into consideration and in particular the question of whether hypovolemia is present, as, for example, in shock, or whether normal amounts of blood are circulating in steady-state conditions. *Motta* and co-workers also found a decrease in the blood viscosity following the infusion of 500 ml LMWD. If, however, the hematocrit value of the blood samples was artificially set to a value of 45% following the infusion, there was an increase of the blood viscosity which probably can be attributed to the increased plasma viscosity.

All sorts of different viscosimeters were used to measure the blood viscosity. Most of the investigators who mentioned the temperature of measurement kept to 37 °C. If rotational viscosimeters were used, all authors found a decrease in the blood viscosity at the shear rates measured; this was most prominent at low shear rates. Contrary to the, in part, considerable changes in the blood viscosity after the infusion with LMWD, the changes in the plasma viscosity were relatively small. Thus *Groth* and co-workers found an increase in the plasma viscosity after the infusion with LMWD; on the other hand, when starting at high initial plasma viscosity values, a slight decrease was found. Just like *Groth, Yao,* and *Shoemaker* only found a decrease in the plasma viscosity when the plasma viscosity values were already primarily raised.

Schülke and *Hartel* found a slight increase in the plasma viscosity after LMWD and a strong increase after MMWD. They pointed out the special significance of plasma viscosity with decreasing vessel radius. In the capillaries there is an apparent decrease in the hematocrit, and the viscosity of the blood is primarily determined by the plasma viscosity [344]. Owing to the axial migration of the erythrocytes, there is a flow of thinned blood from the marginal flow into smaller vessels, while the high-erythrocyte axial flow stays mainly in larger vessels. Thus the intravasal hematocrit decreases in smaller vessels.

Bollinger and co-workers found a slight increase in the plasma viscosity after the infusion with LMWD. Similarly to *Schülke* and co-workers, *Lim* and co-workers found a slight increase in the plasma viscosity after MMWD. In 1974 *Warstat* reported on an increase in plasma viscosity after LMWD infusions. In various series of experiments the author differentiated the behavior of plasma viscosity with regard to a dependency on the erythrocyte sedimentation rate. The smallest increase in the plasma viscosity was found in patients with a highly increased erythrocyte sedimentation rate. Contrary to this, there was a higher increase in plasma viscosity in patients with just a slightly increased erythrocyte sedimentation rate.

Other authors, such as *Burke* and co-workers, *Humphreys* and co-workers, *Landgraf* and *Ehrly, Krepp* and co-workers, and *Haas* and co-workers, also reported on an increase in the plasma viscosity. *Kiesewetter* and co-workers found no significant changes in the plasma viscosity after dextran 40; *Staedt* and co-workers found a decrease in the plasma viscosity values. It may be seen from these studies that, while on the one hand the influence of the plasma viscosity on the whole blood viscosity level is of secondary importance, on the other hand the influence of an increased plasma viscosity per se, as described following infusions of LMWD, can be of much greater importance in the capillary vessels in the microcirculatory region (*Bayliss* 1960). Therefore a therapeutically induced increase in the plasma viscosity must be considered when using LMWD for hemodilution. The decrease in structural viscosity (blood viscosity at low shear rates), found by many authors, points toward a decrease in the erythrocyte aggregation.

Contradictory research results have been published on the question of whether the reduction of the erythrocyte aggregation is primarily caused by the decreasing hematocrit values and thereby by the decreasing concentration of plasma proteins which support aggregation, or whether it is in part an expression of a specific disaggregating effect of LMWD. In their consideration of the question of whether low-molecular weight dextran has a specific effect on the reduction of the erythrocyte aggregation, *Groth* and *Thorsén* are more reticent than *Gelin* and *Ingelmann*. *Groth* and *Thorsén* are of the opinion that the disaggregation observed with microscopy depends considerably on the flow conditions in the vessels. The reduction of the erythrocyte aggregation with LMWD, presumed to be due to the inhibition of the erythrocyte sedimentation rate (corrected for hematocrit), is attributed by *Groth*, on the one hand, partly to the dilution of the plasma and the aggregating proteins contained therein and, on the other hand, to a specific disaggregating effect of LMWD.

It is also assumed that the increased plasma viscosity due to LMWD inhibits the erythrocyte sedimentation rate and thereby feigns a specific disaggregating effect. *Ehrly* also reported on the reduction of the sludge phenomenon that can be observed by means of capillary microscopy. Since no erythrocyte disaggregation could be found on blow-out smears, the decrease in the sludge phenomenon was explained by the improved flow properties of blood, which are attributed solely to hemodilution.

Yao and *Shoemaker* reported on a significant decrease in the blood viscosity in patients in postoperative states of shock and after the infusion of LMWD (standardized hematocrit). The authors explain the reduction of the blood viscosity by a specific disaggregating effect of LMWD. Contrary to this, such changes in the blood viscosity could not be found in a healthy control group.

Bernstein and *Castaneda* postulated a specific disaggregating effect of LMWD owing to the increase in the electronegative charge of the erythrocytes. *Lim* and co-workers found a reduction of the blood viscosity after the infusion with LMWD in postoperative patients (corrected hematocrit). The authors saw the cause of this decrease in blood viscosity in a specific viscosity-reducing effect of LMWD. *Dormandy* found a linear correlation between the logarithm of the blood viscosity and the hematocrit. A specific viscosity-reducing effect of LMWD is therefore not assumed.

Amtoft postulated a possible disaggregating effect of LMWD that corresponds with *Gelin's* statements. *Dawidson* measured a reduction in blood viscosity highly concentrated erythrocyte suspensions after the infusion of LMWD. A specific disaggregating characteristic of LMWD was discussed.

Humphreys and co-workers did not find any reduction in blood viscosity with mathematically corrected hematocrit nor an increase in blood viscosity with LMWD at high shear rates. Like *Ehrly* in 1966 and *Dormandy* in 1971, the authors considered the mode of action of LMWD to be based on hemodilution. *Burke* and co-workers found a decrease in the relative blood viscosity (blood viscosity at standardized hematocrit divided by plasma viscosity) after the infusion of LMWD in postoperative patients. The authors were of the opinion that LMWD exerts a specific disaggregating effect on the erythrocyte aggregates in the blood.

Ehrly and *Landgraf* carried out a quantitative determination of erythrocyte aggregation with the help of a photometric technique after hypervolemic hemodilution with 500 ml 9% LMWD solution. Irrespective of the hematocrit, there was a decrease in erythrocyte aggregation for a short time, which was statistically significant up to 1 h after the end of the 30 min infusion.

Since the plasma viscosity had increased statistically significantly at the same time, the authors did not assume a specific disaggregating effect of LMWD on the erythrocyte aggregates, but rather explained the short-term inhibition of aggregation after the end of infusion as the result of a decreased collision rate of the erythrocytes caused by the increase in the viscosity of the suspension medium plasma. The objective results of the photometric determination of erythrocyte aggregation correspond with the viewpoints of other authors [540, 501, 812] and confirm the opinion of *Groth* [394] and *Ehrly* [230], according to which the disaggregating effect (anti-sludge effect) in vivo described by many authors, as well as the data measured with viscosimeters may be traced back to a dynamic disaggregation as a result of the improved flow rate of the blood after the reduction of the hematocrit values. This means that, owing to the reduction of the peripheral resistance with a decreased hematocrit, especially the flow velocity increases and a dynamic disaggregation of the erythrocyte aggregates occurs.

It should be noted that, according to the results of *Yao* and co-workers, *Lim* and co- workers, and *Burke* and co-workers with corrected hematocrits, a specific disaggregating effect of LMWD may be feigned by a dilution-dependent decrease in the plasma proteins promoting aggregation, e.g., fibrinogen. This is especially the case when the fibrinogen concentration is increased, as in patients just having undergone surgery.

Siekmann and co-workers and *Heilmann* also found a reduction of the photo-optically measured erythrocyte aggregation after a single infusion of 500 ml dextran 40, while *Kiesewetter* and co-workers and *Staedt* and co-workers found no significant changes in these values. In part it must, of course, also be taken into consideration that, especially in patient blood containing proteins with strong aggregation-supporting properties, the concentration of these protein units decreases owing to the dextran-induced hemodilution, and a disaggregating effect can therefore be feigned. The indirect measurement of the disaggregating effect of LMWD with a viscosimeter at low shear rates is much less meaningful compared to the direct photo-optic methods, since these viscosimeters abound with fundamental errors, especially at low shear rates (see Sect. 3.5.2). Accordingly, most authors consider the real source of a decrease in the erythrocyte aggregation to be the hematocrit reduction and a reduction of the plasma proteins promoting aggregation.

There are three research reports available concerning the possible changes in the erythrocyte filtrability (erythrocyte deformability). In 1974 *Warstat* did not find any changes in this parameter, or a minor increase in the perfusion of red cells through the Millipore filters. The author postulated that theoretically a deterioration of the filtration parameters had been expected owing to the increased plasma viscosity. The filtration measurements were made at 23 °C. The erythrocyte flexibility measurements of *Burke* in 1979 (polycarbonate filter, 3-μm pore diameter) showed no changes in this parameter of the flow properties of blood when the plasma viscosity remained unchanged. *Kiesewetter* and co-workers, on the other hand, found a marked decrease in the erythrocyte rigidity using a "single pore" rigidometer.

It may be deduced from these results that with infusions of LMWD the flow rate through the filter system and/or the erythrocyte rigidity is possibly improved, even though a deteriorated filtration rate owing to the increase in the plasma viscosity (at least with healthy subjects) could have been expected. *Warstat* discussed the idea that dextran caused a coating of the filter system and could thus have caused an improved passage of the erythrocytes owing to an improved slip. In addition, the opinion of *Chien* and co-workers [125] has to be discussed according to this the fluidity of the erythrocytes is better the smaller the quotient between the whole blood viscosity and the plasma viscosity becomes (in vitro studies). A final evaluation of the question of whether dextran really influences the erythrocyte filtrability or rigidity and whether these results are not superimposed by other rheologic parameters must remain open.

In a comparison between dextran 40 and hydroxyethyl starch solution (molecular weight 200 000), *Krepp* and co-workers found a higher increase in plasma viscosity after dextran infusion compared to the starch solution. This group, moreover, also measured the urine viscosity and also saw a distinct increase in this parameter after dextran compared to hydroxyethyl starch with a molecular weight of 200 000.

In summary, it may be said that hemodilution with dextran solutions effects a decrease in the blood viscosity. The decrease in the erythrocyte aggregation often described in earlier years is obviously not the result of a specific disaggregating effect of LMWD but mainly an expression of a dynamic disaggregation due to hemodilution. After infusion with LMWD, the plasma viscosity is changed depending on the initial values of the plasma viscosity and, as a rule, is increased. At the present time, no final statement can be made as to the influence of dextran on erythrocyte deformability or filtrability. The improvement of the flow properties of blood leads to an improved blood flow, especially if the microcirculatory flow is found to be impaired. In patients with pains at rest and with peripheral ulcerations in stage IV according to Fontaine, the better flow conditions can lead to a cessation of pains at rest and also to a complete healing of ischemic skin necrosis.

Gelin and co-workers were the first to show in detail that LMWD does not only achieve favorable clinical effects in patients with various circulatory states of shock, but also effectively improves the flow conditions of the blood in chronic occlusive arterial disease.

4.1.8 Penicillin

4.1.8.1 Introduction

Penicillin is the first antibiotic that found world-wide acclaim in the therapy of bacterial infections.

4.1.8.2 Publications

In 1960 **Ritzman** and co-workers [731] studied the influence of 10 million units of penicillin sodium, dissolved in 250 ml 5% glucose solution. The infusion lasted 1 h; the serum viscosity was measured before, immediately after, and 30 and 90 min after the end of the infusion. The authors found a marked decrease in the serum viscosity immediately after the end of infusion; after 90 min the values had readjusted to the initial level.

Note: There was no comparative study as to the influence of the carrier fluid on the serum viscosity.

In 1971 **Harkness** and **Whittington** [418] found that long-term penicillin therapy reduced the plasma viscosity in malaria (capillary viscosimeter).

Note: No further rheologic studies.

4.1.8.3 Discussion

According to these studies, it is improbable that there is a direct viscosity-reducing effect of penicillin.

4.1.9 Hemapheresis (Plasmapheresis/Plasma Exchange, Cytapheresis, Hemoperfusion)

4.1.9.1 Introduction

The word "plasmapheresis" is a combination of "plasma" and the Greek word aphairesis – remove. Strictly speaking, it means the removal of plasma from the circulating blood volume. Cytapheresis accordingly means the removal of corpuscular components from the blood without influencing the plasma. Plasmapheresis is, in the strictly literal sense, only possible in larger amounts if, following the removal of plasma, the patients are reinfused with foreign plasma or plasma substitutes, i.e., when plasma exchange is carried out. Therefore, the term "plasma exchange" is better suited to this type of therapy. Nevertheless, in the literature there are some authors who did not perform plasma exchange but rather pure plasmapheresis. It seemed reasonable not to separate these works from one another in this section. The selective removal of erythrocytes (erythrocyte apheresis) was discussed in Sect. 4.1.2 because of its clinical analogy to venesection. A definition of these terms and their differentiation can be found in ***Kennedy*** *and* ***Domen*** *[448], and* ***Nusbacher*** *[670].*

Plasma exchange, which is most commonly mentioned in literature, is strictly speaking just partial plasma exchange, since only a part of the plasma in the circulating blood is eliminated with different techniques and replaced by foreign plasma or plasma substitutes. In 1914 ***Abel*** *and co-workers [3] reported on a favourable therapeutic effect of plasmapheresis. In 1937* ***Albers*** *was the first to carry out serum viscosity measurements in patients with multiple myeloma, and the increased values that were found were given an important clinical status [9]. The increased viscosity of the plasma and of the serum, and occasionally of the whole blood is the result of an increased protein concentration, in which particularly the large molecular protein units (paraproteins) are increased. Theoretically, plasmapheresis or plasma exchange is pertinent, since with this method various rheologic parameters are improved without, ideally, changing the hematocrit value. It is therefore to be expected that not only the serum or plasma viscosity is clearly decreased, but also the blood viscosity and in particular the erythrocyte aggregation.*

Various technical procedures are used to carry out plasma exchange or plasmapheresis as well as cytapheresis (separation, filtration, etc.); the various procedures are not emphasized in this section.

The clinical importance of plasma exchange or plasmapheresis lies mainly in the treatment of paraproteinemias and cold agglutinin diseases, but also of the most peripheral ischemic disorders such as Raynaud's syndrome. In this connection, Waldenström's disease is particularly important, but so are other paraproteinemias. Patients complain of tiredness, decreasing functional capacity, weight loss, tendency to bleed, and neurologic symptoms. The removal of paraproteins from the circulating blood with subsequent reduction of the highly increased plasma or serum viscosity values often also leads to a favorable influence on the clinical symptoms, even though it must be mentioned that this therapy is not without side effects.

The number of publications on the hemorheologic changes after plasmapheresis or plasma exchange has clearly increased in recent years. In 1983 a short monograph [872] appeared, and at the Third European Congress on Clinical Hemorheology a symposium was held on this topic.

4.1.9.2 Publications

The first viscosimetric studies in connection with plasmapheresis or plasma exchange as a therapeutic measure in various diseases were carried out by **Schwab** and **Fahey** [780, 781] in 1960. Two patients with Waldenström's disease were examined. An Ostwald capillary viscosimeter was used to measure serum viscosity (H_2O = 1) at 37 °C. The normal range of serum viscosity of a healthy control group was between 1.4 and 1.8. With the removal of 1000 ml blood and the reinfusion of the erythrocytes suspended in 25–30 ml isotonic saline solution, there was a decrease in serum viscosity that was synchronous with a distinct reduction in the total protein concentration and accordingly also in the macroglobulins. In another case there was also a clear reduction of the serum protein level as well as of the serum viscosity after plasmapheresis and additional chlorambucil therapy.

In 1963 **Fahey** [342] published a further report on the effect of plasmapheresis on serum viscosity in a patient with macroglobulinemia. The relative serum viscosity (H_2O = 1) was measured with an Ostwald capillary viscosimeter at 37 °C. Over a period of 1 year up to 1000 ml plasma was removed 12 times. A decrease in the total protein level and the macroglobulin level correlated with a decrease in the serum viscosity.

Similar reports on the effect of plasmapheresis on the serum viscosity in ten patients with macroglobulinemia were presented by **Cline, Solomon,** and **Fahey** [135, 780, 781] in 1963. The serum viscosity was also measured with a capillary viscosimeter.

Note: No measurements of further rheologic parameters.

Clinical findings: Within the framework of plasmapheresis treatment carried out over a number of months, the patients in **Fahey's** group showed a cessation of bleedings of the mucosa and an improvement in the retinopathies due to hemorrhage. Existing heart trouble, neurologic symptoms such as lethargy and nausea, as well as a developed hepatomegaly in one patient were mostly relieved or clearly improved and they were interpreted as the result of the decrease in the plasma viscosity.

In 1965 **Fahey** and co-workers [343] again reported on a plasmapheresis therapy in ten patients with macroglobulinemia. As with **Schwab** and **Fahey** [780], blood was sampled, the plasma rejected, and the erythrocytes, suspended in physiologic saline solution, were reinfused. The relative serum viscosity (H_2O = 1) was evaluated with an Ostwald capillary viscosimeter at 37 °C. During intensive plasmapheresis (removal of 2–6 l plasma within 1–2 weeks), there was a decrease in the serum viscosity simultaneously with a drop in macroglobulin concentration.

Note: No measurements of other rheologic parameters.

Clinical findings: The authors observed an improvement in the clinical symptoms after plasmapheresis. The neurologic symptoms, which expressed themselves in a loss of hearing and also in EEG changes, disappeared. In three patients with Raynaud's syndrome, there was an increased tolerance to cold and in one patient a complete healing of chronic ulcers of the foot. Heart trouble was influenced favorably.

Further viscosimetric reports on plasmapheresis in two patients with multiple myeloma were published by **Smith** and coworkers [803] in 1965. In this case 1 liter blood was sampled, the plasma rejected and the erythrocytes, suspended in isotonic saline solution, were reinfused. The relative serum viscosity (0.9% NaCl = 1) was measured with an Ostwald capillary viscosimeter of 37 °C. The normal value in healthy subjects was around 1.5. Within the framework of intensive plasmapheresis (removal of approximately 5 l plasma within 6 days), a strong decrease was observed in serum viscosity and in plasma protein concentration. The treatment lasted in part over a period of 1 year.

Note: No measurements of further rheologic parameters.

Clinical findings: The authors found a clinical improvement in the retinopathies, in the tendency to bleed, as well as in the neurologic symptoms.

In 1965 **Hoffmann** and **Ritzmann** [450] treated ten patients with Waldenström's disease with plasmaphereses which were performed in part a number of years. Within 24 days, 4200 ml plasma was sampled from each patient. A decrease in the serum viscosity was noted at the same time as a decrease in the total protein and the macroglobulin concentration.

Note: No data on the measurement technique; no measurements of further rheologic parameters.

Clinical findings: An improvement was observed in the clinical symptoms (no exact data).

In 1967 **Kopp** and co-workers [524] examined a patient with multiple myeloma. The serum viscosity was measured with an Ostwald capillary viscosimeter (H_2O = 1) at 37 °C. Within the framework of a 10-day plasmapheresis (removal of 4500 ml plasma), there was a decrease in the serum viscosity and in the plasma protein level.
Note: No measurements of further rheologic parameters.

Clinical findings: The investigators found an improvement in nasal bleeding and in vision, as well as a decreased sludge formation in the conjunctival vessels.

In 1968 **MacKenzie** and **Brown** [602, 603] studied six patients with Waldenström's disease and increased viscosity values. The serum viscosity was determined with an Ostwald capillary viscosimeter. Following plasmapheresis there was a drop in the relative serum viscosity, and the previously increased plasma volume of the patients was also statistically significantly decreased.

Note: The temperature of the viscosity measurements was not mentioned; no measurements of further rheologic parameters.

In 1969 **Ritzmann** and co-workers [733] reported on a reduction of the serum viscosity as a result of plasmapheresis in patients with macroglobulinemia. The serum viscosity measurements were carried out with an Ostwald capillary viscosimeter at 37 °C.

Note: No measurements of further rheologic parameters.

Clinical findings: There was an improvement in the tendency to bleed as well as in the neurologic symptoms, which expressed themselves in deafness, visual disturbances, and paralysis of the hands and feet.

In 1970 **Abruzzo** and co-workers [4] treated a patient with polyarthritis, polymyositis, polysynovitis, hyperviscosity, and a highly increased 13 S IgG proportion which made up 40% of the total protein. The relative serum viscosity (H_2O = 1) was measured with an Ostwald capillary viscosimeter at 25 °C. After a 3-day plasmapheresis (removal of 2000 ml plasma) and prednisone therapy, there was a reduction of the serum viscosity, of the gamma globulins, and of the 13 S proportion.

Note: Viscosity measurements were performed below body temperature; no measurements of further rheologic parameters.

Clinical findings: The severe myositis and synovitis symptoms disappeared. Moreover, there was an improvement in the neurologic deficiency symptoms.

Fritz and co-workers [357] treated ten patients with paraproteinemia (nine with plasmyocytoma, one with Waldenström's disease) by means of plasmapheresis in 1970. The plasma viscosity was determined using the Viskowaage manufactured by Haake Brs. (balance viscosimeter) at 20 °C. The patients were divided into two groups, those with particularly severe clinical symptoms showing an increased plasma viscosity. Intensive plasmapheresis leads to a normalization of the plasma viscosity values together with a drop in the paraprotein concentration.

Note: No measurements of further rheologic parameters; the viscosity measurements were performed at 20 °C.

Clinical findings: The bleedings, the retinopathies as well as the cardiac insufficiencies and the neurologic symptoms disappeared.

In 1971 **Benninger** and co-workers [65] tested the therapeutic effect of plasmapheresis in a patient (80 years old) with multiple myeloma and hyperviscosity. The relative serum viscosity (H_2O = 1) was measured with an Ostwald capillary viscosimeter at 22 °C. The normal range of serum viscosity was between 1.4 and 1.9. Within the framework of the plasmapheresis therapy, the serum viscosity decreased from 11 to 5 as did the the macroglobulin level.

Note: No measurements of further rheologic parameters; the blood viscosity was measured at a temperature of 22 °C.

Clinical findings: A clinical improvement was not observed.

In 1971 **Powles** and co-workers [709] reported on the rheologic effects of plasma exchange (NCI-IBM Cell Separator) in 11 patients with paraproteinemia (multiple myeloma and macroglobulinemia) and partially established cryoglobulins. The plasma of healthy subjects was used as a plasma substitute. The viscosimeter, with which measurements were made under constant pressure at 37 °C, had a capillary with a diameter of 1 mm. Following the exchange of 7 l plasma there was a normalization of the plasma viscosity in one patient. A decrease in the macroglobulin concentration was observed in only six patients.

Note: No measurements of further rheologic parameters.

Clinical findings: Owing to a decrease in the retinal exudates and in the hemorrhages, there was an improvement in the retinopathies. Furthermore, the neurologic symptoms, nasopharyngeal bleedings, and comas ceased.

Pruzanski and **Watt** [715] treated five patients with multiple myeloma and increased serum and plasma viscosity with plasmapheresis and additional cytostatic medication. The serum viscosity was measured with a Roger falling ball viscosimeter at 24 °C (H_2O = 1) as well as with a Wells-Brookfield viscosimeter at 37 °C. A decrease in the serum viscosity was observed, together with a decrease in the total protein and the IgM macroglobulin concentration.

Clinical findings: In four of the five cases favorable clinical results such as cessation of the mucosal bleedings, improvement in the retinopathies and in the neurologic symptoms were achieved.

Wolf and co-workers [921] treated patients with multiple myeloma and hyperviscosity syndrome with plasmapheresis by removing up to 4500 ml plasma. A decrease in the serum viscosity was found simultaneously with a reduction of the abnormal plasma protein concentration.

Note: No measurements of further rheologic parameters; no data on the viscosimeter used nor on the temperature of measurement.

Clinical findings: After the plasmapheresis the bleedings ceased and there was a decrease in the neurologic symptoms. In addition, there was an improvement in the retinopathic symptoms, the cardiac insufficiencies, and kidney function.

In 1973 **Reich** and co-workers [719] studied the influence of plasma exchange (IBM Cell Separator) on the rheologic properties of serum in eight patients with Waldenström's disease (two patients with cryoglobulins). After the removal of 2500–3000 ml plasma, the abnormal protein level and the serum viscosity were reduced.

Note: The type and means of viscosimetric measurement were not mentioned; measurements of further rheologic parameters were not carried out.

Clinical findings: The loss of sight and the unconsciousness that had occurred in some of the patients was cured by the treatment.

In 1973 **Lindsley** and co-workers [579] reported on a patient with multiple myeloma. The blood viscosity was measured with a Brookfield micro-cone-plate viscosimeter at 37 °C and shear rates between 11.5 and 230 s^{-1}. After repeated plasmapheresis there was a simultaneous decrease in the macroglobulins and a decrease in the blood viscosity.

Note: A correction for the hematocrit was not performed; no measurements of further rheologic parameters.

Clinical findings: The patient's clinical picture improved and the retinal bleeding ceased.

In 1974 **Tuddenham** and co-workers [865] treated four patients with multiple myeloma and a hyperviscosity of the blood by means of plasmapheresis by removing 2–3.7 l plasma. In three patients the serum viscosity was determined with an Ostwald capillary viscosimeter or a Coulter Electronics capillary viscosimeter (H_2O = 1) and in one patient the blood viscosity was determined using a Wells-Brookfield viscosimeter with a defined shear rate of 230 s^{-1}. The temperature of the viscosity measurements was 37 °C. In the course of the plasmapheresis, a decrease was found in the total protein level in the plasma as well as a decrease in the serum and blood viscosity.

Note: No measurements of further rheologic parameters.

Clinical findings: After plasmapheresis the authors found an improvement in the retinopathies and in the tendency to bleed.

In 1976, within the framework of a plasma exchange with a cell separator (celltrifuge), **Grindle** and co-workers [392] carried out viscosimetric studies in 12 patients with Waldenström's disease. In four of these patients the retinopathies were interpreted as an expression of a so-called hyperviscosity syndrome. During the plasma exchange approximately 5.5 l of the patient's own plasma was removed and substituted with 1000 ml isotonic saline solution, 4000 ml plasma protein fraction, and 150 ml dextran solution. The serum viscosity was measured with an Ostwald capillary viscosimeter at 37 °C ($H_2O = 1$). The blood and plasma viscosity measurements were made with a Wells-Brookfield viscosimeter at a shear rate of 230 s^{-1} and at a temperature of 37 °C. At the end of treatment a reduction was found in the serum, plasma, and blood viscosity, which was in agreement with the drop in the total protein and macroglobulin levels.

Note: Viscosimetric measurements at low shear rates were not performed; no measurements of further rheologic parameters.

Clinical findings: A normalization of the retinal blood flow was observed by means of fluorescence angiography. Without exception there was an improvement in the retinopathies owing to the elimination of retinal hemorrhages, the reduction of papillary edema, and the decrease in venous dilatation in the ocular fundus.

Buskard and co-workers [104] treated two patients with Waldenström's disease, Raynaud symptoms, and increased blood viscosity. The patients' complaints could not be improved with standard chemotherapy. Plasma exchange was performed with a blood cell separator (celltrifuge). In this case approximately 5–6 l plasma was replaced by 1500 ml isotonic saline solution, 3800 ml plasma protein fraction, and 500 ml dextran. This treatment was carried out for up to 3 years. The serum and plasma viscosities were measured using an Ostwald capillary viscosimeter at 37 °C and the blood viscosity with a Wells-Brookfield viscosimeter at shear rates of between 1.15 and 230 s^{-1}. A reduction was found in the blood, plasma, and serum viscosity analogous to the decrease in the paraproteins in the plasma.

Note: Measurements of erythrocyte aggregation and the erythrocyte filtrability (erythrocyte deformability) were not carried out.

Clinical findings: The authors found an improvement in the retinopathic symptoms. The finger pains ceased, and the clinical symptoms of cardiac insufficiency disappeared in one patient.

In 1972 **Milan** and co-workers [646] treated 22 patients with multiple myeloma and increased serum viscosity with plasmapheresis. The serum viscosity was measured with a Wells-Brookfield viscosimeter at 37 °C and at shear rates of 23–230 s^{-1}. A decrease in the serum viscosity was found synchronous to a decrease in the gamma-globulin and the serum protein level.

Note: No further measurements of rheologic parameters.

Clinical findings: The authors found an improvement in the retinopathic symptoms and a cessation of the hemorrhages which were often present.

In 1978 **Preston** and co-workers [711] published a report on the rheologic effect of plasma exchange performed with an Aminco cell separator in 11 patients with multiple myeloma and severely increased blood viscosity. The blood viscosity was determined with a mathematically corrected hematocrit of 45% and shear rates of 23–230 s^{-1}. The measurement temperature was 37 °C. After plasma exchange, a decrease was found in the IgA and IgG_3 macroglobulins synchronous to the decrease in blood viscosity.

Note: No measurements of further rheologic parameters.

Clinical findings: The authors observed a dramatic improvement in the general condition of patients in a coma. The gastrointestinal bleeding and the epistaxis ceased. Occlusions of the retinal veins due to bleeding and the retinopathy caused by papillary edema disappeared. The neurologic symptoms, which included lethargy, nausea, deafness, diplopia, ataxia, etc., could be clearly improved by means of plasma exchange.

In 1978 **Talpos** and co-workers [853] reported on the therapeutic effect of plasma exchange in five patients with Raynaud's disease which had previously been unsuccessfully treated by sympathectomy. Plasma exchange was carried out five times in weekly intervals, whereby the plasma removed using a cell separator (Aminco celltrifuge) was replaced by 2–2.5 liters plasma protein fraction or by normal plasma. The blood viscosity in three patients was particularly reduced at low shear rates.

Note: Details about the temperature of the measurements and the type of viscosimeter used were not given; no measurements of further rheologic parameters.

Clinical findings: Finger ulcers healed in three out of four patients. There was a reduction in the frequency and the intensity of the pains.

In 1979 **Dodds** and co-workers [191] treated eight patients with Raynaud's syndrome by means of a weekly plasma exchange. The 1.9 l plasma removed with a cell separator (Haemonetics 30 blood processor) was substituted with 1.2 l human plasma protein fraction, 0.5 l plasma substitute and 0.5–1 l isotonic saline solution. The blood viscosity was determined with a Contraves Low Shear cylinder viscosimeter with a mathematically corrected hematocrit of 45% at two shear rates (0.77 and 91 s^{-1}). The measurement temperature was 37 °C. The measurements of erythrocyte filtration (erythrocyte deformability) were carried out at 21°C using a polycarbonate filter system with a pore size of 5 μm. Compared to a control group of healthy subjects, the patients showed a decreased erythrocyte deformability. Following plasma exchange, there was a decrease in blood viscosity which was most distinct at low shear rates.

After each plasma exchange the plasma viscosity, the fibrinogen level, and the hematocrit decreased. The erythrocyte deformability, determined via the erythrocyte filtration rate, was improved as well. Six weeks after the end of therapy, a statistically significant reduction in blood viscosity could still be found at a shear rate of 0.77 s^{-1}.

Similar results with a larger number of patients with Raynaud's disease were also published by **O'Reilly** and co-workers in 1979 [671].

Note: The hematocrit was corrected mathematically. Erythrocyte aggregation measurements were not performed. The erythrocyte deformability measurements were carried out below body temperature; the effect of the high plasma viscosity on the erythrocyte filtration probably influenced the results considerably.

Clinical findings: The finger ulcers of three patients healed completely. In addition the authors observed a decrease in the strength and intensity of the periodic attacks of pain.

In 1979 **Kilpatrick** and co-workers [505] reported on plasma exchange performed in seven patients with hyperlipoproteinemia who were suffering from coronary and peripheral vascular diseases. The plasma, which was removed using an Aminco celltrifuge, was replaced with 7.2–4.0 l plasma protein fraction. The blood and plasma viscosity measurements were performed at 37 °C using a Contraves Low Shear 100 viscosimeter which was connected to a Rheomat T 15 (15 shear rates of between 0.77 and 4.6 s^{-1}) for the measurement of low shear rates. In order to avoid surface artifacts in the viscosity measurements, an air-interface supplement was used. At the end of the treatment period a particularly strong reduction of the blood viscosity was found at low shear rates. At the same time as the blood viscosity reduction, a strong decrease in the fibrinogen level, plasma viscosity, cholesterol, and triglyceride concentration were observed, as well as a decrease of 5% in the hematocrit.

Note: A correction of the hematocrit was not made; measurements of erythrocyte aggregation and erythrocyte flexibility were not carried out.

Clinical findings: In the two patients with peripheral arterial vascular disease, the improvement in the flow properties of blood was synchronous with the improvement in the intermittent claudication symptoms.

In 1980 **Yamagata** and co-workers [926] reported on the influence of plasma exchange on the flow properties of blood in a group of patients with various rheumatic symptoms (systemic lupus erythematosus, rheumatoid arthritis, progressive systemic sclerosis, mixed connective tissue disease (MCTD)). 500 ml blood was sampled from the patients, the pathologically changed plasma was centrifuged off, and the erythrocytes gained in this manner were reinfused together with a plasma solution from healthy donors. The serum viscosity was measured with an Ostwald capillary viscosimeter. The authors found a decrease in the serum viscosity and, at the same time, a reduction of the total protein level as well as of the IgG, IgA, and IgM immunoglobulins.

Note: The temperature of the serum viscosity measurements was not given; no measurements of further rheologic parameters.

Clinical findings: The patients reported that the fever attacks disappeared, as did the joint pains, possibly associated with immobility, the exanthema, and the Raynaud's symptoms.

In 1980 **Wrabetz-Wölke** and co-workers [925] reported on a decrease in plasma viscosity following repeated plasmaphereses in patients with Waldenström's disease and multiple myeloma.

Note: No measurements of further rheologic parameters, no details of the type of viscosimeter used.

Clinical findings: At the same time as the decrease in plasma viscosity, the authors found an improvement in the clinical symptoms.

In 1980 **Schoonemann** and co-workers [771] treated insulin-dependent diabetics with increased plasma viscosity values as well as patients with Waldenström's disease and plasmacytoma. Plasma viscosity was measured using a capillary viscosimeter at 37 °C and blood viscosity was evaluated using a coaxial cylinder viscosimeter (Contraves LS 30) at a constant hematocrit and at five shear rates (0.204, 0.512, 0.95, 20.4, and 128.5 s^{-1}). After each plasma exchange (blood cell separator), the authors found a drop in the blood viscosity at low shear rates as well as a decrease in plasma viscosity.

Note: No measurements of further rheologic parameters; the quantity of exchanged plasma was not given.

In 1980 **Iwamoto** and co-workers [469] performed a plasma exchange with a Plasmaflo membrane separator in two patients with multiple myeloma after chemotherapy with prednisolone, melphalan, and cyclophosphamide had not improved the clinical picture. At the same time as the drop in the total proteins and the IgA immunoglobulins there was a decrease in the plasma viscosity.

Note: No data on the type of viscosimeter, nor on the temperature of the measurements; no measurements of further rheologic parameters.

Clinical findings: The patients reported that the mucosal bleeding ceased and that they no longer felt so exhausted.

In 1981 **Dodds** [192] reported on the hemorheologic effects of a single plasma exchange in ten patients with vascular disease. The 2 l plasma removed with a cell separator (Haemonetics 30 blood processor) was exchanged with an infusion of plasma substitute and plasma protein fraction. The blood and plasma viscosity measurements were carried out using a rotational viscosimeter at shear rates of 0.77 and 91 s^{-1}. After plasmapheresis a particularly high reduction of the plasma viscosity was found at the low shear rate. In addition, a synchronous reduction could be observed in plasma viscosity and the fibrinogen level. There was no change in the erythrocyte filtrability (erythrocyte deformability).

Note: Details of the filtration method, the type of viscosimeter, and the temperature of the measurements were not given. A correction of the hematocrit was not carried out.

In 1981 **Allan** [11] treated patients with Waldenström's disease by means of plasma exchange. The 2.5–4 l plasma removed using a cell separator was substituted by a plasma protein fraction. A Brookfield micro-cone-plate viscosimeter was used for the measurement of blood and plasma viscosities at 37 °C. The hematocrit was mathematically corrected to a standard value of 30%. Within the framework of treatment, there was a simultaneous reduction of blood and plasma viscosities corresponding to the reduction in the IgM paraproteins.

Note: No details on the shear rates used.

Clinical findings: The author reported on a favorable improvement in the retinopathies, cardiac insufficiencies, neurologic symptoms, and in dyspnea.

In 1981 **Stoltz** and co-workers [825] reported on a study with patients having various diseases and, at the same time, increased blood viscosity. Various amounts of plasma were either sampled or exchanged. In this case, there was also a rheologic improvement in the control group of patients who did not suffer from hyperviscosity (abstract, no details).

In 1981 **Fazzini** and co-workers [347] studied the effect of plasmapheresis or plasma exchange in 11 patients with multiple myeloma or cryoglobulinemia. Saline solution or albumin solution was used as exchange fluid. Measurements of the blood and plasma viscosities were carried out with a Brookfield LVT cone-plate viscosimeter at shear rates of 23–230 s^{-1}. Plasma viscosity was measured at 230 s^{-1}. Blood filtrability was measured using a modification of the whole blood method according to Reid and co-workers. In most cases there was a decrease in blood and plasma viscosities. In two patients the increase in plasma viscosity was not very distinct. The erythrocyte filtration behavior showed an improvement in flow in most cases, but no parallels were found with plasma viscosity.

Note: No measurements of the erythrocyte aggregation.

Clinical findings: There was also an improvement in the clinical symptoms in most cases, especially concerning renal complications.
The same authors also reported on these results elsewhere [346] (abstract, no details).

In 1981 **Brown** and co-workers [88] studied blood and plasma viscosities in patients with polycythemia before and after plasma exchange. Cerebral blood flow was measured at the same time. The authors found a significant reduction of blood and plasma viscosities at unchanged hemoglobulin values. Nonetheless, cerebral blood flow remained unchanged. The authors concluded that changes in blood viscosity only have a slight effect on blood circulation in the brain (abstract, no details).

In 1981 **Buskard** [105] treated 12 patients with Waldenström's macroglobulinemia using a cell separator and substituted the plasma removed by plasma protein solution and saline solution. The relative blood viscosity of plasma was measured before and after each plasma exchange using an Ostwald capillary viscosimeter at 37 °C. The blood viscosity of heparinized blood was measured with a Wells-Brookfield microviscosimeter at shear rates of 1.5–230 s^{-1}. There was a marked reduction of the blood and plasma viscosities.

Note: No measurement of further hemorheologic parameters.

In 1981 **Somer** and **Ditzel** [810] studied the blood and plasma viscosities in a patient with paraproteinemia using a Wells-Brookfield microviscosimeter before and after intensive plasmapheresis. There was a clear reduction of the blood and plasma viscosities. In addition, the authors studied the oxygen absorption curve and the oxygen binding curve and found no change whatsoever in these parameters. Thus they deduced that the microcirculatory disturbances are mainly the result of the highly increased plasma viscosity, the strong aggregation of the erythrocytes, and existing hypervolemia.

Also in 1981 **Strauer** and co-workers [843] reported on three patients with coronary insufficiency caused by hyperlipidemia and multiple myeloma. The blood and plasma viscosities were determined using a modified Wells-Brookfield viscosimeter and erythrocyte aggregation using photometric measurement (light transmission). The authors found a reduction of the blood and plasma viscosities, as well as a reduction of the erythrocyte aggregation as a result of plasmapheresis.

Clinical findings: The patients' anginal complaints decreased.

In 1982 **Terekhov** and co-workers [855] studied the flow properties of blood before and after hemoperfusion (method for the adsorption of, for exaample, toxins on carbon or synthetic resin whereby plasma protein is lost) in 23 patients with intoxication. They found an improvement in the flow properties of blood (original in Russian, English summary, no details).

In 1982 **Goldberg** and co-workers [381] described the effect of a partial exchange of blood versus plasma in infants with highly raised blood viscosity. The viscosity was measured at shear rates of 11 and 106 s^{-1}; there was an improvement in the blood viscosity values.

Note: No rheologic parameters other than whole blood viscosity were measured.

Clinical findings: A definite improvement of the clinical symptoms was described, in particular of the neurologic abnormalities.

Beck and co-workers [54] also studied the serum viscosity in three patients with paraproteinemia before and after plasma exchange (plasma albumin and saline solution) using an Ostwald capillary viscosimeter. There was a therapy-induced reduction of the viscosity values. The authors found the measurement of the serum viscosity to be a good measure for the control of therapy in such patients.

In 1982 **Baron** [45] and co-workers carried out plasmapheresis in patients with rheumatic hyperviscosity and clinical symptoms. Serum viscosity was measured with an Ostwald capillary viscosimeter at 25 °C. After therapy there was a distinct decrease in the value.

Note: No measurement of further hemorheologic parameters.

In 1982 **Brown** and **Marshall** [89] studied the effect of plasma exchange on blood viscosity and cerebral blood flow in eight patients with various diseases. The endogenous plasma, obtained with an instrument from Haemonetics & Co., was replaced by a 5% human albumin solution and physiologic saline solution. Whole blood viscosity and plasma viscosity were measured with a Contraves Low Shear viscosimeter at 37 °C. In addition, the hematocrit and the plasma protein concentration were determined. There was a statistically significant decrease in whole blood viscosity and plasma viscosity. While the hematocrit did not change statistically significantly, there was, as expected, a clear reduction of the concentration of the plasma proteins.

Note: Other rheologic parameters were not measured. The measurements of the cerebral blood flow performed in the same patients showed no significant change. This was traced back to the compensatory function of cerebral autoregulation. The authors pointed out that, according to their results, the reduction of blood viscosity does not necessarily lead to an improvement in the circulation.

In 1982 **Rosenkranz** and **Oh** [740] reported on the behavior of blood viscosity (Brookfield viscosimeter) within the framework of a partial plasma exchange in children with polycythemia. There was a distinct reduction of the viscosity values.

In 1983 **Strauer** [844] studied individual patients with multiple myeloma or paraproteinemia in whom blood viscosity was reduced and whose coronary venous oxygen saturation was increased following plasmapheresis.

Note: No exact data on the measurement techniques.

Clinical findings: After plasmapheresis and viscosity reduction, there was a major improvement in the clinical symptoms of the patients.

In a short monography **Van den Berg** and co-workers [872] reported in 1983 on the therapy of the so-called hyperviscosity syndrome in patients with paraproteinemia following plasmapheresis using a cell separator. Venous blood was sampled and anticoagulated with EDTA. The viscosity measurements were carried out with a Contraves Low Shear 100 rotational viscosimeter at 37 °C and at low shear rates. In eight patients there was a reduction of the whole blood and plasma viscosities.

Note: Further rheologic parameters were not studied. The authors compared the reduction of the blood viscosity as a result of plasmapheresis with the blood flow at rest in the acral region of the extremities as measured by plethysmography. In some patients there was an improvement in the blood flow at rest.

Clinical findings: In all of the eight plasmapheresis patients studied there was an improvement in the clinical symptoms.

In 1983 in the course of a plasma exchange **Gueguen** and co-workers [399] studied the rheologic parameters of blood in 66 patients with various monoclonal dysglobulinemias. Blood viscosity was measured using a coaxial viscosimeter (Low Shear 30) and the filtrability of the blood was determined according to the method of Reid and co-workers. The blood and plasma viscosities were reduced after plasma exchange and after 48–72 h they increased again. The filtrability of the blood improved during the plasma exchange therapy (abstract, no details).

Schooneman and co-workers [772] found that a major plasma exchange using a cell separator led to a reduction of the blood and plasma viscosities in various paraproteinemias. In the 25 patients studied, the aggregation tendency of the erythrocytes was reduced when 4% human albumin solution was used as an exchange solution (abstract, no details).

In 1983 **Weber** and co-workers [907, 908] carried out plasma filtration in 12 patients with Raynaud's phenomenon and measured erythrocyte aggregation and plasma viscosity. An improvement of these rheologic parameters could be determined if these were pathologically increased before therapy. In the majority of cases there was a clinical improvement (abstract, no details).

In 1983 **Walker** and co-workers [892] studied the flow properties of the blood in ten patients with intermittent claudication following a plasmapheresis substituting 1500 ml plasma substitute and 500 ml isotonic saline solution. Blood viscosity at various shear rates, plasma viscosity, and erythrocyte filtrability were studied. After treatment there was a decrease in plasma viscosity and blood viscosity at all shear rates measured. The functional tests that were carried out at the same time corresponded to the viscosimetrically measured improvement in the flow properties of blood. It was hypothesized that the reduction of plasma viscosity as a result of plasmapheresis could be of clinical value in patients with initially increased plasma viscosity.

Glöckner and co-workers [378, 379] treated one patient with Waldenström's disease using a so-called cascade filtration technique in which only the pathologically high molecular weight proteins were removed and the other protein fractions were reinfused. The viscosity of the serum was clearly reduced after the filtration therapy (abstract, no details).

In 1983 **Gülzow** and co-workers [401] found an increase in plasma viscosity following plasmapheresis in 20 patients with paraproteinemia. The measurements were performed with a capillary viscosmeter at 37 °C (abstract, no details).

In 1983 **Ryabov** and co-workers [746] reported on the flow properties of blood in patients with peritonitis and pancreatitis before and after a hemoperfusion and found that there was no change in plasma viscosity but rather a reduction of the erythrocyte aggregation tendency (English summary, no details).

In 1983 **Derksen** and co-workers [167] carried out plasma exchange in five patients with cryoglobulinemia and studied the serum viscosity before and after therapy. There was a decrease in these values which were measured at 37 °C using an Ostwald capillary viscosimeter and compared to the viscosity of distilled water ($H_2O = 1$).

Note: No measurement of further rheologic parameters.

In 1983 **Di Lollo** and co-workers [171] carried out plasmapheresis and plasma exchange in patients with plasma hyperviscosity of different origin and determined the blood and plasma viscosities (no exact data on the method). There was a reduction of the blood viscosity as well as of the plasma viscosity.

Note: No measurement of further rheologic parameters.

In 1983 **Shirinova** and co-workers [796] carried out hemosorption in patients with severe burns and sepsis and found an improvement in the flow properties of blood (original in Russian).

In 1984 **Walker** and **Dormandy** [89]) treated six patients with intermittent claudication by means of plasma exchange using a cell separator. A total of 2000 ml plasma was sampled and substituted by 1500 ml gelatin preparation and, in addition, by 500 ml physiologic saline solution. The authors found a significant decrease in the whole blood viscosity (measured at three different shear rates using a Contraves LS 3O viscosimeter) and in the plasma viscosity (measured using a capillary viscosimeter).

Note: No improvement in the pain-free walking distance; in addition, there was also no change in the reactive hyperemia (measured plethysmographically) of the lower extremities.

In 1984 **Mehta** and co-workers [632] reported a clear improvement in the whole blood viscosity in three patients with chronic granulocyte leukemia after intensive leukocyte apheresis. The patients were suffering from major visual disturbances with very high leukocyte counts in the blood. The partial removal of leukocytes with the help of a blood cell separator led to a reduction of the whole blood viscosity as measured using a Contraves LS 100 rotational viscosimeter at a temperature of 37 °C and shear rates of 22 s^{-1} and 0.775 s^{-1}. The clinical symptoms of the three patients studied improved in a time-dependent manner with the reduction in viscosity.

Note: No studies of further rheologic parameters.

In 1985 **Crawford** and co-workers [151] treated patients with immunoglobulin diseases and hyperviscosity by means of plasmapheresis and studied plasma viscosity with a Couette-Harkness viscosimeter at 25 °C. They found a marked decrease in the values as a result of this therapy, as well as a simultaneous improvement in the clinical situation.

Note: No measurement of other rheologic parameters.

In 1985 **Stoltz** and co-workers [831] reported on the behavior of hemorheologic parameters before and after plasma exchange in 50 patients with monoclonal dysglobulinemia. Whole blood viscosity as well as plasma viscosity decreased, and erythrocyte aggregation (rheoscope) normalized almost completely (abstract, no details).

In 1985 **Schoonemann** and co-workers [773, 774] reported on hemorheologic changes as a result of plasma exchange in 50 patients with monoclonal dysglobulinemia. Plasma exchange was carried out using a cell separator. Plasma viscosity was measured at 37 °C using a capillary viscosimeter and whole blood viscosity was measured using a Contraves LS 30 viscosimeter after hematocrit correction to 40%. Erythrocyte filtraton was carried out through Nucleopore filters of 5-μm pore diameter. Erythrocyte aggregation was measured using an aggregometer (light transmission). The authors found an improvement in all of the hemorheologic parameters studied.

In 1985 **Litmanovich** [580] reported on the combination treatment of 28 patients with peripheral arterial occlusive disease. He used plasmapheresis in combination with hemodilution and found an improvement in the hemorheologic parameters measured (original in Russian).

In 1985 **Baer** and co-workers [39] described the effect of leukapheresis on the clinical situation of a patient with chronic lymphatic leukemia and a leukocyte count of over 500 000 per mm^3. The authors assumed a hyperviscosity syndrome, without, however, having carried out viscosimetric measurements. After leukapheresis there was a clinical improvement in the patient's symptoms.

In 1985 **von Rhede van der Kloot** and co-workers [889] treated patients with Raynaud's phenomenon by means of plasmapheresis (plasma filtration) and found an improvement in six out of seven cases with increased erythrocyte aggregation. Those patients whose plasma viscosity was already raised before therapy showed a therapy-induced reduction of the values.

Clinical findings: Half of the patients studied reported a subjective improvement of their symptoms.

In 1986 **Bada** and co-workers [38] determined whole blood viscosity using a Wells-Brookfield micro-cone-plate viscosimeter at 37 °C and various shear rates before and after blood exchange versus plasma exchange in newly born infants with polycytemia. In the course of therapy there was a statistically significant reduction of the whole blood viscosity.

4.1.9.3 Discussion

Schwab and *Fahey* were the first to use so-called plasmapheresis as a therapeutic means of decreasing serum or plasma viscosity in patients with strongly increased plasma and serum viscosities as a result of macroglobulinemia or paraproteinemia. The authors were convinced that the increased viscosity of the plasma or the serum and the clinical symptoms of paraproteinemia, which are often present in the form of retinopathies, mucosal bleeding, cardiac insufficiency, neurologic symptoms, and coma, have a causal relationship with the deterioration of the flow properties of blood. One possibility of decreasing the viscosity of serum or plasma was to remove part of the patient's plasma. *Schwab* and *Fahey* reinfused the erythrocytes only; later other authors additionally infused plasma or plasma fractions from healthy subjects. From the publications of some of the authors it cannot be seen whether they performed only plasma removal or whether they also substituted healthy plasma or other infusion solutions for the pathologic plasma that was removed.

The investigators all concurred in finding a decrease in serum or plasma viscosity. The improvement in the clinical symptoms was particularly strongly expressed in patients with hyperviscosity symptoms, i.e., retinopathies, mucosal bleeding, and in those who showed neurologic symptoms. Furthermore, two authors found an improvement in the perfusion of the fundus vessels in the eye (fluorescence angiography).

In a whole series of studies plasmapheresis was combined with chemotherapy so that the component which improves the flow properties of blood as a result of plasmapheresis or plasma exchange cannot easily be separated from the effect of the medications.

The large number of results available do, however, show that this technique is suited to improving the symptoms of this group of diseases with para- or dysproteinemia. This was also the case with some patients with peripheral occlusive diseases (Raynaud's syndrome) who did not really have paraproteinemias or macroglobulinemias.

It may also be seen from these studies that this therapeutic measures was only palliative with malignant paraproteinemia including Waldenström's disease, i.e., that the symptoms could only be improved for a limited time. The patients in fact did die of their basic disease. One further difficulty is the fact that it is, on the whole, quite time consuming to carry out plasmaphereses or plasma exchanges.

Nonetheless, it can be said that all the authors are in agreement that plasmapheresis or plasma exchange leads to an improvement in the clinical symptoms in a whole series of pathologic diseases in which the plasma or serum viscosity is strongly increased as a result of various reasons. Plasmapheresis plays a significant role in therapeutic hemorheology.

The means of action of plasmapheresis is relatively easy to explain: owing to the removal of aggregation-supporting proteins which strongly influence the serum or plasma viscosity and owing to a decrease in the total protein concentration in the plasma there is an improvement in the blood and plasma viscosities and a decrease in erythrocyte aggregation. As far as erythrocyte filtrability (erythrocyte deformability) is concerned, the data are contradictory, even though most authors found an improvement in erythrocyte filtrability.

4.1.10 Chemotherapy, Cytostatic Drugs

4.1.10.1 Introduction

In the case of malignant paraproteinemia (macroglobulinemia, plasmocytoma), a pathologic protein pattern leads to a strong increase in the serum viscosity and to a deterioration of the flow properties of the blood, which manifest themselves clinically in mucosal bleeding, retinopathies, neurologic deficiency symptoms, and a much deteriorated general condition. Chemotherapeutic drugs (cytostatic drugs) have often been used to reduce the increased plasma protein concentration in this group of diseases.
The chemotherapeutic treatment of paraproteinemias together with rheologic measures was first reported by **Ritzmann** *and co-workers [732] in 1960. For the treatment of macroglobulinemia, they used d-penicillamine which is supposed to have a depolymerizing effect on the macroglobulins and which suppresses immunoglobulin synthesis.*
d-penicillamine is a d-βdimethylcysteine and is used for the treatment of chronic polyarthritis and heavy-metal poisoning. Furthermore, some investigators have used more aggressive chemotherapeutic substances such as cyclophosphamide as well as chlorambucil and melphalan for immunoglobulin suppression.

4.1.10.2 Publications

The first rheologic studies on the improvement of the flow properties of blood using chemotherapeutic drugs in patients with Waldenström's disease were published by **Ritzmann** and co-workers [731] in 1960. A patient whose macroglobulins made up 38% of the total serum proteins was treated with d-penicillamine for a period of 19 days. The patient received 250 mg d-penicillamine orally every 6 h. At the same time as the reduction of the serum macroglobulin level (depolymerization of the macroglobulins), there was a decrease in the serum viscosity (Ostwald capillary viscosimeter, H_2O = 1, 20 °C) which still remained at a lowered level 30 days after the end of therapy.

Note: The serum viscosity measurements were carried out below body temperature; no measurements of further rheologic parameters.

Clinical findings: The tendency to bleeding was normalized and the attacks of pain at exposure to cold ceased.

In 1960 similar findings were reported by **Schwab** and **Fahey** [779, 780] concerning a reduction of the serum viscosity (Ostwald capillary viscosimeter, H_2O = 1, 37 °C) following plasmapheresis and intermittent oral chlorambucil administration in a patient with Waldenström's disease. On the other hand, penicillamine and corticosteroids showed no effect.

Note: The dosage level of chlorambucil was not mentioned. An assessment of the specific effect of chlorambucil is difficult because plasmapheresis was performed at the same time.

Clinical findings: The authors observed a cessation of the mucosal bleedings and an improvement in the retinopathies caused by hemorrhaging.

In 1965 **Smith** and co-workers [803] treated a patient with multiple myeloma both chemotherapeutically with melphalan (0.1–0.8 mg per day orally) and with weekly plasmaphereses. There was a simultaneous decrease in the paraprotein concentration and the serum viscosity (Ostwald capillary viscosimeter, 37 °C, 0.9% NaCl = 1).

Note: No measurements of further rheologic parameters; an assessment of the specific effect of melphalan is difficult because plasmaphereses were carried out at the same time.

Clinical findings: The authors found a clinical improvement in the retinopathies, in the increased tendency to bleeding, and in the neurologic symptoms.

In 1967 **Kopp** and co-workers [524] published a report on the cytostatic treatment of a patient with multiple myeloma. After oral administration of 100 g cyclophosphamide, which was given for a period of 1 year, a normalization of the pathologic protein pattern and a reduction of the serum viscosity occurred simultaneously (Ostwald capillary viscosimeter, 37 °C, $H_2O = 1$).

Note: No measurements of further rheologic parameters.

Clinical findings: The investigators found an improvement in nasal bleeding and in vision, as well as reduced sludge formation in the conjunctival vessels.

In 1968 **Azzena** and co-workers [36] reported on the rheologic effects of plasmapheresis and on chemotherapy with steroids and cytostatic drugs in three patients with multiple myeloma. The relative serum viscosity ($H_2O = 1$) was measured using an Ostwald capillary viscosimeter at 24 °C. A reduction of the serum viscosity was observed at the same time as a normalization of the pathologic protein pattern.

Note: No measurements of further rheologic parameters; type and quantity of steroids and cytostatic drugs used were not defined.

Clinical findings: The authors interpreted the improvement in the retinopathies, the cessation of nasal bleeding, the absence of neurologic deficiency symptoms as well as of various pain conditions (e.g., back pains) as an expression of a real improvement in the flow properties of blood.

In 1968 **Levine** [570] published a short article according to which serum viscosity was not reduced in two patients with macroglobulinemia after oral administration of penicillamine (1–2 g per day).

Note: The type of viscosimeter used was not mentioned.

Clinical findings: The author did not observe an improvement in the clinical symptoms.

In 1969 **Pronk** and co-workers [714] treated four patients with Waldenström's disease and one patient with multiple myeloma by means of oral administration of chlorambucil and prednisolone. After the oral administration of chlorambucil the authors only found a reduction of the serum viscosity and the macroglobulins (Ostwald capillary viscosimeter, 37 °C, $H_2O = 1$) in one patient with Waldenström's disease.

Note: No measurements of further rheologic parameters.

In 1970 **Abruzzo** and co-workers [4] treated a patient with polyarthritis, polymyositis, polysynovitis, raised serum viscosity, and a highly increased 13S IgG component which was 40% of the total protein. The relative serum viscosity (H_2O = 1) was measured using an Ostwald capillary viscosimeter at 25 °C. After the oral administration of prednisone (60 mg per day), there was a reduction in the serum viscosity, the gamma-globulins, and the 13S paraprotein component of the serum.

Note: The viscosity measurements were performed below body temperature. Measurements of further rheologic parameters were not carried out.

Clinical findings: The extreme symptoms of myositis and synovitis disappeared. In addition, there was an improvement in the neurologic deficiency symptoms.

In 1970 **Ritzmann** and co-workers [733] reported on the treatment of a patient with Waldenström's disease by means of oral administration of chlorambucil. At the same time as the reduction of the paraproteins, there was a decrease in the serum viscosity (Ostwald capillary viscosimeter, H_2O = 1, 37 °C).

Note: No measurements of further rheologic parameters.

Clinical findings: The clinical picture of dyspnea, lethargy, and mucosal bleeding was improved.

In 1972 **Mackenzie** and **Fudenberg** [604] treated patients with Waldenström's disease, a number of whom had a so-called hyperviscosity syndrome. Chlorambucil, cyclophosphamide, and melphalan were given as alkalyzing medication, and therapeutic plasmapheresis was performed. The relative serum viscosity (H_2O = 1) was measured at room temperature using an Ostwald capillary viscosimeter. If cryoglobulins were found, the viscosity measurements were performed at 37 °C. The authors found a decrease in the IgM globulins and in serum viscosity.

Note: No measurements of further rheologic parameters; the viscosity measurements were performed below body temperature. The dosage of the cytostatic drugs used was not mentioned. An evaluation of the rheologic effects of the individual substances on the flow properties of blood is difficult as plasmapheresis was performed at the same time.

Clinical findings: There was a cessation of the mucosal bleeding, the retinopathies, and the symptoms of acquired cardiac disease. Good results were also achieved in coma patients.

In 1972 **Sugai** [850] treated a patient with multiple myeloma, cryoglobulins, and increased serum viscosity. In addition to plasmapheresis therapy, 100 mg cyclophosphamide and 30 mg prednisone were given daily. With this combined therapy the serum viscosity (Ostwald capillary viscosimeter, 37 °C, H_2O = 1), the macroglobulin level, and the cryoglobulins were decreased.

Note: No measurements of further rheologic parameters. Owing to the fact that plasmapheresis was performed at the same time, an assessment of the specific effect of the cytostatic medications is difficult.

Clinical findings: A reduction of the tendency to bleeding, the visual disturbances, and the neurologic symptoms such as nausea and headaches was observed. The improvement in the flow properties of blood was also expressed in a reduction of the parasthesias.

In 1973 **Bloch** and **Maki** [71] reported on a reduction of serum viscosity in patients with macroglobulinemia after oral administration of 250 mg penicillamine (every 6 h). After the end of therapy, the serum viscosity remained at a low level for several weeks.

Note: The type of viscosimeter used and the temperature of the viscosity measurements were not given.

Clinical findings: The authors found a normalization of the increased tendency to bleeding and a reduction of the pain attacks at exposure to cold.

In 1980 **Iwamoto** and co-workers [469] treated two patients with multiple myeloma by means of chemotherapy using prednisolone, melphalan, and cyclophosphamide. The authors found no reduction of the pathologic proteins and the plasma viscosity as a result of chemotherapy.

Note: Type of viscosimeter and how it was used were not mentioned.

In 1982 **Case** [117] reported on combined chemotherapeutic treatment of patients with Waldenström's macroglobulinemia and hyperviscosity who were suffering from the corresponding symptoms such as a tendency to bleeding and peripheral neuropathy. Serum viscosity was determined using a Wright and Jenkins pipette viscosimeter and reported as relative viscosity compared to distilled water. The combined chemotherapeutic therapy, which consisted of cyclophosphamide, vincristine, melphalan, and prednisone, led to a normalization of the serum viscosity values.

Note: Further rheologic tests were not performed.

Clinical findings: For most of the patients with clinical symptoms resulting from hyperviscosity there was an improvement in these symptoms. The author spoke of "symptomatic hyperviscosity" and thereby characterized those patients who not only had a strongly increased serum or plasma viscosity in Waldenström's disease, but who also had clinical symptoms.

In 1985 **Miller** and **Heilmann** [647, 648] reported on the effect of chemotherapy with cyclophosphamide, 5-fluorouracil, and methotrexate in patients with malignant tumors. While there were no statistically significant changes in the whole blood viscosity, there was a statistically significant decrease in erythrocyte aggregation (abstract, no details).

4.1.10.3 Discussion

In multiple myeloma there is a hyperproteinemia of IgG and IgA immunoglobulins and in Waldenström's disease a hyperproteinemia of IgM immunoglobulins. The immunoglobulins can aggregate among themselves to form macroglobulins. Owing to the type and size of the circulating immunoglobulins in the blood, the serum viscosity is pathologically raised. The deteriorated flow properties of blood mainly determine the clinical symptoms of these diseases which are expressed as bleeding, visual disturbances, neurologic symptoms, cardiac and circulatory problems, and vascular disease. The production of immunoglobulins can be suppressed by chemotherapeutic drugs (cytostatic drugs).

Most of the authors found that viscosity values decreased with this type of therapy. Since chemotherapy was usually carried out in combination with plasmapheresis, no definite statement can be made about the specific viscosity-reducing effect of the different cytostatic drugs.

On the whole, as far as the improvement in the flow properties of blood and a consecutive improvement in the clinical symptoms are concerned, the medicinal therapy with cytostatic drugs is less effective than well-performed plasmapheresis according to the results of the majority of the authors. Theoretically serum viscosity should decrease after cytostatic therapy with simultaneous depression of the pathologic protein synthesis, which in turn also leads to a decrease in the whole blood viscosity and the plasma viscosity. In addition, a considerable decrease in the erythrocyte aggregation is to be expected, since the macromolecular long chain proteins show a well-known, strong erythrocyte-aggregating effect. However, up to now this aspect has not been tested by investigators. When measuring the whole blood viscosity, it must be taken into account that the aggressive therapeutic technique with chemotherapeutic drugs may also alter blood cell formation, which then leads to a decrease in the hematocrit or in the number of erythrocytes in the whole blood. As a result, the hematocrit-dependent whole blood viscosity would be decreased as well.

4.1.11 Cardio-active Glycosides

4.1.11.1 Introduction

The main effect of cardiac glycosides (digitalis preparations) is an increase in the contractile force of the heart resulting from an improvement in the electromechanical coupling.

4.1.11.2 Publications

In 1962 **Wasilewski** [901] studied the effect of intravenous administration of lanatoside (a genuine glycoside from *Digitalis lanata*) on blood viscosity in patients with cardiac decompensation. The capillary of an Ostwald viscosimeter was directly connected to the circulation, and the blood was drawn up at a constant negative pressure (200–300 mmHg). The aspiration time required for the quantity of blood to flow through the instrument was taken as a relative measure of blood viscosity. Contrary to the effect in healthy subjects, there was a decrease in blood viscosity after intravenous administration of lanatoside (no dosage given) in the decompensated patients. If an accompanying medication with a polycarbonic acid derivative was given in addition, then there was an even stronger reduction of the blood viscosity.

Note: Measurements of further rheologic parameters were not performed, and the dose of the digitalis preparation was not given.

In 1964 **Eisenberg** [317] studied patients with myocardial insufficiency, cardiomegaly, edemas, and dyspnea. Blood viscosity was measured using a Brookfield micro-cone-plate viscosimeter at 37 °C and at shear rates of 12–230 s^{-1}. After compensation (treatment with digitalis, salt restriction, bed rest, and diuretics), there was an increase in blood viscosity, in the hematocrit value, and in the fibrinogen level.

Note: No measurements of further rheologic parameters.

4.1.11.3 Discussion

Wasilewki and *Eisenberg* had different results regarding the effect of digitalis on the blood flow properties. This may be associated with the fact that both authors used completely different viscosimeters and that *Eisenberg* used diuretics as well as digitalis and prescribed a salt restriction. Studies on this problem would be useful because a decrease in viscosity would ease the work of the heart as the result of a reduction in the peripheral resistance. The fact that after successful digitalization and a resultant strenghtening of the heart patients, as a rule, have very strong diuresis (this is especially true for patients with edema) could lead to the theoretical conclusion that, on the whole, blood viscosity rises or normalizes again from previously lowered values as a result of a relative hemoconcentration.

4.1.12 Diuretics

4.1.12.1 Introduction

***Kuschinsky** and **Lüllmann** [530], **Lyons** [601], **Völker** and **Eichler** [882], **Gormsen** [383], and others have reported on the increased frequency of thromboembolic disorders following the application of diuretic drugs. The cause of this complication is considered to be an increased diuresis which, owing to a decrease in the plasma volume, leads to a thickening of the blood (hemoconcentration) and thereby supports the occurrence of thromboembolism. In a diuretic-induced hemoconcentration, there is an increase in the hematocrit and in the total protein concentration, which should necessarily lead to an increase in the blood and plasma viscosities. Therefore, rheologic studies have been carried out by different authors, especially in patients in whom the use of diuretic drugs was indicated owing to various types of edemas. In addition, diuretic drugs are useful to a considerable extent in the basic therapy of essential hypertension. Here there is natriuresis as well as diuresis, which, owing to a reduction of the intravasal blood volume, could lead to a thickening of the blood and to an increase in the viscosity. This would particularly be the case when there is a discrepancy between the urinary discharge on the one hand and fluid intake on the other. In rheologic studies, the following diuretics were used for the therapy of edemas and for baseline therapy of hypertension: furosemide; bemetizide; triamteren, amiloride, hydrochlorothiazide, xipamide, and various combinations of these substances.*

4.1.12.2 Publications

In 1964 **Eisenberg** [317] determined the blood viscosity in 30 patients with cardiac insufficiency, cardiomegaly, edemas, and dyspnea using a Brookfield micro-cone-plate viscosimeter at 37 °C and shear rates of $12-123\ s^{-1}$. In a decompensated condition, the blood viscosity was in most cases normal or just slightly raised. After compensation (treatment with digitalis, salt restriction and diuretic drugs), there was an increase in the hematocrit values, the fibrinogen values, and the blood viscosity values.

Note: No measurements of further rheologic parameters. The diuretic drugs and their dosages were not given. Since a parallel therapy with digitalis was carried out during the treatment period, the rheologic changes in the flow properties of the blood cannot be definitely traced back to the effect of the diuretic drugs used.

A further publication, by **Völker** and **Eichler** [882], on the deterioration of the flow properties of the blood in heart and vascular diseases following the administration of diuretics followed in the year 1965. Blood viscosity was measured using a capillary viscosimeter constructed for this purpose. Simultaneous to the increase in the hematocrit values, the authors found an increase in blood viscosity of up to 100%, independently of the type of diuretic and the form of application.

Note: The temperature for the viscosity measurements was not given and measurements of further rheologic parameters were not carried out. The different types of diuretic drugs used and their dosages were not listed separately.

In 1972 **Ehrly** [247] reported on changes in the flow properties of blood after intravenous administration of 20 mg furosemide in nine patients with cardiac edema and ten patients with ascites (liver cirrhosis). The relative blood and plasma viscosities ($H_2O = 1$) was measured using a capillary viscosimeter, and the apparent blood viscosity was measured at shear rates of 5.8, 11.5, 23.0, and 46.0 s^{-1} using a Brookfield micro-cone-plate viscosimeter. The patients with myocardial edemas showed no major changes in the flow properties after injections of furosemide in comparison with a placebo group, while, on the other hand, patients with ascites showed a slight hemoconcentration during the same period. The hemoconcentration presented itself as an increase in the blood and plasma viscosities and in the hematocrit value.
Similar results were published by **Köhler** [518] in 1973 and by **Ehrly** in 1975 [264].

Note: Measurements of the erythrocyte aggregation and the erythrocyte deformability were not carried out; the temperature of the viscosity measurements was 23 °C.

Clinical findings: Both groups of patients showed a significant increase in the urinary discharge.

In 1975 **Strauss** [845] from the Ehrly team reported in his dissertation on the 4-week treatment of eight patients with essential hypertension by means of an oral application of an antihypertensive containing a diuretic. The patients received one tablet containing 60 mg furosemide daily. Measurements of the flow properties of the blood were carried out using an Ostwald capillary viscosimeter. The investigators found a significant decrease in the blood and plasma viscosities along with decreased total protein levels and unchanged hematocrit values.

Note: Measurements of the erythrocyte aggregation and of the erythrocyte deformability were not performed; the temperature of the viscosity measurements was 23 °C.

In 1976 and 1977 **Bös** [76] and **Ehrly** [276] studied patients with chronic venous edemas of the lower extremities due to former deep vein thrombosis. The patients received a single injection of 20 mg furosemide. In a second experimental series, a so-called venodiuretic (20 mg triamteren and 10 mg bemitizide daily) was administered orally to another group of patients for a period of 3 days. The relative blood and plasma viscosity ($H_2O = 1$) was measured using an Ostwald capillary viscosimeter at 23 °C. The erythrocyte deformability measurements were performed using a filtration method (Millipore filter 12.9 mm H_2O, hematocrit 10%) according to Ehrly and Rosbach. After administration of furosemide, there was a significant increase in the blood and plasma viscosities, the hematocrit, and also the total protein concentration compared to the placebo-patient group. Sixty minutes after the injection, the definite deterioration of the flow properties of the blood reached a maximum. After the oral administration of the diuretic combination triamteren/bemitizide, which has a long-lasting effect and needs a comparatively low dosage, there was a slight, nonsignificant increase in the protein concentration. Only the hematocrit increased significantly. A decrease in erythrocyte deformability (erythrocyte flexibility) was also registered.

Note: No quantitative measurement of erythrocyte aggregation; the temperature of the viscosity measurements was 23 °C.

Clinical findings: With diuretic drugs there was a significant increase in the urinary discharge and a reduction of the edemas (e.g., diameter of lower leg).

In 1979 **Ehrly** and co-workers [280] studied the question of how much the diuretic xipamide influences the flow properties of blood. In two experimental series, 10 and 20 mg xipamide daily were administered orally to patients with essential hypertension over a period of 10 days. The blood and plasma viscosities of the heparinized blood samples were determined using an Ostwald capillary viscosimeter, erythrocyte filtrability was determined according to the method of Ehrly and Rosbach using Millipore filters with a pore diameter of 8 µm; and the erythrocyte aggregation was determined quantitatively according to the method of Ehrly and Schmitt. The measurement temperature was 37 °C. After the administration of 10 mg xipamide daily there was a minimal but significant increase in the hematocrit value on the 1st day. Significant changes in the blood and plasma viscosities, erythrocyte aggregation and erythrocyte filtration were not measured. Significant increases in the blood and plasma viscosities, the hematocrit value, and the total protein concentration were found upon administration of 20 mg xipamide daily.

Note: No blood viscosity measurements at low shear rates.

Clinical findings: There was a significant decrease in the raised blood pressure values and an increase in the urinary discharge both with the low and the high dosage.

In 1980 **Mann** [608] studied the effect of various saluretic drugs on the flow properties of the blood in patients with essential hypertension. The patients received oral doses of the following combination of substances within the framework of three parallel series of experiments performed over a treatment period of 10 days: half a tablet of a combination of bemitizide and 20 mg triameteren per day (A), half a tablet (B) and one tablet (C) of a combination of 50 mg hydrochlorothiazide with 5.677 mg *N*-amino-3,5-diamino-6-chlorpyrazine-2-carboximide-hydrochloride per day. The blood and plasma viscosities were determined using an Ostwald capillary viscosimeter; erythrocyte flexibility was determined according to the method of Ehrly and Rosbach through a Millipore filter with a pore size of 8 µm; and erythrocyte aggregation was dtermined quantitatively according to the method of Ehrly and Schmitt (hematocrit of 40%). The measurement temperature was 37 °C. During the low dosage with (A), the flow properties of the blood were hardly affected. With the second substance, there was a significant increase in the blood and plasma viscosit only on the 2nd day with a dosage of half a tablet per day (B). On the other hand, with the higher dosage (C) there was a significant increase in the hematocrit values from the 1st to the 3rd day, a significant increase in the blood viscosity on the 1st day, a significant increase in the plasma viscosity on the 1st and 2nd day and there was a significant increase in the total protein concentration in the plasma.

In 1981 **Wodniok** [920] reported similar results with still higher dosages of the saluretics bemitizide, triamteren, and hydrochlorothiazide in patients with hypertension. There were no significant changes in erythrocyte aggregation and deformability.

Note: Blood viscosity measurements were not performed at low shear rates.

Clinical findings: Even with lower dosages of the saluretic drugs there was a significant reduction of hypertension and an increase in the urinary discharge.

In 1981 **Palareti** and co-workers [682] reported on the effect of a combination of 5 mg amyloride plus 50 mg hydrochlorothiazide per day in patients with simple essential hypertension. Blood viscosity was measured at 23 and 230 s^{-1}, as were the plasma viscosity and the erythrocyte filtration. After 7 days of treatment, there was a statistically significant increase in the blood and plasma viscosities and a decrease in the erythrocyte filtrability. After 30 days the values were normal again (abstract, no details).

In 1982 **Karasawa** and co-workers [485] studied the whole blood viscosity of preeclamptic female patients being treated with a daily dose of 80 mg furosemide and found an increase in this parameter. The measurements were made with a cone-plate viscosimeter at shear rates of 150 and 375 s^{-1}.

In 1983 **Jahnsen** and co-workers [471] reported on the effect of a high dosage of intravenously administered furosemide in patients with lung edema caused by cardiac disease. At the same time, the patients received pethidine intravenously and oxygen via a nasal catheter. Some patients were additionally given other medications. Blood, anticoagulated with EDTA, was removed from the antecubital vein just before treatment and at intervals of up to 2 h thereafter. The whole blood was measured using a Brookfield cone-plate viscosimeter at shear rates of 5.75–230 s^{-1} at 37 °C. In accordance with a slight decrease in the hematocrit value, there was a statistically significant reduction of the whole blood viscosity in patients 60 and 120 min after the injection. In spite of the fact that there is a close correlation between the changes in blood viscosity and in the hematocrit values, the authors could not find a statistically significant relationship between these two parameters and the urinary discharge.

Note: No studies of further rheologic parameters.

In 1984 **Kiesewetter** and co-workers [495] studied the effect of a 7-day therapy using a combination of 50 mg triamteren and 25 mg hydrochlorothiazide in patients with chronic venous edema. On the 1st, 4th, and 7th day after the beginning of therapy, plasma viscosity, erythrocyte rigidity, erythrocyte aggregation and the yield stress were measured. Plasma viscosity increased continuously up to the 4th day after the beginning of therapy and then remained constant; the erythrocyte rigidity first increased, but then returned to its initial state. Erythrocyte aggregation showed a significant increase on the 7th day.

In 1985 **Jahnsen** and co-workers [472] found that a daily dose of, on average, 6.5 mg bendroflumethiazide led to an increase in the hematocrit and blood viscosity during an 8-week treatment of patients with arterial hypertension and/or unspecific edema. The measurements were carried out with a cone-plate viscosimeter at different shear rates. This increase, however, did not surpass the variations of these two parameters, as found, for example, in the course of a normal circadian rhythm.

4.1.12.3 Discussion

In the patients with myocardial edemas [247, 471] it was shown that, despite a considerable loss of fluid via the kidneys after the intravenous administration of a short-acting and very effective diuretic (furosemide), there was no increase in the blood and plasma viscosities. The body is capable of mobilizing the fluid contained in the edemas just as quickly as the urine is discharged. This is different in patients with ascites [247, 518] in whom the ascites fluid cannot be mobilized so quickly, resulting in a hemoconcentration with a deterioration of the flow properties of blood.

Patients with chronic venous edemas [76, 276, 495] also showed a thickening of the blood (similar to that of patients with ascites) with a significant increase in the blood and plasma viscosities, in the hematocrit values, and the total protein concentration. On the other hand, the oral application of a comparatively low-dosage diuretic combination [76, 276] did not lead to a deterioration of the flow properties of the blood. After the oral administration of xipamide in patients with essential hypertension [280, 920], there was a deterioration of the flow properties of the blood only at higher doses. Further studies confirm that the level of the oral dosage of saluretics is of primary importance for the flow properties of the blood in patients with essential hypertension.

Eisenberg's study, in which he was able to show an increase in the blood viscosity after the administration of diuretic drugs in patients with cardiac edema, has to be interpreted from the point of view that these were patients whose edemas were already considerably flooded out owing to simultaneous digitalis therapy, so that there was, as it were, an iatrogenic diuretic-induced exsiccosis. A similar explanation would be indicated for the results of *Völker* and *Eichler* as well. An increase of 100% in the viscosity values is, however, not very likely and can probably be traced back to methodological problems.

The studies of *Palareti* and co-workers support the observations that the administration of high dosages of diuretic drugs can lead to a hemoconcentration but only in patients with hypertension without edemas. The studies of *Jahnsen* and co-workers, who found a decrease in the viscosity values after the administration of high dosages of the fast-acting diuretic furosemide, are difficult to explain. On the one hand, this could be due to an accumulation of fluid around the lungs, which works against a hemoconcentration even with increased diuresis. On the other hand, the simultaneous administration of pethidine in these patients makes the explanation of the effect found by these authors more difficult. The results of *Kiesewetter* and co-workers as well as of *Jahnsen* and co-workers confirm the fact that an incorrect dosage of certain diuretic drugs can lead to hemoconcentration.

In summary, it may be said that the elimination of edema with fast- and strong-acting diuretics (e.g., furosemide) in patients with ascites and chronic venous edemas leads to a marked hemoconcentration. This results in a deterioration of the flow properties of blood and in an increase in the concentration of coagulation factors in the blood. Thus, the risk of thrombosis is raised by two mechanisms, namely the increase in the concentration of coagulation factors like fibrinogen and the slowing-down of the blood flow.

The hemoconcentration may be explained by the fact that, after the administration of the fast- and strong-acting drug furosemide in the diseases mentioned, the influx of fluid from the interstitium or intercellular space into the intravasal space is slower than the fluid loss via the kidneys. On the other hand, the studies in patients with cardiac edemas show no increase in blood viscosity after administration of furosemide, despite considerable quantities of urinary discharge. In this case, the body is capable of mobilizing the fluid accumulated in the edemas just as quickly as the urine is discharged. In this clinical picture, there is obviously no or only a limited temporal phase shift between the urine production of the kidneys and the influx of edema fluid into the intravasal space [264]. With low-dosage, delayed-action diuretics, no significant deterioration of the flow properties of the blood is to be expected in the therapy of arterial hypertension and post-thrombotic insufficiency.

4.1.13 Adenyl Compounds

4.1.13.1 Introduction

The intra-arterial infusion of so-called high-energy phosphates (mostly mixtures of adenosinetriphosphate, adenosinediphosphate, and others) as a therapeutic measure still stands high in the estimation of a small number of angiologists. In chronic arterial occlusive disease an intra-arterial infusion is necessary because of the rapid disintegration of adenosin-triphosphate in particular. The mode of action can probably be explained by vasodilation.

4.1.13.2 Publications

In 1974 **Franke** [356] from the Hess team reported on the treatment of patients with vascular diseases by means of adenyl compounds. During an infusion time of 30 min the patients were infused via the femoral artery with 20 ml of a mixture of adenosinetriphosphate, adenosinediphosphoric acid, adenosinemonophosphoric acid, adenosine, guanosine, guanosine monophosphoric acid, inosine and uridine. The blood viscosity measurements of the blood samples, anticoagulated with heparin, were carried out using a Brookfield micro-cone-plate viscosimeter at four shear rates of 23, 46, 115, and 230 s^{-1} and at a temperature of 37 °C. At the end of infusion, the authors found a decrease in blood viscosity which normalized again 10 min after the end of infusion. In addition, there was a slight decrease of 1% in the hematocrit.

Note: Since no hematocrit correction was carried out for the viscosity measurements in these studies, the decrease in viscosity could also have been caused by the hematocrit reduction which occurred. No measurements of further rheologic parameters.

In 1982 **Artale** and co-workers [31] found that an intra-arterial application of 200 mg *S*-adenosyl-l-methionine was able to reduce the increase in whole blood viscosity and the decrease in the filtrability following muscular exercise. While there was an increase in the blood viscosity and a reduction of the filtration values of the femoral venous blood without this therapeutic agent, after therapy this increase could no longer be shown with corresponding muscular exercise. Blood viscosity was measured using a Wells-Brookfield viscosimeter and the erythrocyte filtrability using the method of Reid and co-workers. The authors found no change in the plasma viscosity values, the hematocrit, and the fibrinogen concentration.

4.1.13.3 Discussion

Franke as well as *Artale* and co-workers found a reduction of the whole blood viscosity after the intra-arterial administration of adenyl compounds. A complete rheologic analysis was not performed by either of the teams; the assessment of the erythrocyte deformability based on the filtration method of Reid and co-workers must be relativized. Thus the question must remain open as to whether the reduction of blood viscosity that was found may be considered to be reliable and how the mechanism of action of this viscosity reduction can be explained.

4.1.14 Glucose Solutions/Levulose Solutions

4.1.14.1 Introduction

Glucose (dextrose) solutions and levulose solutions are used in low concentrations both as infusion solutions per se and as carrier solutions for various drugs. At low concentrations these solutions are used to replenish the blood volume; at high concentrations they are used for osmotherapy in edema of the brain and the lungs.

4.1.14.2 Publications

In 1965 **Groth** and co-workers [395] reported on the improvement in the flow properties of blood in patients with various diseases as a result of administering 500 ml 5.5% glucose solution intravenously in an infusion time of 25–35 min. Plasma viscosity was measured using an Ostwald capillary viscosimeter at 37 °C. Owing to the glucose infusion there was a short-term but highly significant reduction of plasma viscosity and a significant decrease in the hematocrit values independently of the initial plasma viscosity values in the patients. On the other hand, infusions of low-molecular dextran caused no significant changes in the plasma viscosity. Similar results were also reported elsewhere [394, 398].

Note: No measurements of further rheologic parameters.

In 1966 **Langsjoen** and co-workers [543] treated patients with acute myocardial infarction using glucose infusions. The viscosity of the blood samples anticoagulated with heparin was measured with a Brookfield micro-cone-plate viscosimeter at shear rates of 2–230 s^{-1} and at a temperature of 37 °C. The raised blood viscosity and hematocrit values were reduced with repeated infusions of 500 ml 5% glucose solution. The blood viscosity decrease was particularly marked at low shear rates. Similar results were reported elsewhere [545].

Note: No measurements of further rheologic parameters.

In 1968 **Bernstein** and **Castaneda** [68] treated 25 patients, who were connected to a heart-lung machine during heart surgery, with 10% low-molecular dextran solution and 5% dextrose solution. The temperature of the blood viscosity measurements was 30°–31 °C. Simultaneous to the decrease in the hematocrit, the blood viscosity (Brookfield micro-cone-plate viscosimeter, 1.3–130 s^{-1}) was reduced to a greater degree and for a longer time with dextran than with the 5% dextrose solution. The reduction of blood viscosity was particularly marked at low shear rates.

Note: No measurements of further rheologic parameters.

In 1969 **Ditzel** and co-workers [189] treated patients with acute myocardial infarction using daily infusions (500 ml 5% glucose solution, 500 ml LMWD – no percentage data) of 4 h each for a treatment period of 5 days. Blood viscosity was measured using a Brookfield micro-cone-plate viscosimeter at 12 revs/min. At the same time as the reduction of the hematocrit values, the investigators found a decrease in blood viscosity owing to the infusion solutions. However, the blood viscosity-reducing effect of dextran was sigificantly stronger than that of the glucose solution.

Note: No measurements of further rheologic parameters; no data on the temperature of the measurements.

In 1974 **Heidrich** and **Ott** [421] studied patients with and without vascular diseases and found, using a falling ball viscosimeter, that the infusion of 250 ml 5% levulose produced a reduction of the blood viscosity with a slight decrease in the hematocrit value, which, however, lasted only a limited time. The authors attributed the reduction of blood viscosity to a thinning effect of the plasma.

Note: No measurements of further rheologic parameters.

In 1985 **Gottschalk** [384] determined the whole blood viscosity using a Höppler rheoviscosimeter at 38 °C in children who were infused with both a 10% invertose solution and an amino acid solution. Owing to the low viscosity of the amino solution itself, there was a marked reduction of the viscosity; the effect was somewhat less with the 10% invertose solution.

Note: No measurement of further hemorheologic parameters.

4.1.14.3 Discussion

Groth and co-workers found a highly significant reduction of plasma viscosity after the infusion of 500 ml 5.5% glucose solution compared to nonsignificant changes in plasma viscosity after infusions of low-molecular dextran. This is due to the low specific viscosity of the glucose solution itself compared to the high viscosity of the dextran solution.

Groth and co-workers discussed the importance of a strong plasma viscosity-reducing effect in particular in the region of the microcirculation, because there the whole blood viscosity is much more strongly influenced by plasma viscosity than is the case in the vessels with a larger diameter. The authors stated that the effect of the glucose solution on the flow properties of blood is of very short duration.

Langsjoen and co-workers found a parallel reduction of the blood viscosity and the hematocrit after repeated infusions of 500 ml 5% glucose solution in patients with myocardial infarction. A much stronger hemodilution effect could be achieved with the same quantity of 10% low molecular dextran solution.

Ditzel and co-workers reported on a simultaneous reduction of the blood viscosity and the hematocrit value after infusions of 500 ml 5% glucose solution. Compared to an infusion of 500 ml low molecular dextran, the viscosity and hematocrit reductions with glucose solutions were much less significant because with, the infusion of isotonic glucose solution, no additional fluid is mobilized from the interstitial and intracellular space, as is the case with dextran.

For levulose solutions *Heidrich* and *Ott* found similar results to those for glucose solutions, which – compared to the total quantity infused – are interpreted as a nonspecific, induced hemodilution. According to the studies of *Gottschalk,* there is an expected decrease in the whole blood viscosity due to dilution after the infusion of both invertose solutions and amino acid solutions.

In summary, it may be said that a sugar solution, just as an isotonic saline solution, leads to a relatively rapid but short-term hemodilution, which all the authors ascribe to a reduction of the blood and plasma viscosities and of the hematocrit. The viscosity reduction with isotonic glucose solutions is determined by the flow velocity and by the total quantity of the fluid infused. As expected with noncolloidal solutions, the fluid leaves the intravasal space relatively quickly, so that the improvement in the flow properties of blood due to hemodilution does not last very long.

4.1.15 Human Albumin and Human Plasma Solutions

4.1.15.1 Introduction

Human serum albumin constitutes 52%–62% of the total plasma proteins, and with its pronounced water-binding capacity it is to considerable extent involved in the regulation of the colloidosmotic pressure. In addition, albumin is able to bind many substances reversably and it is a carrier molecule for low molecular, water-insoluble substances (e.g., bilirubin, salicylic acid). It has an average molecular weight of 69 000. Among other things, human serum albumin solutions and human plasma solutions are used rheologically for the thinning of blood (hemodilution). After the infusion of such solutions, there is a decrease in the hematocrit value and – in particular with pathologic protein constellations such as plasmocytomas – a decrease in the total protein concentration. The infusion of albumin or human plasma solutions is either carried out without previous blood removal (hypervolemic hemodilution) or as a so-called isovolemic hemodilution (blood sampling and subsequent infusion of plasma or albumin solution). The various forms of hemodilution have already been considered elsewhere (see Sect. 3.4). The following publications refer to hypervolemic hemodilution without previous venesection. Studies in which blood was first withdrawn with subsequent infusion of plasma protein solutions are considered in Sect. 4.2.1.2.

4.1.15.2 Publications

The first article on the influence of albumin solutions on the flow properties of blood in patients with increased plasma viscosity values and various types of diseases was published by *Groth* in 1966 [396]. The patients received 10 ml 10% albumin solution per kilogram body weight within an infusion time of 1 h. A control group received an equivalent infusion of low molecular weight dextran. The blood and plasma viscosities of the heparinized blood samples were measured using a Brookfield micro-cone-plate viscosimeter (Model LVT) at 37 °C and at four shear rates of 23, 46, 115, and 230 s^{-1}. Contrary to the dextran infusion, the author found a significant reduction of the plasma viscosity with albumin infusions independently of the initial plasma viscosity values. A significant decrease in the blood viscosity was achieved with both infusion solutions at all shear rates; at low shear rates the blood viscosity reduction was particularly marked. Four hours after the end of infusion, the plasma viscosity values and the serum albumin concentration were almost back to the initial value. On the other hand, blood viscosity, the hematocrit, and the erythrocyte sedimentation rate (measured with corrected hematocrit) were still reduced by one half.

In 1968 and 1969 **Groth** [397, 398] published further studies with similar results. Once again the author found no significant difference between the hemorheologic effects of albumin solution and low molecular weight dextran solution with the exception of plasma viscosity.

Note: Measurements of further rheologic parameters were not performed.

In 1984 **Ehrly** and co-workers [301] reported on the effect of a 30-min intravenous infusion with 500 ml 5% human albumin solution in patients with intermittent claudication. There was a reduction of the whole blood and the plasma viscosities, as well as a decrease in the erythrocyte aggregation.

4.1.15.3 Discussion

The first studies on the effect of albumin (hypervolemic hemodilution) came from *Groth*. As expected, the administration of human albumin solution at a concentration of 10% had a hemodiluting effect resulting in a reduction of the hematocrit and the blood viscosity. Since the viscosity of the solution itself is low, the plasma viscosity is also markedly decreased after infusion of a 10% albumin solution. The decrease in the erythrocyte sedimentation rate with corrected hematocrit indicates a possible disaggregating effect of the albumin solution on erythrocyte aggregates. The studies of *Ehrly* and co-workers confirm the results of the *Groth* team. The possible disaggregating effect discussed by *Groth* could be confirmed by *Ehrly* and co-workers using the light transmission method. According to this, albumin possesses a special affinity for improving the flow properties of blood. However, its high price and the occasional side effects limit the utilization of this solution for hemorheologic indications.

4.1.16 Anesthetics

4.1.16.1 Introduction

Substances which either by themselves or in combination with others produce a reversible state of unconsciousness, which make the organism insensitive to pain, and which reduce its defense reflexes are called anesthetics. They are the prerequisites for carrying out operations and painful surgical treatment without putting too much strain on the patient.

4.1.16.2 Publications

In 1966 a publication by **Boyan** and coworkers [85] appeared on the effect of inhalation anesthetics on the flow properties of blood in patients having undergone tumor operations. The patients were anesthesized with diethylether, halothane, and cyclopropane. Blood viscosity was measured using a Brookfield micro-cone-plate-viscosimeter at shear rates of 46, 115, and 230 s^{-1}. At the same time as a slight reduction of the hematocrit of about 1%, there was a minimal, tendential, nonsignificant reduction of blood viscosity.

Note: A correction of the hematocrit was not carried out, no measurements of further rheologic parameters, e.g., of plasma viscosity.

In 1970 **Hess** [442] found a decrease in blood viscosity as measured using a falling ball viscosimeter (no time given) with anesthetics containing thiopenthal and halothane.

Note: No further rheologic measurements.

In 1979 **Gramstad** and **Stovner** [390] reported on the effect of anesthetics containing cremophores (althesine, propanidide) in 29 female patients undergoing exploratory curettage. Blood samples were taken and examined at various times. Plasma viscosity was measured with a Wells-Brookfield microviscosimeter at 37 °C. There was a reduction of plasma viscosity in all patients, the degree of the changes being different from patient to patient.

Note: The considerable decrease in plasma viscosity of 40% of the initial value was probably artificial and may be traced back to the surface activity of cremophore. From the available data it cannot be determined whether there was a plasma dilution, since other rheologic parameters were not studied.

In 1980 **Drummond** and co-workers [220] reported on the effect of inhalation anesthesia with halothane and on the effect of spinal anesthesia with cinchocain hydrochloride on the flow properties of the blood of patients who were being operated on as a result of a collum femoris fracture. Blood viscosity was determined using a Contraves LS 30 viscosimeter at shear rates of 0.94 and 94.5 s^{-1} and with a mathematically corrected hematocrit. Erythrocyte deformability was measured according to the method of Reid and co-workers. After the operation, each patient had, on average, lost 260 ml blood. The deficient blood volume was compensated for with an infusion of plasma expanders, during the course of which the patients undergoing spinal anesthesia treatment were always infused with double the quantity of fluid. The type of plasma expander was not mentioned. At the end of the operation, the authors found a significant increase in blood viscosity at high and low shear rates, as well as a significant reduction of erythrocyte deformability in the patients who were anesthesized with halothane. Significant changes in the rheologic parameters were not found during anesthesia.

Note: The hematocrit correction was only performed mathematically. It is quite probable that there was an interaction between the plasma expanders given and the anesthetics, so that a definite statement concerning the rheologic evaluation of halothane cannot be made. Measurements of further rheologic parameters, e.g., of the plasma viscosity, were not carried out.

In 1982 **Orr** and co-workers [672] studied the effect of an intravenous infusion of althesine (0.5ml/min) on the flow properties of the blood in 11 patients undergoing minor surgical operations. Blood samples were taken before and 5–10 min after the beginning of anesthesia. The blood viscosity was measured using a Contraves Low Shear 30 viscosimeter at 37 °C at a standardized hematocrit of 45%. Plasma viscosity was determined at 25 °C using a capillary viscosimeter. In addition, the blood was filtered according to the method of Reid and co-workers. The authors found a slight decrease in the hematocrit and in the plasma viscosity. The whole blood viscosity was statistically significantly decreased at only one shear rate. There were no changes in the erythrocyte filtrability values. The authors assumed that the minor changes in the rheologic parameters were an effect of the sedation and that they did not point to a direct rheologic effect of the anesthetic.

Note: No measurements of the erythrocyte aggregation.

In 1984 **Gibbs** and co-workers [375] studied the effect of althesine in 12 patients during a long-term infusion and found an increase in the plasma viscosity (letter to the editor, no details).

In 1985 **Gonzalez de Zarate** and co-workers [382] reported on the effect of propanididecremophor EL in eight patients and studied the whole blood viscosity at different shear rates using a Brookfield LVT viscosimeter at a temperature of 37 °C. The dosage was 7–10 mg per kilogram body weight. There was a statistically significant decrease in the whole blood viscosity just a few minutes after administration of the medication. The hematocrit decreased on the average from 43.5 to 41.0, which, however, could not be proven statistically. Thus, the authors came to the conclusion that hemodilution was not the cause of the viscosity reduction.

Note: No measurement of the plasma viscosity.

4.1.16.3 Discussion

Based on these contradictory results it is impossible to make any statements on a possible effect of anesthetics on the flow properties of blood. This is, to a great extent, due to the very small number of articles published and due to methodologic difficulties. In the individual studies there are, however, indications that it is not the anesthetics themselves which have an effect on the blood viscosity, but rather that the slight reduction of the blood and plasma viscosities described by some authors is to be traced back to an effect of vasodilation leading to endogenous hemodilution. Furthermore, the possibility should not be ruled out that a decrease in the blood viscosity can also be the result of the hematocrit reduction resulting from the mandatory blood loss during operations. A short review on this topic was presented by *Dormandy* [213] in 1984.

4.1.17 Gelatine Solutions

4.1.17.1 Introduction

Gelatin solutions are colloidal plasma substitutes that are the product of collagen degradation. In 1951 oxypolygelatin was developed; in 1952 a polymerization from decomposed succinyl gelatin followed, as did a cross-linked polypeptide mixture in 1962. Owing to their low molecular weight of 23 000 to 27 000 gelatin solutions are eliminated from the circulation faster than comparable dextran substances. Gelatin substances are used for shock therapy, in all types of blood and plasma loss, as well as in extracorporal circulation and in hemodialysis.

4.1.17.2 Publications

In 1966 **Schülke** and **Hartel** [776] reported on the improvement in the flow properties of blood after the infusion of 1000 ml 3.5% gelatin solution in patients undergoing cholecystectomy. Plasma viscosity was measured using a capillary viscosimeter at 37 °C. The authors found a slight increase in the plasma viscosity after the infusion of the gelatin solution and a comparatively marked rise in the plasma viscosity after the infusion of a 10% solution of low molecular weight dextran.

Note: Measurements of further rheologic parameters were not carried out.

In 1980 **Motta** and co-workers [658] treated patients with stage II, according to Fontaine, chronic occlusive arterial disease. The blood and plasma viscosities of the heparinized blood samples were measured using a capillary viscosimeter at 37 °C. For blood viscosity measurements, the hematocrit was partially corrected to a value of 45%. After the infusion of 1250 ml 3.5% gelatin solution, there was a strong decrease in the blood viscosity, a slight increase in the blood viscosity with corrected hematocrit, and a 5.6% rise in the plasma viscosity. Similar changes were observed in the flow properties of blood after the infusion of 10% solutions of low molecular dextran.

Note: Measurements of further rheologic parameters were not carried out.

4.1.17.3 Discussion

Motta and co-workers reported on a reduction of blood viscosity after the infusion of a 3.5% gelatin solution in patients with peripheral arterial vascular disease. The authors attributed this blood viscosity reduction to hemodilution (thinning of the blood) which is expressed as a hematocrit reduction. If, on the other hand, the hematocrit was set to a standardized value for the blood viscosity measurements, then the authors found a slight increase of the blood viscosity at a standardized hematocrit value of 45%. Using a standardized hematocrit, the blood viscosity results are independent of the number of the erythrocytes present. This increase in blood viscosity at an adjusted hematocrit was traced back to an increase in the plasma viscosity. In addition, an increase in erythrocyte aggregation and a deterioration of the erythrocyte filtration is conceivable; however, corresponding studies were not performed. Measurements of the

plasma viscosity by *Schülke* and *Hartel* after infusions of 3.5% gelatin solutions also showed a slight increase in the values. *Schülke* and *Hartel* pointed out that the increase in plasma viscosity after gelatin infusions is due to the specific viscosity of the infused gelatin substance itself as human plasma has lower viscosity values.
There are only a few publications on plasma substitutes based on gelatine, so that a proper evaluation of the hemorheologic properties of these substances is not possible.

4.1.18 Streptokinase/Urokinase

4.1.18.1 Introduction

Streptokinase is an enzyme with a molecular weight of 47 000 which is won from a β-hemolyzing streptococcus; it splits fibrin and fibrinogen in the blood plasma. Clinical indications for therapy with streptokinase are mainly deep venous thrombosis, pulmonary embolism, as well as acute and chronic arterial occlusions. It was already determined at an early stage that this fibrinolytic enzyme not only splits fibrin and thereby dissolves the thrombus, but also splits the soluble fibrinogen present in the plasma into split products. This is also the reason for the most significant side effect of streptokinase, i.e., the complications of bleeding due to a major reduction of the plasma fibrinogen concentration accompanied by an increase in the fibrinogen split products in the plasma.
***Ehrly** [235] was able to show in 1969 that the therapeutic reduction of the fibrinogen concentration as a result of fibrinogen splitting with streptokinase in vitro and in an analog manner also in patients in vivo leads to an improvement in the flow properties of the blood. The in vitro studies of **Merrill** and co-workers [636, 637] from 1963 and 1966 already showed that the flow properties of blood could be influenced by the addition or removal of fibrinogen. In in vitro studies in 1966 **Chien** and co-workers [124] also found that the non-Newtonian behavior of the blood is mainly dependent on the plasma proteins, especially fibrinogen.*
In recent years, another thrombolytic enzyme, urokinase, has been used in addition to streptokinase. Urokinase is a highly purified polypeptide made from human urine with a molecular weight of 45 000. This endogenous substance is also able to split fibrin and thereby dissolve thrombi. Since it also splits the soluble fibrinogen at the same time, there is a decrease in the fibrinogen concentration, which in turn has an effect on the flow properties of blood. The indications for urokinase correspond to a large degree to those for streptokinase.

4.1.18.2 Publications

In 1969 **Ehrly** and **Lange** [235] reported for the first time on an improvement in the flow properties of blood during the streptokinase treatment of patients with arterial occlusive diseases. The dosage was 250 000 and 500 000 units per hour, which corresponds to a streptokinase concentration of 50–100 units per milliliter of blood. The results of these experiments with patients correspond to in vitro studies performed at the same time, in which it was shown that the total blood viscosity could be reduced especially at low shear rates. Blood viscosity was measured using a Brookfield micro-cone-plate viscosimeter and an Ostwald capillary viscosimeter; plasma viscosity was only measured using an Ostwald capillary viscosimeter at 23 °C. There was a clear reduction of the viscosity values during streptokinase treatment and at the same time a considerable decrease in the fibrinogen concentration in the plasma. A decrease in erythrocyte aggregation was observed in so-called blow-out smears; a decrease in the blood cell sedimentation rate was also interpreted in this sense.

Note: Quantitative measurements of the erythrocyte aggregation and of the erythrocyte flexibility were not carried out. The temperature of the measurements was 23 °C. There are no statistical studies on a larger number of patients.

In a further publication **Ehrly** and **Lange** [243] discussed their results with regard to the therapy of acute myocardial infarction. The idea was postulated that the endangered peripheral regions around the infarction are better perfused as a result of an improvement in the flow properties of blood, thus possibly limiting the size of the infarct region. The possible clinical importance of the improvement of the flow properties by streptokinase has also been considered elsewhere in connection with the therapy of venous thrombosis and certain forms of shock [238, 240, 244, 252].

In 1971 **Benda** and co-workers [60] reported on fibrinolytic therapy with streptokinase in 28 patients with severe angina pectoris. In 19 patients there was an improvement in the nitroglycerine-resistant stenoçardic pains. A Brookfield micro-cone-plate viscosimeter was used for the measurement of blood viscosity. Following a dose of 500 000 units streptokinase within 45 min followed by a dose of 37 000 units streptokinase in the next 4–8 h, a decrease in the blood viscosity and the plasma fibrinogen concentration was noted. After completion of the streptokinase treatment, an anticoagulant therapy was carried out. The authors explained the favorable clinical effect on an imminent heart infarction as being the result of the reduction of blood viscosity, the inhibition of thrombocyte adhesion and lysis of the fibrin platelet microthrombi in the coronary arteries.

Note: No details of the temperature and the shear rates during the blood viscosity measurements; no evaluation of further rheologic parameters; no correction for hematocrit.

In 1972 **Bapistella** [42] tested the effect of streptokinase on the flow properties of blood in 12 patients with arterial and venous thromboses. After an intial dose of 250 000 units streptokinase, within 30 min the patients received a maintenance dose of 150 000 units streptokinase per hour. Blood viscosity was measured using a Brookfield micro-cone-plate viscosimeter ($11.5\ s^{-1}$) as well as using an Ostwald capillary viscosimeter at a constant hematocrit of 38%. Plasma viscosity was measured using an Ostwald capillary viscosimeter at 23 °C. There was a synchronous drop in the blood viscosity at high and low shear rates, as well as in the plasma viscosity and in the fibrinogen concentration in the plasma, which reached an optimum 3 h after the beginning of infusion. The erythrocyte sedimentation rate and the decreased erythrocyte aggregates in the blow-out smears indicate a disaggregation of the erythrocytes.

A study with similar results was later published by **Ehrly** [255].

Note: The viscosity measurements were made at 23 °C. A quantitative registration of the aggregation behavior of the erythrocytes was not carried out. No measurements of the erythrocyte flexibility.

In 1972 **Benda** [61] reported on reductions of blood viscosity and fibrinogen during fibrinolytic therapy with streptokinase in a patient with acute myocardial infarction. Blood viscosity was measured using a Brookfield micro-cone-plate viscosimeter at shear rates of 5.6–115 s^{-1}.

Note: No measurements of further rheologic parameters. No data on the temperature.

In 1972 **Ehringer** [228] described the hemorheologic effects of streptokinase infusions on patients with peripheral arterial occlusions. The initial dose of 400 000 units streptokinase within 15 min was followed by a maintenance dose of 100 units streptokinase per hour. The blood viscosity measurements were carried out using a Brookfield micro-cone-plate viscosimeter at a shear rate of 11.5 s^{-1} and at 37 °C. There was a reduction of the blood viscosity with a simultaneous decrease in the fibrinogen concentration in the plasma.

Note: The viscosity measurements were carried out without hematocrit correction; other rheologic parameters were not studied.
Similar results were also reported elsewhere [229].

In 1973 **Vogeler** [883] reported on the improvement in the flow properties of blood as a result of streptokinase infusions in patients with peripheral arterial vascular disease. Erythrocyte deformability was determined using an erythrocyte filtration method according to Ehrly and Rossbach (Millipore filter, diameter 8-μm, hematocrit 10%, 12.9 mm H_2O). Blood and plasma viscosities of the heparinized blood samples were measured using an Ostwald capillary viscosimeter at 23 °C. There was a decrease in the blood viscosity and the plasma viscosity; the deformability of the erythrocytes was not influenced.

Note: Measurements of the erythrocyte aggregation were not carried out. There was no hematocrit correction for the blood viscosity measurements. The viscosity measurements were carried out below body temperature.

In 1973 **Benda** and co-workers [62] studied blood viscosity in 40 patients with acute cardiac infarction at a defined shear rate of 5.8 s^{-1} in the course of fibrinolytic therapy with streptokinase. The dosage of the streptokinase was 500 000 units in the first 30 min, 750 000 units in the next 4 h and 750 000 units in the following 8 h. There was a comparable decrease in blood viscosity, in the fibrinogen level, as well as in erythrocyte aggregation as measured using a Fleisch and Perren aggregometer (no exact details).

Note: The blood viscosity measurements were carried out at only one shear rate; no temperature data for the viscosity measurements; no correction of the hematocrit.

Clinical findings: In comparison to an untreated control group, the authors found no reduction in mortality from acute cardiac infarction following treatment with streptokinase.

Also in 1973, **Breddin** and co-workers [86] reported on the short-term fibrinolysis in acute myocardial infarction. Six patients were infused with 750 000 units streptokinase each within 3 h and maintained on a continuous heparin infusion afterwards. Blood viscosity was measured using a Brookfield micro-cone-plate viscosimeter and an Ostwald capillary viscosimeter; plasma viscosity was only measured using an Ostwald capillary viscosimeter. After the administration of streptokinase, the blood and plasma viscosities and the fibrinogen level decreased clearly. The reduction of blood viscosity was strongest at low shear rates. As in earlier studies [244], the decrease in erythrocyte aggregation was determined with the help of the blood cell sedimentation rate and microscopic blow-out smears.

Note: No correction of the hematocrit; no data on the temperature of the measurements. The blood viscosity reduction was not controlled statistically.

Clinical findings: Compared to a control group (heparin), the investigators found a mortality decrease in the cardiac infarction patients after streptokinase treatment.

In the year 1973 **Jauch** [475] studied the effects of 18 thrombolysis treatments with streptokinase on blood viscosity. The patients received 250 000 units streptokinase as an initial dose within 30 min, followed by a maintenance dose of 100000 units streptokinase per hour. Blood viscosity was measured using a falling ball viscosimeter at 37 °C. During fibrinogenolysis there was a simultaneous decrease in the whole blood viscosity and in the fibrinogen level.

Note: No correction for hematocrit; no measurements of further rheologic parameters.

In 1974 **Franke** [356] treated 18 patients with vascular occlusions using an initial infusion of 250 000 units streptokinase in 50 ml isotonic saline solution, which was followed by a maintenance dose of 100000 units per hour. The blood viscosity changes (Brookfield micro-cone-plate viscosimeter at 37 °C and a defined shear rate of 230 s^{-1}) were compared before and after the intravenous streptokinase infusions. The strongest reduction of the blood viscosity was found 2 h after the start of treatment. A comparable reduction of the fibrinogen level and of the hematocrit values was observed.

Note: No correction for hematocrit; no measurements of further rheologic parameters; no viscosity measurements at low shear rates.

In 1976 **Benda** and co-workers [63] tested the behavior of blood viscosity in 57 patients with acute or impending myocardial infarction before and after fibrinolytic treatment with streptokinase. The initial streptokinase dose of 500 000 units within 30 min was followed by various maintenance dosage levels. Blood viscosity was measured using a Brookfield micro-cone-plate viscosimeter at 37 °C and a shear rate of 5.8 s^{-1}. Four hours after the beginning of infusion, a maximal drop was observed in the blood viscosity and the plasma fibrinogen level.

A similar report on the behavior of blood viscosity following streptokinase treatment was published by **Benda** and co-workers in 1977 [64].

Note: No correction for hematocrit; no measurements of further rheologic parameters.

Clinical findings: The authors observed a reduction in the intensity of the previously therapy-resistant stenocardic pains.

In 1977 **Schmid-Schönbein** and co-workers [759] studied 11 patients with chronic occlusive arterial disease. After an initial dose of 250 000–750 000 units streptokinase followed by a maintenance dose of 100 000 units streptokinase per hour, the blood and plasma viscosities were measured using a Brookfield micro-cone-plate viscosimeter with a Mooney adaptor at a temperature of 37 °C and at shear rates of 2.3–160 s^{-1}. During streptokinase therapy, a normalization of the increased plasma viscosity values was observed in comparison to the controls, as well as a decrease in the apparent relative blood viscosity (blood viscosity values divided by plasma viscosity values), which was most marked at a shear rate of 2.3 s^{-1}. The reduction of the apparent relative blood viscosity could not, however, be statistically proven in any shear rate range as compared to the control values. In accordance with the lowered fibrinogen concentration during fibrinolytic therapy, there was also a decrease in erythrocyte aggregation, which could be verified for the shear resistance and formation rate of erythrocyte aggregation using a light-transmission method. The hematocrit varied around 2%.

Note: No measurements of the erythrocyte filtrability (erythrocyte deformability).

In 198O **Theiss** and co-workers [856] carried out a randomized study of patients with acute myocardial infarction. During fibrinogenolytic therapy with streptokinase, the blood and plasma viscosities values were measured using a Brookfield micro-cone-plate viscosimeter with a Mooney adapter at shear rates of 8–80 s^{-1} at 37 °C. The initial dose of 250 000 units streptokinase and the subsequent maintenance dose of 100 000 units streptokinase per hour in 50 ml 15% glucose solution led to a marked decrease in the fibrinogen concentration in the treatment group compared to the placebo group. The plasma and blood viscosity values of the patients, which were statistically raised compared to healthy subjects, normalized. On the other hand, even the viscosity values of the placebo group rose.

Note: Measurements of the erythrocyte aggregation and filtration were not carried out.
Similar results were also reported elsewhere [888].

In 1985 **Arntz** and co-workers [27] treated patients with acute myocardial infarction with intravenous high-dosage streptokinase and found a marked decrease in the whole blood viscosity at low shear rates measured using a Wells-Brookfield cone-plate viscosimeter, while at high shear rates there were only slight changes. Plasma viscosity also decreased statistically significantly. Erythrocyte aggregation, measured using a light-transmission instrument, also showed a distinct decrease, while the erythrocyte filtrability (apparatus according to Teitel) decreased in the streptokinase group as well as in the placebo group (abstract, no details).

In 1985 **Ehrly** and co-workers [307] studied the influence of systemically applied urokinase in patients with thrombotic disease. Synchronous to the decrease in the fibrinogen concentration, there was a reduction of the whole blood viscosity, the plasma viscosity, and the erythrocyte aggregation, while the erythrocyte filtrability remained unchanged (abstract, no details).

Kiesewetter and co-workers [497] studied the influence of a 15-min intravenous administration of 250 000 units urokinase on the flow properties of the blood in ten patients with deep venous thrombosis. In all patients the flow properties of the blood were improved during therapy (yield shear stress, erythrocyte aggregation, and plasma viscosity), while the erythrocyte deformability remained unchanged.

4.1.18.3 Discussion

All the authors found a reduction of blood viscosity in the course of fibrinolysis or fibrinogenolysis with streptokinase, even though the effect of the changing hematocrit on the blood viscosity was disregarded in some studies. It could be shown by several authors that at low shear rates there was an especially marked decrease in blood viscosity both at a temperature of 23 °C as well as at 37 °C.
Therefore, theoretically, a decrease in erythrocyte aggregation should have been expected. There was, in fact, a considerable decrease in the erythrocyte aggregation in corresponding semiquantitative or quantitative measurements. Since the reduction of the

fibrinogen concentration also caused a reduction of the plasma viscosity, it may be said that streptokinase causes a multifactorial improvement in the flow properties of the blood [263]. No changes were found in erythrocyte filtrability i.e., erythrocyte deformability. In summary, it may be said that, in addition to the reduction of the blood and the plasma viscosities, streptokinase effects in particular a decrease in the erythrocyte aggregation. The reduction of the blood viscosity at low shear rates (structural viscosity) is of significance for those vascular areas in which the shear rates are pathophysiologically very low. The improvements in the flow properties of blood may explain the clinical effects of streptokinase found by some authors, as, for example, the decrease in the frequency of angina pectoris attacks in patients with impeding myocardial infarction [63, 64].
Compared to conventional heparin therapy, the reduction in the mortality of acute myocardial infarctions by the use of streptokinase may be mainly attributed to the improvement in the flow properties of blood. It is probable that by the use of streptokinase the perfusion in the peripheral areas of the infarctions (penumbra) is improved and a dissemination of intravasal microthrombi takes place [86]. This could prevent a further expansion of the infarction area.
It is also of interest that treatment with streptokinase already leads to a subjective improvement in acute deep venous thrombosis with livid discoloration and pains in the legs, even when the occlusion still remains phlebographically unchanged. An improvement in the hemorheologic parameters in the microcirculation may also be made responsible for this [240].
According to the studies of *Ehrly* and co-workers and *Kiesewetter* and co-workers the rheologic response to urokinase is similar to that of streptokinase.

4.1.19 Ancrod/Batroxobin

4.1.19.1 Introduction

In 1963 ***Reid*** *and co-workers [720] published initial reports on the anticoagulatory activity of a poisonous snake enzyme from the Malaysian pit viper* (Agkistrodon rhodostoma). *They observed that after a nonlethal snake bite the blood of humans was incoagulable for several days.* ***Reid*** *believed to have found a new anticoagulant. Shortly thereafter intensive research efforts led to the separation of a fibrinogen-reducing fraction (generic name: ancrod) from the total snake poison. Chemically ancrod is a glycoprotein with a molecular weight of 30 000. The means of action of the reduction of the fibrinogen concentration by ancrod is caused by a splitting of the fibrinopeptides A, AP, and AY from the fibrinogen molecule, so that the fibrin monomeres which are produced polymerize to fibrin filaments and are split up by the lytic activity of the endogenous fibrinolysis.*
After intravenous administration, which was standard in the early days, there was a rapid decrease, almost to zero, in the fibrinogen concentration in the blood. The induced anticoagulant action of the drug was first used for the treatment of deep venous thrombosis, thrombophlebitis, myocardial infarction, priapism, and thrombosis of the central vein of the retina. This substance could not, however, assert itself as an anticoagulant ver-

sus the other standard coagulation-inhibiting substances. In 1970 **Ehrly** *[237] recommended the use of ancrod for the therapy of chronic arterial vascular disorders. The theoretical basis for this recommendation was based on the assumption that, under unchanged hemodynamic and anatomic conditions, there should be an increase in the microcirculatory flow rate resulting from an improvement in the flow properties of blood. In the following period therapeutic defibrinogenation was confirmed as a new principle for the treatment of peripheral arterial occlusive diseases [263]. Batroxobin is, in analogy to ancrod, a purified fraction of the poison from the snake* Bothrops atrox.

4.1.19.2 Publications

In 1971 **Ehrly** [253], **Ehringer** and co-workers [223], and **Vinazzer** [875] presented results on a reduction of the blood viscosity in the course of therapy with ancrod.

Ehrly [253] studied the flow properties of the blood over the first 3 h after intravenous administration of one to three ampules of ancrod (70–210 units) in five patients with peripheral arterial and venous vascular disorders. The measurements of the blood, plasma, and serum viscosities were carried out using an Ostwald capillary viscosimeter, and that of the blood viscosity using a Brookfield micro-cone-plate viscosimeter. During the therapy with ancrod, there was a simultaneous reduction of the fibrinogen concentration and of the blood and plasma viscosity values, whereby the plasma viscosity almost reached the values of the serum viscosity. The reduction of the blood viscosity was marked at all shear rates, but particularly so at low shear rates. The normalization of the raised erythrocyte sedimentation rate indicated the reduction of erythrocyte aggregation. This result also corresponds to the microscopically observed reduction in the number and size of the erythrocyte aggregates present in the so-called blow-out smears.

Note: The temperature of the viscosity measurements was not given; no hematocrit correction. Measurements of further rheologic parameters were not carried out.

Clinical findings: Within the framework of therapy with ancrod and in the course of the improvement in the flow properties, there was an improvement in the ischemic symptoms, in particular a reduction of the pains at rest.

Also in 1971 **Ehringer** and co-workers [223] reported on a reduction of the blood viscosity within the framework of intravenous treatment of patients with chronic occlusive arterial disease. The patients received 134 units ancrod twice daily within a treatment period of 10 days. At the end of treatment there was a clear reduction of the blood viscosity parallel to the reduction of the fibrinogen level to values of approximately 40 mg%.

Note: No precise data as to the technique of measurement and the type of viscosimeter used; no measurements of further rheologic parameters; no correction for hematocrit.

Clinical findings: A correlation was assumed between the improved flow conditions of the blood and the disappearance of pains at rest in four of the five patients with multiple vascular occlusions.

In 1971 **Vinazzer** [875] studied the blood and plasma viscosities in five patients before and after a 5-day treatment with ancrod. As an initial dose the patients received a continuous slow drip infusion of 150–200 units ancrod in 500 ml isotonic saline solution over an infusion time of 4 h and thereafter a maintenance dose via short infusions of 75–150 units ancrod in 100 ml isotonic saline solution twice daily.

A viscosity scale by Haake and Co. (ball balance viscosimeter) was used for the viscosity measurements at 20 °C and at three different shear rates. The viscosity values were expressed as relative viscosity (H_2O = 1). The therapy with ancrod resulted in a clear decrease in the blood and plasma viscosities.

Note: No measurements of the erythrocyte aggregation and the erythrocyte filtration; no data on the shear rates used.

In the following period **Ehrly** [238, 244] and **Ehringer** and co-workers [224] published further studies which basically contained the results already mentioned.

In the year 1972 **Ehrly** [245, 255, 246] presented more detailed and statistical studies on the improvement of the flow properties of blood after therapeutic defibrinogenation using ancrod in 12 patients with peripheral arterial occlusive diseases. With an initial dose of 490–560 units ancrod per day and a maintenance dose of 350–420 units ancrod (administered intravenously with a perfusor), there was an extensive reduction of the plasma fibrinogen level within a few days. The blood viscosity values (Brookfield micro-cone-plate viscosimeter) decreased statistically significantly at shear rates of 5.8–46 s^{-1} and at a temperature of 23 °C, while the hematocrit values remained unchanged. The viscosity reduction was particularly marked at low shear rates. The plasma viscosity (Ostwald capillary viscosimeter, 23 °C) also decreased statistically significantly.
The viscosity values remained low during the entire period of therapy and became normal again after cessation of treatment. Parallel to the reduction of the fibrinogen level, the erythrocyte sedimentation rate reached physiologic values, and no pathologic erythrocyte aggregates could be shown in the blow-out smears. The deformability of the erythrocytes, measured using a filtration method according to Ehrly and Rossbach at 23 °C and 37 °C, was not affected by ancrod.
These rheologic and clinical results of the therapy with ancrod were presented elsewhere as a new rheologicotherapeutic concept for the treatment of chronic peripheral arterial occlusive diseases [249, 248].

Note: Except for the erythrocyte filtration the measurements were carried out at 23 °C. A quantitative measurement of the erythrocyte aggregation was not performed.

Clinical findings: The improvement in the microcirculatory flow and supply of the extremities correlated in a time-dependent manner with the disappearance of pains at rest.

In 1972 **Ehringer** and co-workers [225] reported on patients with rest pains due to chronic arterial occlusive diseases. On the 1st day the patients received 134 units ancrod over 6 h and on the 2nd–10th day 134 units ancrod every 12 h as a short intravenous infusion. Blood viscosity was measured using a Brookfield micro-cone-plate viscosimeter at shear rates of 11.5–230 s^{-1} and at a temperature of 37 °C. During treatment there was a simultaneous decrease in blood viscosity, the plasma fibrinogen level, and the hematocrit.
Similar studies were also published by this team in 1973 [227].

Note: No hematocrit correction; no measurements of further rheologic parameters.

Clinical findings: The authors found a slight but statistically nonsignificant increase in the ultrasound Doppler pressure values distal to the occlusions in the arteries in the extremities and an increase in the rest flow and in the reactive hyperemia of the lower leg. In five of the nine patients the symptoms improved and remained so even after the end of treatment.

In 1973 **Ehrly** and **Jung** [251] studied the effect of ancrod on the flow properties of human blood with the subcutaneous application of the substance in healthy subjects and in patients with chronic arterial occlusive diseases and angina pectoris. Blood viscosity was determined using an Ostwald capillary viscosimeter and a Brookfield micro-cone-plate viscosimeter; the plasma and serum viscosities were measured using an Ostwald capillary viscosimeter. The aggregation behavior of the erythrocytes was evaluated microscopically, and the erythrocyte filtrability was tested according to Ehrly and Rossach. In both cases the temperature of the measurements was 23 °C.
Compared to the intravenous application, the subcutaneous administration of ancrod effected a sustained reduction of the blood and plasma viscosities and of the erythrocyte aggregation. The fibrinogen level gradually decreased within 2–3 days to the desired low concentration of about 80 mg%. The simplicity of application and the easily controllable dosage made it possible to maintain the fibrinogen level at the desired low level during the entire period of treatment. Similar changes were observed in the flow properties of the blood of healthy subjects.
These results were presented elsewhere as well [254, 263, 254].

Note: All measurements were made at 23 °C, and a quantitative determination of the erythrocyte aggregation was not carried out.

Clinical findings: During treatment there was a decrease in the foot pains at rest, and in some cases there was also a reduction of the anginal complaints.
In 1973 the theoretical and experimental situation concerning the correlation between the improvement of the flow properties of blood and the therapy of angiologic diseases was treated in a monography using previously published measurement data [252]. The previously stated concept concerning the therapy of peripheral arterial occlusive diseases by improving the flow properties of blood [245] was described in more detail.

In 1973 **Völker** reported [879] on rheologic changes after medicinal therapy with batroxobin, a highly purified snake poison enzyme from the Brazilian viper *Bothrops atrox*. Eleven patients with chronic arteriopathies in stages III and IV according to Fontaine were treated for a period of 5 days. The patients received a short daily infusion of 75 µl batroxobin per kilogram body weight. A Brookfield micro-cone-plate viscosimeter was used for the measurement of the blood and plasma viscosities. The aggregation tendency of the erythrocytes was observed using a rheoscope. The result was a reduction of the blood and plasma viscosities; the fibrinogen level which was initially around 300 fell to 19 mg%. Erythrocyte aggregation was decreased by batroxobin. There was a 3% change in the hematocrit.

Note: No data on the temperature of the measurements; no hematocrit correction.

The study by **Völker** and **Martin** [877] in the year 1973 on the behavior of the blood viscosity after batroxobin corresponds to a large degree to the above-mentioned study.

In 1975 **Ehrly** [252, 263] studied the effect of a subcutaneous dose of ancrod in large number of patients with severe chronic occlusive arterial disease in stages III and IV according to Fontaine. Using an Ostwald capillary viscosimeter, the blood, plasma, and serum viscosities ($H_2O = 1$) were measured at 23 °C. The dosage used here was 1 unit ancrod per kilogram body weight daily for 4 days and on the 5th day a depot of 4 units per kilogram body weight. As a follow-up a depot injection of 4 units ancrod per kilogram body weight was given twice a week. During the treatment there was a decrease in blood and plasma viscosities (no data on the significance).

Note: All measurements were made at 23 °C. Quantitative erythrocyte aggregation measurements were not carried out.

Clinical findings: There was a decrease in the pains at rest and a more rapid healing of ulcers in patients at stage IV according to Fontaine.

In 1975 **Tilsner** and **Greul** [860] reported on the behavior of the blood, plasma, and serum viscosities following the subcutatenous application of ancrod in 53 patients with chronic arterial occlusive diseases in stages III and IV according to Fontaine. The patients received 1 unit ancrod per kilogram body weight daily until they reached a therapeutically effective low fibrinogen level and then, as a maintenance dose, 3 units ancrod per kilogram body weight twice a week. The therapy was continued for 4–8 weeks. At the same time as the reduction of the fibrinogen level, there was a decrease in the blood and plasma viscosities with almost unchanged serum viscosity.

Note: No information on the viscosimeter used or other viscosimetric data. Further rheologic parameters were not measured.

Clinical findings: During treatment there was a cessation of the pains at rest in 43 patients; in some patients the necroses also healed. Six weeks after the end of therapy only 33 patients were still free of pains at rest. Of the 26 patients who could be observed for 1 year, 15 still showed favorable clinical results 1 year after the end of therapy.

In 1976 **Dormandy** and co-workers [203] treated 18 patients who were suffering from intermittent claudication, pains at rest, or gangrene using ancrod. The dosage of 8 units ancrod per kilogram body weight was given subcutaneously twice a week. The whole blood viscosity was measured at a shear rate of 0.77 s^{-1}. As the fibrinogen level decreased from an average of 510 mg% to values of about 50 mg%, the blood viscosity fell clearly in all patients.

Similar data were presented in a review article in 1976 [205].

Note: No data on the type of viscosimeter and the temperature of the measurements. Further rheologic parameters, such as, for example, plasma viscosity, were not measured; no hematocrit correction.

Clinical findings: In 14 of the 18 patients the pains at rest stopped and some of the necroses healed. The patients with intermittent claudication reported an increase in the pain-free walking distance.

In 1976 **Ehrly** and **Köhler** [269] tested the flow properties of blood within the framework of a modified dosage scheme for the subcutaneous application of ancrod in 30 patients with severe chronic arterial occlusive diseases, in whom the usual conservative and physical methods of treatment available had not brought any clinical improvement. In these cases 1 ampule ancrod (70 units) daily was injected subcutaneously, independently of the body weight and initial fibrinogen level. In ten patients the relative blood, plasma, and serum viscosities (23 °C, H_2O = 1) were determined using an Ostwald capillary viscosimeter.

Synchronous to the prolonged reduction of the plasma fibrinogen concentration to values of about 100 mg%, there was a reduction of the blood and plasma viscosities while the serum viscosity remained constant. The erythrocyte sedimentation rate decreased with corresponding hematocrit values, which was attributed to a decrease in the erythrocyte aggregation. The extent of the viscosimetric changes is comparable to the dosage scheme described earlier [260].

Note: All measurements were carried out below body temperature at 23 °C. Quantitative studies of the erythrocyte aggregation were not carried out. There are no viscosimetric measurements at low shear rates.

Clinical findings: In the patients with severe intermittent claudication there was an increase in the pain-free walking distance, and in some of the other patients the ischemic rest pains disappeared as well.

In 1976 **Forbes** and co-workers [352] treated 13 patients with thromboembolic occlusive diseases using a 12 h continuous drip infusion with 2 units ancrod per kilogram body weight for 3–6 days. The blood samples, anticoagulated with EDTA, were measured within 30 min after sampling. The blood viscosity measurements (Dintenfass viscosimeter) and the plasma viscosity measurements (U-shaped capillary viscosimeter) were performed at 37 °C. With the reduction of the plasma fibrinogen level, the blood and plasma viscosities also decreased.

Note: No hematocrit corrections; no data on the shear rates.

In 1977 **Schmitt** [766] reported on the improvement of the flow properties of blood as a result of ancrod in patients with peripheral chronic arterial disease. Erythrocyte aggregation was measured using a light-transmission aggregometer at a constant hematocrit and at 37 °C. During the therapy with ancrod, the author found a decrease in the erythrocyte aggregation simultaneous to the decrease in the fibrinogen level.

Note: No measurements of further rheologic parameters.

In 1977 **Dormandy** and co-workers [206] reported on rheologic studies during subcutaneous administration of ancrod in 15 patients with intermittent claudication (introductory phase: 4 units per kilogram body weight daily for 6 days; and a maintenance phase: 8 units ancrod per kg body weight twice a week for 4 weeks). A rotational viscosimeter (no details) was used at shear rates of 0.77–91 s^{-1}. After only a few days there was, parallel to the reduction of the fibrinogen level, a statistically significant reduction of the blood viscosity at all shear rates, which was most marked at low shear rates.

Note: No hematocrit correction. Further rheologic parameters were not measured. No temperature data for the viscosity measurements.

Clinical findings: All patients reported an increase in the pain-free walking distance.

Similar results were also published by **Dormandy** elsewhere [202, 207].

In 1977 **Humphreys** and co-workers [465] studied the blood viscosity within the course of intravenous treatment with ancrod in 16 patients with peripheral arteriopathies, with pains at rest and gangrene of the skin less than 3 cm^2 in size. The whole blood viscosity was measured using a Contraves Low Shear 2 rotational viscosimeter at the following shear rates: 124 s^{-1}, 0.675, and 0.786 s^{-1}. The hematocrit was corrected mathematically to a value of 45%. After an initial dose of 4 units ancrod per kilogram body weight in 500 ml NaCl over the first 8 h, a short infusion of 2 units ancrod per kilogram body weight was given twice a day on each of the following 4 days. Simultaneously with the reduction of the fibrinogen level from 4.24 g/l–0.455 g/l, there was a statistically significant reduction of the blood viscosity which was particularly prominent at low shear rates.

Note: The temperature of the measurements was not given and other rheologic parameters were not measured.

Clinical findings: Subjectively, the pains disappeared in only one patient and in five patients there was a slight reduction of the pains at rest. After the end of therapy the pains at rest returned again.

In 1978 **Lowe** and co-workers [584] studied the behavior of the viscosity values during antithrombotic therapy with ancrod in 110 patients with fractures of the femur of the neck (double-blind study). Blood viscosity was measured using a Dintenfass coaxial rhombospheric rotational viscosimeter at a temperature of 37 °C and at shear rates of $12\ s^{-1}$–$280\ s^{-1}$. The hematocrit was corrected mathematically to a value of 40%. Plasma viscosity was determined using a capillary viscosimeter at 37 °C. At the end of surgery 53 patients received four ampules ancrod of 70 units each, subcutaneously and one ampule each over the following 4 days. Unlike the increase in the fibrinogen values observed in the control group, there was a clear reduction of the fibrinogen level to 0.9 g/l in the ancrod group accompanied by a statistically significant decrease in the blood and plasma viscosities. Similar results were published elsewhere [585].

Note: The correction of the hematocrit values was only carried out mathematically; no measurement of the erythrocyte deformability and the erythrocyte aggregation.

In 1979 **Lowe** and co-workers [586] tested the viscosity-reducing effect of subcutaneous ancrod injections in nine patients with intermittent claudication and pains at rest. Blood viscosity was measured using a Dintenfass coaxial rhombospheric rotational viscosimeter at shear rates of 12–280 s^{-1}; plasma viscosity was determined using a capillary viscosimeter at 37 °C. After an initial dose of 4 units ancrod per kilogram body weight, all subjects received an injection of 1–4 units ancrod per kilogram body weight subcutaneously over a period of 1–4 days. At the same time as the decrease in the fibrinogen level from an initial 3.76 g/l–0.31 g/l on the 5th day of treatment, a statistically significant decrease was observed in the blood and plasma viscosities. The duration of treatment was 3 weeks.

Note: No measurement of the erythrocyte deformability and the erythrocyte aggregation. The hematocrit correction was only carried out mathematically. The improved flow properties of blood had no effect on the values obtained by venous occlusion plethysmography; however, an improvement in the circulation in the skin was measured using the ^{133}xenon clearance method.

Clinical findings: In three of the six patients there was an increase in the pain-free walking distance; two of the three patients with pains at rest were completely relieved of their complaints.

In 1979 **Yasunaga** and co-workers [929] studied the effect of batroxobin on the flow properties of blood in eight patients with chronic occlusive arterial disease. Batroxobin was administered intravenously as a drop infusion (20–40 μl per kilogram body weight) in order to maintain the fibrinogen concentration at a range of 80–150 mg%. Blood viscosity was measured using a Brookfield micro-cone-plate viscosimeter at a shear rate of 115 s^{-1} and at 37 °C. Within hours there was a reduction of the blood viscosity, the lowest being measured 24 h following the start of therapy.

Note: No hematocrit correction; no measurements of further rheologic parameters.

The effect of ancrod on the filtrability of the blood in patients after a hip joint operation was studied by **Ernst** and **Dormandy** in 1981 [319]. These patients first received an intravenous injection as an initial dosage; in the course of another 10 days the medication was applied subcutaneously. Erythrocyte filtration was determined according to the method of Dodds and co-workers. In these patients treated with ancrod, the authors found a longer-lasting decrease in the filtrability of the blood compared to an operated group without ancrod.

Note: No data on the anticoagulants used for blood sampling. In in vitro studies there was no direct effect of ancrod on the erythrocyte filtrability. The authors deduce from their studies a correlation between fibrinogen split products and erythrocyte filtrability.

In 1982 **Lowe** and co-workers [591] studied the effect of an 8-day therapy with the intravenous administration of ancrod in a double-blind study with 27 patients with ischemic pains at rest. The blood viscosity was measured using a Dintenfass coaxial rhombospheric rotational viscosimeter; plasma viscosity was determined using a capillary viscosimeter at 37 °C. The blood and plasma viscosities decreased statistically significantly.

Note: No measurements of further rheologic parameters.

Clinical findings: The authors found no improvement in the pains at rest in the ancrod group compared to a placebo group. However, instead of the usual 4-week therapy, only an 8-day therapy was carried out.

In 1984 **Kindl-Mascher** [506] studied the effect of batroxobin on the flow properties of the blood of patients with intermittent claudication. The blood as well as the plasma viscosities were measured using a capillary viscosimeter at 37 °C. Erythrocyte aggregation was studied photo-optically. Erythrocyte filtrability was determined according to the method of Ehrly and Rossbach. Following the subcutaneous injection of batroxobin, there was a statistically significant reduction of the whole blood and plasma viscosities as well as a considerable decrease in erythrocyte aggregation. There were no changes in erythrocyte filtration.

In 1985 **Kollár** and co-workers [520] studied the influence of ancrod on the whole blood viscosity, the plasma viscosity, and the erythrocyte aggregation in a double-blind study. During therapy they found a reduction of these parameters (abstract, no details).

Cappelli and co-workers [115] studied the influence of defibrotide on the flow properties of the blood of ten patients with intermittent claudication. Defibrotide is a poly-deoxynucleotide extracted from lung tissue. It was infused with a 100 ml saline solution over about 30 min. There was a reduction of the fibrinogen concentration, a decrease in the whole blood and plasma viscosities, as well as an improvement in the erythrocyte filtrability.

4.1.19.3 Discussion

Although from the present point of view many of the hemorheologic methods and studies during the early years of therapy with defibrinogenating enzymes (ancrod, batroxobin) showed methodologic deficits, nonetheless, the decrease in the whole blood viscosity correlating with the decrease in the fibrinogen level was in accordance with the findings of all authors and independent of the dosage and the type of viscosimeter used. Some viscosity measurements were not made at constant hematocrit values; in others the viscosity values were only mathematically corrected to a particular standard hematocrit.

The few viscosometric studies in which the whole blood viscosity was measured at various shear rates and particularly at low shear rates showed a clear decrease

in the whole blood viscosity at low shear rates of below 50 s^{-1}. The plasma viscosity measurements verify a decrease in the values with reduced fibrinogen concentrations in the plasma both at 23 °C and at 37 °C.
Indirect measurements of erythrocyte aggregation, which were only performed in some cases, showed a clear erythrocyte disaggregating effect during therapy with ancrod or batroxobin. In three studies the erythrocyte aggregation was also measured directly in a quantitative manner using a photometric technique (light transmission). The evident decrease in the formation of erythrocyte aggregates is also an explanation for the particularly strongly expressed reduction of the blood viscosity at low shear rates (structural viscosity). At high shear rates (as measured using an Ostwald capillary viscosimeter), the decrease in the plasma fibrinogen concentration also contributes to a decrease in the whole blood viscosity due to a reduction of the total protein concentration.
Studies by *Ehrly* and co-workers [245, 255] on the effect of ancrod on the erythrocyte filtrability did not show any changes in this parameter. These findings are in agreement with the results of *Sirs* and co-workers, which were obtained using a different methodology in animal experiments [800]. On the other hand, *Ernst* and *Dormandy* found a decrease in the filtrability of the blood in hip-operated patients, which was even more marked with a high dose of ancrod. The authors could not find a direct effect of ancrod on the filtrability of the blood in vitro, but they did find a positive correlation to the concentration of fibrinogen split products. Possibly the decrease in the filtrability of blood in the studies of these authors may be the result of a partial coagulation of the blood in vitro, which cannot be prevented with the standard anticoagulant additives.
With batroxibin there was a decrease in the blood and plasma viscosities as well as in erythrocyte aggregation [879, 929]. From the studies of *Kindl-Mascher* it follows that the filtrability of the blood is not affected by the subcutaneous administration of batroxobin.
Positive clinical improvements were reported by almost all authors. In addition, *Lowe* [586] determined an improvement in the circulation in the skin. In a double-blind study *Lowe* and co-workers reported on the improvement of the flow properties of blood after the administration of ancrod, while the ischemic pains at rest did not decrease to a statistically significant degree. Since this therapy only lasted 8 days, a clinical improvement was hardly to be expected in this short time.
In summary, it may be said that, owing to defibrinogenating substances, there is a multifactorial improvement in the flow properties of the blood. There is a reduction of the plasma viscosity and the blood viscosity at low shear rates (structural viscosity) and a decrease in the aggregation tendency of the erythrocytes without any effect on the number of oxygen transporting erythrocytes. The opinions on the effect of defibrinogenating substances on the erythrocyte filtrability (erythrocyte deformability) diverge; these studies were carried out with various patient collectives and with different substances and various methodologic approaches.
The reduction of the blood viscosity at low shear rates should be of particular importance for clinical needs. These low shear rates are found in the entire venous system as well as in the arterial system distal to occlusions. It may be assumed that, especially in post-steno-

tic ischemic areas (low-flow areas), the dynamic aggregation of the erythrocytes as well as the exponential rise in the structural viscosity of the blood can be prevented to a large degree by defibrinogenating enzymes (therapeutic effect on the circulus vitiosus of erythrocyte aggregation [252, 263]).

4.1.20 Electrically Induced Sleep

4.1.20.1 Introduction

It may be necessary to put patients to sleep for therapeutic reasons. One method is the induction of sleep by means of electrical stimulus.

4.1.20.2 Publications

In 1970 **Magora** and co-workers [606] studied the effect of a 1-h long, electrically induced sleep on the blood viscosity and the hematocrit in 33 patients. Blood viscosity was measured using a Brookfield LVT cone-plate viscosimeter at 37 °C and at various shear rates. The blood samples, which were removed from the veins without applying a tourniquet, were anticoagulated with heparin. After 1 h of electrically induced sleep, there was a reduction of the blood viscosity compared to the initial values in 27 of the 33 patients. The viscosity decrease corresponded to a reduction of the hematocrit. Surprisingly, the blood viscosity also decreased in those patients who served as controls and in whom the electrically induced sleep was only simulated. The authors suspect that the reduction of the blood viscosity is associated with a decrease in the sympathicotonic activity.

Note: No measurements of further hemorheologic parameters.

4.1.21 Contraceptives

4.1.21.1 Introduction

Varying dosages of various, in part chemically modified, female sexual hormones are used as contraceptives. For some years now reports have been available on thrombo-embolic complications following the oral intake of contraceptives. These causal connections and the question of the pathogenetic mechanism of these thrombo-embolic occurrences induced several investigators to carry out measurements on the flow properties of blood. It is conceivable that thrombi are produced with raised blood viscosity and correspondingly decreased blood flow velocity in the sense of Virchow's trias. In addition, it is supposed that a unique hypercoagulability occurs when the flow slows down.

4.1.21.2 Publications

In 1971 **Aronson** and co-workers [29] studied the effect of two oral contraceptives containing ethinylestradiol on the flow properties of the blood of 47 healthy women for an observation period of 6 months. The authors found an increase in the hematocrit and blood viscosity values in eight women (17%) (Brookfield micro-cone-plate viscosimeter model LTV; 37 °, 11.5, 23.0, 46.0, 115.0, and 230.0 s^{-1}). The values returned to normal 2 months after discontinuation of the contraceptives.

Note: No correction of the hematocrit; no measurements of further rheologic parameters.

Clinical findings: In the women with the highest hematocrit and viscosity values the experiment had to be stopped after 3 months owing to thrombophlebitis.

In 1972 **Oski** and co-workers [674] studied the erythrocyte deformability through strips of filter paper according to the method of Teitel in 70 women (36 women were the control group) who were taking various contraceptives (norethindrone and mestranol, ethynodiol and mestranol, norgestrel and ethinylestradiol). The temperature at measurement was 37 °C. Compared to the control group, the investigators observed a significant decrease in erythrocyte deformability. Among the various contraceptives there were no significant differences in the intensity of action in regard to the erythrocyte deformability. Similar changes could be determined in four volunteer male subjects. It could not be determined whether the changes in the erythrocyte flexibility were caused by estrogens or gestagens.

Note: No measurements of further rheologic parameters.

In 1975 **Durocher** and co-workers [221] compared the effect of contraceptives of different composition on the erythrocyte filtrability (polycarbonate filters) after at least 2 months of intake with the data before the start of the investigation. Erythrocyte filtrability was measured using washed erythrocytes. The authors could not find any influence of the contraceptives on erythrocyte deformability.

Note: No measurements of further rheologic parameters.

In 1979 **Leonhardt** and co-workers [565] reported on changes in the flow properties of the blood as a result of oral contraceptives. Thirty-three women who had been taking contraceptives continuously for 2 years were studied. The measurement of the blood viscosity, the plasma viscosity, and the relative blood viscosity (blood viscosity divided by plasma viscosity) was carried out with a corrected hematocrit of 40% and at two shear rates (46 and 115 s^{-1}) using a Brookfield micro-cone-plate viscosimeter. In addition, the plasma viscosity was determined using a capillary viscosimeter (37 °C). Compared to a control group, the authors found a significant increase of the relative blood viscosity as the result of the oral intake of contraceptives only at a shear rate of 46 s^{-1}. The plasma viscosity did not change.

Note: No measurements of further rheologic parameters.

In 1980 **Lowe** and co-workers [587] reported on hemorheologic studies in 25 women who had taken oral contraceptives for 3 months. The measurement of the blood and plasma viscosities was carried out with a rotational viscosimeter (100 s^{-1}). The hematocrit was set to a value of 45% for the blood viscosity measurements. The authors found an increased blood viscosity after the intake of oral contraceptives, which they attributed to an increased plasma fibrinogen level and an increased hematocrit level. Changes in the plasma viscosity were not observed.

Note: No data on the temperature of the measurements; no measurement of further rheologic parameters.

In 1980 **Buchan** and co-workers [92] studied the blood and plasma viscosities in 38 women who had been taking a contraceptive (ethinylestradiol and norgestrel) for six months. A Deer rheometer (37 °C) was used, and the erythrocyte deformability (erythrocyte flexibility) was measured according to the method of Buchan (no detailed data given). After these women had taken contraceptives for 6 months, the authors found a significant increase in the blood and plasma viscosities, in the fibrinogen level, and in the hematocrit, which was not the case with a control group who did not take contraceptives. Furthermore, erythrocyte deformability was reduced.

Note: No correction for the hematocrit; no measurements of the erythrocyte aggregation.

In 1980 **Easthope** and **Brooks** [222] found no statistically significant changes in blood viscosity as measured with a rotational viscosimeter at 25 °C in 31 healthy subjects, 11 of whom were taking oral contraceptives. However, there was a tendency to higher values in the fibrinogen concentration and the blood and plasma viscosities in the group of women taking oral contraceptives; the differences were not statistically significant.

In 1981 **Heilmann** and co-workers [432] reported on the effect of oral contraceptives on the whole blood viscosity (Wells-Brookfield viscosimeter at different shear rates and constant hematocrit) and the eythrocyte deformability (Nucleopore filters, 5-μm pore diameter). In comparison to a control group, the authors found a deterioration of the erythrocyte filtrability in the women taking ovulation inhibitors. In addition, the whole blood viscosity was raised at high shear velocities, while the plasma viscosity remained unchanged.

In 1982 **Palareti** and co-workers [683] reported on hemorheologic changes caused by oral contraceptives. In the course of a 1-year intake of a low-dose oral contraceptive, there was a slight deterioration of the filtration values of the blood compared to a control group. However, no difference could be determined in many women. Blood viscosity, measured at 230 s^{-1}, and plasma viscosity were on the average slightly increased as well, while the plasma fibrinogen concentration was statistically no different from that of the control group (abstract, no details).

In 1982 **Buchan** [93] described raised whole blood viscosity values and a decreased erythrocyte deformability in women taking oral contraceptives who, in addition, had increased blood pressure values and/or were heavy smokers (abstract, no details).

4.1.21.3 Discussion

The first publication on a deterioration of the flow properties of blood as a result of the use of oral contraceptives was published in 1971 by *Aronson* and co-workers. In 17% of the women examined, the authors found an increase in the hematocrit and the blood viscosity at all the shear rates measured. They believed that the flow rate was in a reciprocal relationship to the viscosity and that an increase in the blood viscosity led to a considerably stronger reduction of the flow rate. The authors discussed the idea that this slowing down of the flow led to an increase in the structural viscosity (blood viscosity at low shear rates), which in turn would favor the formation of thrombo-embolic complications. In a further publication *Leonhardt* and co-workers reported on an increase in blood viscosity at low shear rates and with a corrected hematocrit. The authors pointed out that the intake of oral contraceptives increased the reversible erythrocyte aggregation and thereby caused an increase in the structural viscosity.

Lowe and co-workers, *Buchan* and co-workers, and *Heilmann* and co-workers also reported on an increase in blood viscosity and the hematocrit. *Buchan* discusses the problem that up to now no comparative studies have been carried out on changes in the flow properties of blood during the menstrual cycle and while taking oral contraceptives and emphasizes that corresponding deteriorations of the flow properties of the blood were also observed during the menstrual cycle. *Buchan* therefore does not attribute the formation of thrombo-embolic complications to the deterioration of the flow properties of blood as a result of oral contraceptives but rather to other mechanisms.

Leonhardt and co-workers, *Lowe* and co-workers, and *Heilmann* and co-workers did not report any significant changes in plasma viscosity. Nevertheless, *Lowe* and co-workers surprisingly found an increase in the plasma fibrinogen level. *Buchan* found both an increase in the fibrinogen level and in the plasma viscosity.

While *Easthope* and *Brooks* found an increase in the fibrinogen level, a corresponding change was not described by *Palareti* and co-workers. *Oski* and co-workers, as well as *Buchan* and co-workers and also *Heilmann* and co-workers found a decrease in the erythrocyte deformability. The filtration studies of *Palareti* and co-workers, on the other hand, showed that, despite the use of contraceptives, most patients did not have deteriorated filtration values. *Durocher* and co-workers also did not find any influence on the erythrocyte deformability. The methods used by the various authors are so different that no definite statement can be made concerning the question of the effect of oral contraceptives on the filtrability.

In summary, it may be said that most authors reported an increase in the blood viscosity and hematocrit values which were independent of the type of contraceptive used. A hematocrit correction was carried out in some reports for the blood viscosity measurements, after which the increase in the blood viscosity values was less marked. The conclusions regarding the behavior of plasma viscosity differ; most investigators found a slight increase in plasma viscosity. The conclusions on erythrocyte filtrability also diverge.

Many investigators attribute the observed thrombo-embolic complications to an increase in erythrocyte aggregation and in the structural viscosity due to increased hematocrit and fibrinogen values. However, quantitative measurements of erythrocyte aggregation were not performed by any group. A complete evaluation is made more difficult by the fact that, apart from the different methods of measurement, noncomparable combinations of contraceptives were used and that the rheologic studies were partly carried out at different phases in the menstrual cycle.

4.1.22 X-Ray Contrast Media

4.1.22.1 Introduction

Diagnostically necesssary X-ray presentations of arterial and venous vessels are made possible by the intravasal administration of iodine combinations. With the use of intravasally applied X-ray contrast media, there are known side effects that cannot solely be attributed to allergohyperergic mechanisms. For example, the pain which is known to occur with the intra-arterial injection of contrast media cannot yet be explained as far as its mechanism of development is concerned. A vessel intima that is affected by hyperoncotic conditions could be assumed, as could a temporary microcirculatory ischemia due to rigid erythrocytes. The question of whether rheologic factors play a role has been studied in a series of publications. Different X-ray contrast media were used for this purpose (iothalaminic acid, diatrizoic acid, adipiodon and metrizoic acid).

4.1.22.2 Publications

The first in vivo studies on changes in the flow properties of blood in intravenous and intra-arterial contrast media infusions were performed by **Köhler** [516] in the year 1971. Urograms, cholecystangiographies, and arteriographies were made of the patients. Vasopertonic, high viscosity contrast media were injected intravenously for the urograms and the cholecystangiographies and intra-arterially for the arteriographs. Iothalaminic acid (40 ml), diatrizoic acid (40 ml), adipiodone (20 ml), and metrizoic acid (45 ml) were used. The relative blood and plasma viscosities of the blood samples, anticoagulated with sodiumcitrate, were measured using an Ostwald capillary viscosimeter (H_2O = 1); further blood viscosity measurements were performed with a Brookfield micro-cone-plate viscosimeter (5, 8, 11.5, 23.0, 46 s^{-1}). Measurement of erythrocyte deformability was carried out according to the method of Ehrly and Rossbach (hematocrit 10%, Millipore filter diameter 8-µm, 37 °C, driving force 12.9 mm H_2O). All measurements were made at 37 °C. There was a reduction of the hematocrit and the total ptotein level that coincided with the reduction of the blood and plasma viscosities 5, 15, and 30 min after the intravenous injections of the contrast medium. Significant changes did not occur in erythrocyte deformability. Already 1 min after the intra-arterial injection, there was a decrease in the relative blood and plasma viscosities as well as in the hematocrit and the protein levels. There were no significant changes in erythrocyte deformability. In order to get as high a concentration of the contrast medium as possible when the hemodilution is still low and just beginning, not only was an injection catheter used for the kidney angiographies, but also a blood removal catheter which was placed approximately 30 cm behind the injection site in the aorta abdominalis. Measurements of the flow properties of blood showed a slight increase in blood viscosity and a decrease in erythrocyte deformability, while the hematocrit and the plasma viscosity did not change, and the plasma protein level remained constant.

Note: No measurements of the erythrocyte aggregation.

In 1973 **Jauch** [475] reported on the effect of contrast media infusions (100 ml diatrizoic acid) on the flow properties of blood. The relative blood viscosity (H_2O = 1) was measured at 37 °C using a Reiss and Müller falling ball viscosimeter. The authors found a simultaneous decrease in the blood viscosity and the hematocrit values.

Note: No measurements of further rheologic parameters.

In 1976 **Störmer** and co-workers [821] studied the effect two water-soluble X-ray contrast media on the whole blood viscosity, as measured using a Wells-Brookfield viscosimeter and a Contraves Low Shear viscosimeter. Within the framework of angiographies of patients with chronic peripheral occlusive disease, up to 150 ml were injected intra-arterially. The blood samples necessary for the viscosity measurements were removed directly from the positioned catheter every 10, 60, and 180 min after the injection of the contrast medium. There was a reduction of the whole blood viscosity 10 min after injection; later the viscosity values were raised compared to the initial values. These changes correspond to the behavior of the hematocrit values.

Note: No measurements of further hemorheologic parameters.

In 1981 **Everts** and **Böhme** [338, 339] studied the changes in the flow properties of blood after angiographies (femoral artery) using diatrizoic acid in 21 patients with peripheral arterial occlusive diseases and cerebrovascular disorders. The blood viscosity measurements (constant hematocrit at 40%) and the plasma viscosity measurements were performed with a Contraves LS 30 viscosimeter. Thirty minutes after the injection of the contrast medium, the investigators found a decrease in the plasma viscosity and an increase in the blood viscosity at low shear rates.

Note: No measurements of further rheologic parameters; no temperature data.

In 1982 **Schäbitz** and co-workers [753] reported on the effect of an ionic X-ray contrast medium consisting of sodium amidotrizoate and methylglucamine-amidotrizoate on blood viscosity in patients who, during angiocardiography, were given an injection of approximately 50 ml of the contrast medium into the left ventricle or the aorta ascendens. The viscosity was measured using a Rheotest 2 rotational viscosimeter at a blood temperature of 37 °C and at shear rates of 218.7–1312 s^{-1}. The blood samples for viscosity determination were removed before, as well as 3 and 10 min after the injection of the contrast medium and then anticoagulated with heparin. There was a decrease in the whole blood viscosity with a simultaneous reduction of the hematocrit value, the effect of which was most evident 3 min after the injection of the contrast medium. The authors explain this decrease in the whole blood viscosity at high shear rates as being the result of the decreased hematocrit; the cause is claimed to be the hyperosmolar contrast medium solution which draws the fluid out of the interstitial space into the intravasal space. In addition, a slight dilution of the saline solution used for wash-out is still being discussed.

Note: Further rheologic parameters were not studied.
In 1983 these results were also presented elsewhere [754].

In 1983 **Everts** and **Böhme** [340] studied the effect of a nonionic contrast medium on the whole blood viscosity in connection with a phlebography in 20 patients with post-thrombotic syndrome or deep vein thrombosis. The whole blood viscosity was determined with a Contraves Low Shear LS 30 viscosimeter. Blood samples were measured before injection of the contrast medium, immediately thereafter, and 30 and 60 min after the examination. The authors found a considerable reduction of the whole blood viscosity, in particular at low shear rates (abstract, no details).

4.1.22.3 Discussion

Köhler, Jauch, Störmer and co-workers, *Schäbitz* and co-workers, and *Everts* and *Böhme* reported on a reduction of the blood viscosity following injections of highly viscous, hyperosmolar X-ray contrast media. The authors feel the decrease in the blood viscosity is due to hemodilution. As a result of injections with hyperosmolar, water-soluble contrast media, there is an influx of fluid from the interstitial and intracellular spaces into the intravasal space in order to thus compensate for the differences in osmotic pressure. This hemodilution may also be seen in the reduction of the hematocrit value, the plasma viscosity, and the total protein concentration. Close to the injection site the concentration of the X-ray contrast media is very high, thus causing the changes in the flow properties of blood. These changes are reversible if the osmotic difference in pressure is compensated for. With the intra-arterial injection into the aorta abdominalis, there was also a reduction of the blood viscosity and the plasma viscosity 1 min after the injection because most of the contrast medium had already gone through the circulation once, causing an osmotic compensation in the periphery. If, however, the removal site was only 30 cm from the injection catheter, a slight increase in the blood viscosity and a decrease in the erythrocyte deformability occurred, while plasma viscosity, hematocrit values, and the plasma protein concentration remained unchanged. At the injection site itself, the osmotic difference in pressure cannot be reduced so quickly and therefore, owing to the reduction of erythrocyte deformability, also leads to an increase in the blood viscosity.
In patients with peripheral arterial occlusive diseases in whom the osmotic difference in pressure can only be compensated for slowly, the infusion of larger quantities of contrast medium could cause an increase in the blood viscosity via a decrease in erythrocyte flexibility [516].
Some authors also found a reduction in the whole blood viscosity when using nonionic contrast media for phlebographies. Obviously the effects and the scope of these effects on the hemorheologic parameters in vivo are highly dependent on the type of contrast medium used (ionic, nonionic), on the concentration, the quantity, the blood sampling conditions, etc.

4.1.23 Rutosides

4.1.23.1 Introduction

Rutosides are chemically defined substances which are primarily used for the treatment of chronic venous insufficiency. A number of publications are available on O-(β-hydroxyethyl)-rutoside as well as one study on tri-(hydroxyethyl)-rutoside.

4.1.23.2 Publications

The first publication on the effect of intravenous administration of *O*-(β-hydroxyethyl)- rutosides on the flow properties of the blood was published by **Barras** and **Maibach** in 1971 [47]. Five patients suffering from chronic venous insufficiency and having an increased plasma viscosity were infused intravenously for 4 days with a daily dose of 500 mg of the drug. The measurements of the blood and plasma viscosities of the heparinized blood samples were carried out with a capillary viscosimeter. At the end of the infusion, there was a decrease in the plasma viscosity, which was more marked the higher the initial values of the plasma viscosity were. In addition to the reduction of the plasma viscosity, there was also a decrease in the blood viscosity. At a dose of hydroxyethylrutoside which was three times higher the authors found no reduction of the viscosity (abstract, no details).

Note: No measurements of further rheologic parameters; no temperature data; no hematocrit correction.

In a study by **Thulesius** and **Gjöres** [859] 1 year later, the viscosity of the blood samples, anticoagulated with EDTA, was measured before and after the oral administration of different doses of *O*-(β-hydroxyethyl)-rutoside in patients with primary and post-traumatic venous thrombosis (Brookfield-micro-cone-plate viscosimeter, 37 °C, 2.3–115 s^{-1}). With a high daily dosage of 4.8 g per day over a treatment period of 15 days the investigators found a slight but nonsignificant decrease in the blood viscosity.

Note: No measurements of further rheologic parameters; no hematocrit correction.

In 1978 **Schmid-Schönbein** [859] studied the properties of erythrocyte aggregation (photometric technique) following the intravenous injection of 3 g *O*-(β-hydroxyethyl)-rutoside within 10 min in patients with venous leg ulcers and in normal volunteers. At the end of infusion, the author found a much more distinct aggregation-reducing effect of the hydroxyethylrutoside in blood samples with normal aggregating tendency than in the blood samples of patients with primarily increased aggregation.

Note: No measurements of further rheologic parameters; no temperature data.

In 1983 **Barnes** and co-workers [44] reported on the effect of an intravenous administration of 2.5 g *O*-(β-hydroxyethyl)-rutoside in patients with insulin-dependent diabetes mellitus. Compared to normal subjects, who showed hardly any rheologic changes, the diabetics showed a drug-induced reduction of the whole blood viscosity at high shear rates, while the plasma viscosity remained unchanged. The filtrability of the whole blood was also improved in the diabetics after the administration of the drug. The authors conclude from these results that the substance may possibly improve erythrocyte deformability if this parameter is a priori deteriorated in the patient (abstract, no further details).

Lukjan and co-workers [597] studied the effect of 500 mg *O*-(β-hydroxyethyl)-rutoside on the filtrability of whole blood in patients with peripheral vascular disorders. They used the filtration tests according to Reid and co-workers and found an improvement in the values measured (abstract, no details).

In 1985 **Le Devehat** and **Lemoine** [560] reported on the effect of troxerutine on the flow properties of the blood of patients with chronic venous insufficiency of the lower extremities. The filtrability of the blood, the plasma viscosity, and the fibrinogen concentration were measured in 22 patients. After a 10-day therapy with 3 g troxerutine orally per day, there was an improvement in the whole blood filtrability as well as a decrease in the plasma viscosity and the fibrinogen concentration (abstract, no details).

In 1985 **Gallasch** and co-workers [364] tested the effect of tri(hydroxyethyl)-rutoside for various hemorheologic parameters in a controlled double-blind study. The patients received either placebo or 1.89 g for the first 7 days of the experiment and then 0.90 g of the substance to be tested. The measurement of the whole blood viscosity was carried out with a Wells-Brookfield viscosimeter and that of the serum and plasma viscosity with a capillary viscosimeter. Erythrocyte aggregation was studied photometrically, and the erythrocyte filtrability with a Teitel polymicroviscosimeter. After the oral administration of the drug at this dosage, there was no effect on the flow properties of the blood.

4.1.23.3 Discussion

Barras and *Maibach* reported on a decrease in blood viscosity following oral and intravenous administration of *O*-(β-hydroxyethyl)-rutoside in patients with chronic venous insufficiency and raised plasma viscosity alues. Since no correction of the hematocrit was performed for the blood viscosity measurements and since the temperature of the measurements was not reported, the results of the measurements are difficult to interpret, especially since the statistical validity of the results was not tested either.

The report by *Schmid-Schönbein* on the decrease in quantitatively measured erythrocyte aggregation indicates a possible mode of action for the improvement in the flow properties of blood. On the other hand, the decrease in the erythrocyte aggregation of patients with peripheral venous vascular disorders was clearly less than that of healthy subjects.

Studies by the *Barnes* and co-workers on patients with diabetes mellitus showed a statistically significant reduction of the whole blood viscosity at high shear rates without changes in the plasma viscosity. The authors attribute this to an improvement in erythrocyte deformability. *Le Devehat* and *Lemoine* found an improvement in whole blood filtration and a reduction of the plasma viscosity using troxerutine. According to the results of *Gallasch* and co-workers tri-(hydroxyethyl)-rutoside has no hemorheologic effect.

In summary, it may be said that the results presented by the various research groups up until now do not allow a uniform interpretation, nor can a reliable conclusion be derived from these results as to a possible rheologic means of action. (For publications on further plant extracts see Sect. 4.1.56).

4.1.24 Protein-Free Calves' Blood Extract

4.1.24.1 Introduction

Protein-free calves' blood extract is produced by various companies for the therapy of ischemic diseases. This solution is rich in electrolytes and contains a series of amino acids, peptides of unknown structure, and metabolites from the citric acid cycle. According to the manufacturers, it is standardized according to the requirements of the cellular oxygen consumption. They claim that the oxygen absorption is raised, the uptake of glucose and its utilization is promoted and the energy-requiring processes are supported in the sense of an anabolic effect. The function of the cells is also supposed to be retained under hypoxic con-

ditions for a long time. Processes with a high energy requirement, such as, for example, the healing of wounds, the regeneration of ferment systems, etc., are supposed to be stimulated and speeded up.

4.1.24.2 Publications

In 1973 **Ott** [676], from the team of Heidrich and Schlichting reported on the effect of protein-free calves' blood extract on the blood viscosity. In addition to healthy subjects, there were patients with vascular disorders and other diseases. A falling ball viscosimeter was used at 37 °C for the viscosity measurements. Within 1 h five subjects received a single infusion of 30–40 ml protein-free calves' blood extract in 250 ml 5% levulose solution. The viscosity measurements were performed 4–5 h after the end of the infusion. In the meantime, the blood samples were stored at 4 °C. At the end of the infusion, there was a hematocrit reduction of 2.77% but no reduction of the blood viscosity. In comparison, there was a 1.21% reduction of the hematocrit and a 6.61% reduction of the blood viscosity at the end of infusion with a single administration of the carrier solution (250 ml 5% levulose solution).

Note: The blood viscosity was measured with an inadequate viscosimeter; no correction of the hematocrit; no measurements of further rheologic parameters.

In 1976 **Becker** [55] reported on the effect of a single intravenous injection of 10 ml protein-free calves' blood extract on patients with polyglobulism. The blood viscosity was measured using a Brookfield micro-cone-plate viscosimeter at four shear rates of 5.8, 11.5, 23.0, and 46 s^{-1}. An Ostwald capillary viscosimeter was used for the determination of plasma viscosity (H_2O = 1). Erythrocyte deformability was determined according to the technique of Ehrly and Rossbach (hematocrit 10%, Millipore filter diameter 8 μm, driving pressure 1.29 cm H_2O, 23 °C). Contrary to the healthy subjects, the patients with polyglobulism showed a slight but significant reduction of the blood viscosity. Changes in erythrocyte deformability were not found.

Note: No hematocrit correction; no quantitative measurement of the erythrocyte aggregation; no viscosity measurements at body temperature.

In 1978 **Watzlaw** [904] treated eight patients with chronic peripheral arterial disorders for a period of 3 days with daily injections of 20 ml protein-free calves' blood extract. The determination of the relative blood viscosity and plasma viscosity (H_2O = 1) was performed with an Ostwald capillary viscosimeter at 37 °C. Measurement of erythrocyte filtration was carried out according to Ehrly and Rossbach (Millipore filter 14 μm). The author did not find any effect on the rheologic parameters measured.

Note: No hematocrit correction; no measurement of the erythrocyte aggregation.

4.1.24.3 Discussion

The publications available at this time do not indicate an improvement in the flow properties of blood in the experimental situation with intravenous injections of protein-free calves' blood extract in patients with vascular disorders. Thus, clinically favorable observations must be explained by other mechanisms.

4.1.25 Nicotinic Acid Derivatives

4.1.25.1 Introduction

Nicotinic acid preparations have been used as vasodilators for a long time in the treatment of cerebral and peripheral vascular disorders. Xantinol nicotinate, a nicotinic acid salt, is used for the treatment of chronic arterial occlusive diseases with pains at rest and gangrene, as well as in cases of skin damage as a result of cold and in cerebrovascular disorders. Studies on the effect of xantinol nicotinate on the flow properties of the blood have been available since 1972.
Inositol nicotinate, chemically an inosithexanicotinic acid ester, is mainly used for the treatment of cerebral and peripheral vascular disorders and for lowering a raised cholesterol level.
Nicofuranose is also used for the treatment of peripheral arterial and venous vascular disorders such as thrombangitis obliterans, Raynaud's syndrome, intermittent claudication, ulcus cruris, and hyperlipoproteinemia. Niceritrol and ciclonicate have also been studied.

4.1.25.2 Publications

In 1972 **Davis** and **Rozov** [155] reported on the improvement of the flow properties of blood as a result of xantinol nicotinate in patients with peripheral arterial vascular disorders. A Brookfield micro-cone-plate viscosimeter was used for the blood viscosity measurements at shear rates of 3–60 s^{-1}. The patients received 1.8–2.4 g xantinol nicotinate orally each day. The average period of treatment was 6.5 months. At the end of treatment, the investigators found a significant decrease in the blood viscosity at low shear rates, as well as a decrease in the fibrinogen level from 477.6 mg% – 378.8 mg%. A decrease in the cholesterol level was determined, and, using a capillary microscope, a reduction of sludge formation was found in the retinal blood vessels.

Note: No correction of the hematocrit; no data on the temperature of the measurements; no measurements of further rheologic parameters.

Clinical findings: There was a clinical improvement in patients with trophic and necrotic changes of the skin. In addition, the skin temperature increased. All in all, the authors found a clinical improvement in 39 of the 60 patients.

In 1974 **Schmittner** [767] from the Ehrly team studied the effect of a single intravenous application of 300 mg xantinol nicotinate on the flow properties of the blood of patients with chronic occlusive arterial disease. The blood and plasma viscosities were measured using an Ostwald capillary viscosimeter ($H_2O = 1$) and the blood viscosity with a Brookfield micro-cone-plate viscosimeter at shear rates of 5.8–46 s^{-1}. The temperature at the measurements was 23 °C. Measurements of erythrocyte filtration were performed according to the method of Ehrly and Rossbach. At the end of the infusion, there was no statistically significant improvement in the flow properties of blood as compared to the initial values.

Note: No correction of the hematocrit; no measurement of the erythrocyte aggregation; blood viscosity measurements were not performed at 37 °C.

In 1975 **Davis** and **Rozov** [156] published the results of a cross-over double-blind study with xantinol nicotinate. The authors found similar results as in their 1972 study. In addition, the authors reported an improvement in pains at rest and an increase in the pain-free walking distance in patients with intermittent claudication.

In 1975 **Chmiel** and co-workers [131] reported on a decrease in the blood viscosity following the administration of xantinol nicotinate in smokers (abstract, no details).

In 1976 **Djurdjevic** and co-workers [190] studied the effect of inositol nicotinate on the flow properties of blood in patients with hyperlipoproteinemia types IIb and IV (diagnosed by lipid electropheresis). The patients received 900 mg inositol nicotinate orally per day. The treatment period was 3 months. Blood viscosity was measured using a capillary viscosimeter (20 °C). The investigators found a significant decrease in blood viscosity as well as in the cholesterol and triglyceride levels.

Note: No correction for hematocrit; no measurement of blood viscosity at body temperature; no measurements of further rheologic parameters.

In 1978 **Mashiah** and co-workers [620] studied the effect of an oral administration of nicofuranose on the flow properties of blood in patients with intermittent claudication and increased blood viscosity values compared to a healthy control group. The viscosity of the blood samples, anticoagulated with EDTA, was measured using a coaxial cylindrical viscosimeter (Contraves Low Shear 2) at shear rates of 1.24–124 s^{-1}. The temperature at measurementss was 21°–25 °C. After the 6-week treatment period, the investigators found a significant decrease in the blood viscosity values at low shear rates.

Note: The hematocrit was corrected mathematically to 45%. The temperature for the blood viscosity measurements varied between 21° and 25 °C. No measurements of other rheologic parameters. The exact nicofuranose dosage was not given.

In 1979 **Holti** [451] treated patients with Raynaud's syndrome, other peripheral vascular disorders, systemic sclerosis, and subacute lupus erythematodes. Within the framework of a simple blind study the patients received 4 g inositol nicotinate orally every day for 16 weeks. The authors did not find a significant reduction of the plasma viscosity, but they did find a significant decrease in the cholesterol, triglyceride, and plasma fibrinogen levels.

Note: Exact viscosimetric data are missing; no measurement of further rheologic parameters such as, for example, blood viscosity; no temperature data.

Clinical findings: The patients reported an increase in skin temperature, a decrease in the ischemic symptoms, and a less painful exposure to cold.

Strano and co-workers [835] studied the effect of d-glucitol hexanicotinate in one group of patients with vascular disorders and in a second group who also had diabetes mellitus. Nicotinic acid was applied orally at a dose of 100 mg and d-glucitol hexanicotinate at a dose of 400 mg. Erythrocyte deformability was measured according to the whole blood method of Reid. Blood viscosity measurement was carried out with a Wells-Brookfield microviscosimeter at 37 °C and at various shear rates. Plasma viscosity was measured using the same instrument at 230 s^{-1}. Blood was sampled 60 min after the oral administration of nicotinic acid and 150 min after the oral administration of d-glucitol hexanicotinate. With both substances there was an increase in the filtration rate, a reduction of blood viscosity at different shear rates, and a decrease in plasma viscosity. Both groups of patients showed an improvement in the filtration values and a decrease in the viscosity values.

Note: Other rheologic parameters were not measured; hematocrit values were not given.

In 1983 **Pretolani** and co-workers [713] described the effect of a 90-day treatment with ciclonicate (a nicotinic acid derivative) on erythrocyte filtrability, measured according to the method of Reid and co-workers. They found a statistically significant improvement of the measurement values.

Note: No measurement of other hemorheologic parameters.

In 1983 **Annoni** and co-workers [22] studied the effect of ciclonicate on the flow properties of the blood of patients with chronic arterial occlusive disease. A single dose of 400 mg of the drug led to a reduction of the whole blood viscosity, measured using a Wells-Brookfield LVT viscosimeter at various shear rates.

Note: No data on the temperature of the measurements; no measurement of further rheologic parameters.

In 1985 **Hamazaki** and co-workers [415] studied the effect of 1500 mg of a nicotinic acid derivative (niceritol) on the whole blood viscosity of patients with diabetes mellitus and hypertension. Blood viscosity was measured without corrected hematocrit at shear rates of 37.6–376 s^{-1}. The authors found no changes in the viscosity values with this therapy.

4.1.25.3 Discussion

Davis and *Rozov* as well as *Chmiel* and co-workers reported on a reduction of the blood viscosity using xantinol nicotinate in patients with peripheral arterial vascular disorders and in heavy smokers. This reduction was especially marked at low shear rates. The studies of *Jurdjevic* and co-workers on the effect of inositol nicotinate also showed an improvement in the blood viscosity values, as did the studies of *Holti* who found a reduction of the plasma viscosity. *Schmittner,* on the other hand, did not find a decrease of the blood and plasma viscosities. *Strano* and co-workers found a reduction of the blood viscosity as well as of the plasma viscosity with both nicotinic acid and d-glucitol hexanicotinate after an oral administration of these substances over a prolonged treatment period.

With the exception of *Schmittner* most authors did not correct the hematocrit, and with the exception of *Strano* and co-workers, most authors did not measure at 37 °C. Measurements of the erythrocyte aggregation were not carried out. *Davis* and *Rozov,* however, found a decrease in erythrocyte aggregation using capillary microscopy and traced the favorable clinical effect back to a rheologic mode of action. Erythrocyte deformability was only measured by *Strano* and co-workers and by *Schmittner. Strano* carried out his filtration measurements using the method of Reid and co-workers which is nowadays no longer adequate.

There is only one publication available on the changes in the flow properties of the blood as a result of nicofuranose. *Mashiah* and co-workers could not find any change in the blood viscosity with mathematically corrected hematocrit at high and low shear rates.

According to *Annoni* and co-workers and *Pretolani,* ciclonicate effects a decrease in the blood viscosity and an improvement in the erythrocyte filtrability. There is only one study available on niceritol.

In summary, it may be said that the various authors have presented contradictory results. They do not make a clear statement on whether nicotinic acid and its derivatives have a favorable influence on the flow properties of the blood as a whole or on the individual subparameters.

4.1.26 Bencyclane

4.1.26.1 Introduction

Pharmacologically bencyclane has been classified as a musculotropic vasodilator. Since 1970 it has been used for the treatment of cerebral and peripheral vascular disorders.

4.1.26.2 Publications

The first report on the hemorheologic effects of the vasoactive substance bencyclane was published by **Ott** [676] from the Heidrich team, in 1973. Among his subjects there were patients with vascular disorders and other diseases. A Haake falling ball viscosimeter was used; the measurement of the blood viscosity was carried out at 37 °C. Five patients received a single infusion of 150 mg bencyclane in 250 ml 5% levulose solution intravenously over 1 h. The viscosity measurements were performed within 8 h after the end of the infusion. In the meantime, the heparinized blood samples were stored at 4 °C. At the end of infusion, there was a 3.96% reduction of the blood viscosity along with a 6.36% decrease in the hematocrit. In a control group of five subjects, the administration of the carrier solution alone led to a 6.61% decrease in the blood viscosity along with a hematocrit reduction of 1.12%. Accordingly, the use of levulose with bencyclane resulted in a reduction of the blood viscosity that was clearly less than with the infusion of the carrier solution levulose alone.

In 1974 and 1975 **Heidrich** and and co-workers [421, 422, 424] published major parts of this study.

Note: The viscosometric findings with a falling ball viscosimeter are of little value for methodologic reasons. A correction of the hematocrit was not carried out; no measurement of further rheologic parameters.

In 1974 **Ehrly** and co-workers [257] reported on the effect of bencyclane on the flow properties of the blood in six patients with peripheral arterial occlusive diseases in stages III and IV according to Fontaine. Bencyclane was administered intravenously as a single infusion of 500 mg in 60 ml physiologic saline solution over 20 min. A Brookfield micro-cone-plate viscosimeter was used for the blood viscosity measurements at shear rates of 5.8–46 s^{-1}; the relative blood and plasma viscosities (H_2O = 1) were measured using an Ostwald capillary viscosimeter (23 °C and 37 °C). The erythrocyte filtrability measurement was carried out according to the method of Ehrly and Rossbach. Compared to two control measurements, at the end of infusion there was a slight but nonsignificant decrease in the plasma viscosity (Brookfield and Ostwald viscosimeter) that correlated with the minimal decrease in the hematocrit. The plasma viscosity did not change. The measurement of erythrocyte filtrability showed a slight but nonsignificant increase in the values after the administration of bencyclane.

In 1975 *Blendin* [70] reported on similar results in his dissertation.

Note: No measurement of the erythrocyte aggregation.

In 1974 **Hess** and co-workers [445] reported on the effect of bencyclane on the flow properties of the blood in patients with arterial occlusions of the limbs in stages II, III, and IV according to Fontaine. The patients received 100 or 200 mg bencyclane in 30 ml 5% levulose solution intravenously during an infusion time of 20 min. The blood viscosity of the native blood was measured using a Brookfield micro-cone-plate viscosimeter at a defined shear rate of 230 s^{-1}. At the end of the infusion as well as later there was a significant decrease in the blood viscosity, especially with low doses. The viscosity-reducing effect of bencyclane was stronger 150 min after the end of the infusion than that of a 90-min infusion of 500 ml low molecular weight dextran. In control experiments with the application of only the carrier solution, there were slight variations in the blood viscosity.

Note: No data on the temperature of the measurements; no hematocrit correction; no measurements of further rheologic parameters.

In 1974 **Völker** [878] studied the rheologic changes during a long-term oral therapy with bencyclane within the framework of a randomized double-blind study in 26 patients with chronic arteriopathies in stage II according to Fontaine. During an 8-week treatment period, the patients received 600 mg bencyclane daily. The apparent whole blood viscosity was measured using a Brookfield micro-cone-plate viscosimeter at shear rates of 5.75–230 s^{-1} and the plasma viscosity at shear rates of 230–115 s^{-1}. Erythrocyte aggregation was determined with a rheoscope. At the end of treatment, the whole blood viscosity showed a significant decrease of 20% at a shear rate of 11.5 s^{-1}. In the majority of cases a slight, nonsignificant increase in the erythrocyte aggregation was determined. No major changes were found in plasma viscosity, the total protein concentration, as well as in the cholesterol and triglyceride levels. The variations in the hematocrit values were minimal.

Note: No details on the temperature of the measurements; no hematocrit correction; no measurements of the erythrocyte deformability.

In 1977 **Störmer** and co-workers [818] reported on the rheologic effects of an infusion of bencyclane in 250 ml 5% glucose solution (10 mg bencyclane per kilogram body weight) in 15 patients with chronic arterial occlusive diseases. The blood samples for the viscosity measurements were anticoagulated with EDTA and stored for up to 2 h at 37 °C. Blood viscosity was measured using a Contraves Low Shear Couette rheometer at shear rates of 1.0–4.6 s^{-1}. The plasma and serum viscosities were measured using the same viscosimeter at a shear rate of 4.6 s^{-1}. The measurement temperature was 37 °C. A hematocrit correction was performed mathematically. At the end of the 9-day treatment, no uniform changes were observed in the blood, plasma, and serum viscosities after therapy with bencyclane. Analog studies by **Störmer** and co-workers [819] were published in 1978.

Note: No measurements of further rheologic parameters; changes in the blood samples due to storage (2 h at 37 °C) cannot be excluded.

In 1979 **Völker** [881] reported on the effect of bencyclane on the flow properties of the blood of patients with chronic obliterative vascular disorders in stage II according to Fontaine. During a treatment period of 2–3 weeks the patients received 600 mg bencyclane orally every day. Blood viscosity was determined at a corrected hematocrit of 45% and at shear rates of 5.75–230 s^{-1}. The plasma viscosity measurement was carried out with the same viscosimeter at shear rates of 115–230 s^{-1}. The temperature of the measurements was 37 °C. Erythrocyte deformability was determined according to Schmid-Schönbein (Nucleopore filters 5 μm, whole blood, driving force 50 mm H_2O). At the end of treatment a significant decrease in the whole blood viscosity of, on average, 16.6% was found at all shear rates. Furthermore, the author found a significant improvement in erythrocyte deformability. No changes were observed in plasma viscosity, the cholesterol, serum triglyceride level, as well as in the plasma fibrinogen level.

Note: No measurements of further rheologic parameters.

In 1980 **Le Devehat** and co-workers [555] reported on the hemorheologic effect of an oral administration of bencyclane (400 mg per day) on erythrocyte deformability. The measurements were carried out according to the method of Reid and co-workers (whole blood, Nucleopore filters 5 µm, driving force 20 cm H_2O). The reduced erythrocyte deformability found in patients with diabetes mellitus could be normalized with the administration of bencyclane.

Note: No measurements of further rheologic parameters; no data on the temperature used.

4.1.26.3 Discussion

Both *Hess* and co-workers and *Völker* reported a significant decrease in blood viscosity after the oral and intravenous administration of bencyclane. The work of *Völker* should receive particular mention here: using a standardized hematocrit he found a reduction of the blood viscosity at low and high shear rates as the result of the long term oral treatment with bencyclane. On the other hand, the teams of *Heidrich, Ehrly,* and *Störmer* did not find a significant decrease in blood viscosity after the intravenous administration of bencyclane. The observation of *Ott* is interesting in that a stronger viscosity-reducing effect could be achieved with an infusion of the carrier solution alone than with a corresponding bencyclane additive to the levulose solution. There were no changes in the plasma viscosity, the plasma fibrinogen, and the whole protein concentration. *Ehrly* and co-workers reported on a slight but nonsignificant improvement in the erythrocyte deformability after intravenous administration of bencyclane in patients with peripheral occlusive diseases. On the other hand, *Völker* as well as *Le Devehat* and co-workers found a significant improvement in erythrocyte deformability as the result of oral medication. In one publication, semiquantitative measurements of erythrocyte aggregation were described [878], showing in most cases a decrease in this parameter.

Seen as a whole, the results published by the individual investigators are contradictory. This is the case for the blood viscosity measurements as well as for erythrocyte filtrability. It cannot be assumed that only the different form of application of the substances is the cause of these discrepancies. Some of these studies were performed with methods which can no longer be accepted today.

The stronger viscosity reduction with bencyclane than with low molecular weight dextran, which was found by one author, seems improbable. From these results, no conclusion may be drawn about a possible rheologic mode of action of this substance for the therapy of vascular disorders.

4.1.27 Pentoxifylline

4.1.27.1 Introduction

Pentoxifylline is a derivative of xanthine. The drug was first deemed to be a vasoactive substance, especially since a certain vasodilating effect could be shown after intravenous administration of high dosages. Nowadays pentoxifylline is used worldwide as a therapeutic substance for vascular disorders. After ***Hess*** *[444] had described a viscosity-reducing effect, the studies of* ***Ehrly*** *and others [564, 476, 787] showed that pentoxifylline improved the impaired erythrocyte deformability. This discovery was the beginning of an intensive period of study in the field of hemorheology. Apart from the studies on hemodilution with dextran solutions, most of the publications on the effect of "hemorheologic" drugs have been on pentoxifylline.*

4.1.27.2 Publications

Initial studies on the rheologic effect of pentoxifylline on the flow properties of blood in patients came from the Hess team [475, 444, 356] in the years 1973 and 1974. In 1973 *Jauch* [475] infused 200 mg pentoxifylline in 100 ml 5% levulose over 90 min in patients with vascular disorders. Blood viscosity was measured using a Reiss and Müller falling ball viscosimeter (data in falling time units). At the end of the infusion, the author found a reduction of the whole blood viscosity. Despite the simultaneous decrease in the hematocrit values, the author presumed that a substance-specific effect had contributed to the decrease in viscosity.

Note: The measurements were carried out with an inadequate viscosimeter; no measurements of further rheologic parameters.

In a further publication by this group [444] the viscosity of the native blood was studied using a Brookfield cone-plate viscosimeter at a shear rate of 230 s^{-1} after a 90-minute intravenous administration of mg pentoxifylline per kilogram body weight in 20 ml 5% levulose. A reduction of the blood viscosity in the 15 patients with peripheral, obliterative arteriopathies was found at the end of the infusion as well as some time later. On the other hand, there was no significant decrease in the whole blood viscosity in the controls who received just a single dose of the carrier solution (20 ml 5% levulose). Compared to the decrease in viscosity achieved with pentoxifylline, the viscosity reduction with a 90-min infusion of 500 ml low molecular weight dextran was not much greater. Daily variations in the viscosity in ten resting patients who did not receive any medication could not be statistically corroborated.

Note: No data on the temperature of the measurements; no measurements of further rheologic parameters; no hematocrit correction.

Clinical findings: There was a cessation of the pains at rest in patients with vascular disorders in stage III according to Fontaine.

In the same year results were published by **Ott** [676], from the team of Heidrich and Schlichting, on the effect of vasoactive substances on the blood viscosity. Healthy subjects as well as patients with vascular disorders were studied. A Haake falling ball viscosimeter was used; the measurements were carried out at 37 °C. Ten subjects received a single infusion of 150 mg pentoxifylline in 250 ml 5% levulose solution over a period of 1 h. The viscosity measurements were carried out 4–8 h after the end of infusion. In the meantime, the blood samples were stored at 4 °C. At the end of infusion there was a 13% decrease in the blood viscosity, with a 3.8% reduction of the hematocrit. In comparison with the infusion of low molecular weight dextran (500 ml), only a 15.2% reduction of the viscosity could be achieved with a simultaneous 13% decrease in the hematocrit.

Note: The viscosimetric results determined with a falling ball viscosimeter must be relativized owing to technical reasons; no hematocrit correction.

In 1974 **Heidrich** and **Ott** published major parts of this work elsewhere [421].

Analog studies, which corresponded with the previous ones using a falling ball viscosimeter, were published in 1975 and 1976 by **Heidrich** and co-workers [423, 424, 425]. Contrary to the earlier studies, blood viscosity was measured using a cone-plate viscosimeter (Haake Rotovisco RV 2) at 37 °C and at a defined shear rate of 330 s^{-1}. After a 30-min infusion of 200–300 mg pentoxifylline in patients with peripheral arterial occlusive diseases, the blood and serum viscosities were measured at the end of the infusion as well as 90 min later. After a parenteral dose of 300 mg pentoxifylline the authors were able to observe a significant decrease in the blood viscosity (approximately 8.92%) at the end of infusion. The serum viscosity and the hematocrit did not change.

Note: No measurements of further rheologic parameters.

In 1975 **Ehrly** [262, 266] was the first to find changes in erythrocyte filtrability after intravenous infusion of 200 mg pentoxifylline. The method used determined the filtrability of the erythrocyte suspensions at a constant hematocrit of 10% through a Millipore filter with a standardized pore size of 8 μm using a pressure of 1.32 cm H_2O and a constant temperature of 37 °C. In these studies there was a statistically significant reduction of the erythrocyte filtrability both 15 and 60 min after the end of infusion.

Note: No measurements of further rheologic parameters.

Smud and co-workers [804] studied the effect of oral (300–1600 mg per day) administrations of pentoxifylline as well as single intravenous applications of pentoxifylline of between 100 and 300 mg suspended in 100 ml physiologic saline solution in patients with multiple myeloma, hyperlipidemia, and polyglobulism. The time of infusion for the parenteral administration of pentoxifylline was 60 min. The blood samples were taken 60, 120, and 240 min after the end of infusion and anticoagulated with heparin. A Harkness capillary viscosimeter was used for the measurement of blood viscosity; the temperature of the measurements was 25 °C. The intravenous administration of the drug in the above-mentioned dosage led to a reduction of the viscosity, whereas with oral administration at least 600 mg pentoxifylline had to be given in order to achieve a dose-dependent viscosity reduction. There were no changes in the hematocrit, the serum viscosity, and the plasma viscosity.

Note: The measurements were performed at 25 °C; no measurements of further rheologic parameters.

In 1976 **Völker** [880] carried out filtration studies with 10% erythrocyte suspensions (Nucleopore filter, pore diameter 5 μm, driving force 50 mm H_2O). Blood viscosity was measured using a Brookfield micro-cone-plate viscosimeter with a Mooney adapter at a corrected hematocrit of 45% (37°C) and at various shear rates of 1.96–78.36 s^{-1}. The oral administration of 3 × 3 tablets (= 900 mg) of pentoxifylline daily was studied in 19 patients with chronic occlusive arterial disease and was compared to a placebo group. The duration of treatment was 10 days. At the end of treatment both groups showed no significant change in erythrocyte filtrability; however, a slight but nonsignificant decrease in the blood viscosity was observed in the pentoxifylline group. Changes in plasma viscosity, the erythrocyte sedimentation rate, the fibrinogen level, and the total protein concentration could not be determined.

After oral and intravenous administration of pentoxifylline in patients with various types of diseases, **Deguchi** and co-workers [164] carried out blood viscosity measurements using a rotational viscosimeter (736.12 s^{-1}). Viscosity reductions of up to 8.2% were registered.

Note: An evaluation of these results is difficult since concrete data concerning the type of viscosimeter, the temperature of the measurements and the hematocrit are missing.

Knarr [514] studied the blood and the plasma viscosities as well as erythrocyte flexibility according to the method of Ehrly and Rossbach in eight patients with chronic arterial occlusive diseases. After two control measurements before the injection, 200 mg pentoxifylline suspended in 10 ml physiologic saline solution were infused over a period of 10 min. The viscosity measurements were performed 15–16 min after the end of infusion. The blood samples were anticoagulated with heparin. There was a slight rise in the relative blood viscosity (Ostwald viscosimeter H_2O = 1, 23 °C), whereas plasma viscosity remained unchanged. After the intravenous injection of the corresponding amount of physiologic NaCl solution in a placebo group, there were no significant changes in the flow properties of the blood. During the experiment the hematocrit and the number of erythrocytes per milliliter blood increased slightly but significantly. The erythrocyte sedimentation rate showed no change. As far as the erythrocyte filtration is concerned (Millipore filter, pore diameter µm, driving force 1.29 mm H_2O, hematocrit 10%, 23 °C), there was no change in the filtration rate 15 min after the end of infusion; however, 60 min later there was a significant increase in the number of erythrocytes flowing through the capillary system per unit of time.

Note: Measurements of the blood viscosity and erythrocyte filtration were carried out at 23 °C; no measurements of further rheologic parameters.

In 1977 **Schubotz** and **Mühlfellner** [775] compared erythrocyte deformability before and after 2 weeks of pentoxifylline medication in ten elderly diabetics with chronic arterial occlusive diseases. The daily dose was 3 × 400 mg pentoxifylline orally. The filtration of the blood samples anticoagulated with sodium citrate were carried out 30 min after removal of the blood according to the method of Reid and co-workers (whole blood, Nucleopore filter, pore size µm, driving force 20 cm H_2O). The filtration temperature was 25–30 °C. At the end of treatment the erythrocyte filtration rates rose significantly from 0.336 ml/min to 0.702 ml/min.

Using rheologic and audiologic methods, **Schulz** and co-workers [778] tested the effect of pentoxifylline in patients with acute and chronic impaired hearing in the inner ear. Twenty patients were infused with 400 mg pentoxifylline in 500 ml crystalloid solution (Ringer lactate) over 2 h. Ten other patients received the corresponding solution but without medication. The only hemorheologic parameter measured was blood viscosity using a Schott capillary viscosimeter at 37 °C. At the end of the infusion there was a stronger reduction of the blood viscosity in the pentoxifylline group than in the placebo group.

Störmer and co-workers [818, 819] studied the rheologic effects of an infusion of pentoxifylline (10 mg per kilogram body weight in 250 ml 5% glucose solution) in 14 patients with chronic arterial occlusive disease. The blood samples were anticoagulated with EDTA for the viscosity measurements and stored for up to 2 h at 37 °C. The blood viscosity was measured using a Contraves Low Shear rheometer at shear rates of $1.0-4.6\ s^{-1}$, and at high shear rates with a Brookfield cone-plate viscosimeter. The temperature of the measurements was 37 °C. An attempt was made to consider changes in the hematocrit values mathematically. At all of the shear rates measured, especially at low shear rates, there was a significant decrease in the blood viscosity. During the 9-day treatment the plasma and serum viscosity values were also reduced; this behavior was, however, inconsistent.

Note: No measurements of further rheologic parameters (see Sect. 3.3.4 for the validity of low shear measurements using rotational viscosimeters).

In 1978 **Ehrly** [279, 283] reported on the results already described in Knarr's dissertation regarding the improvement of erythrocyte filtrability after an intravenous administration of 200 mg pentoxifylline in patients with chronic arterial occlusive disease.

Following initial intravenous therapy, **Lusov** and co-workers [598] treated 40 patients with peripheral and coronary vascular disorders using 250–750 mg pentoxifylline orally daily. The treatment period varied between 1.5 and 3 months. Blood viscosity was measured using a rotational viscosimeter of an unknown type with shear stresses of 0.7 dyn/cm^2 and 5.0 dyn/cm^2 (no details on temperature). The initial viscosity values were increased in 60% of the patients compared to a healthy control group. At the end of treatment, the blood viscosity normalized, especially in those patients with raised viscosity. No hematocrit changes were observed.

Note: Apart from the fact that there is no data on the temperature of the measurements, the authors could only register a decrease in the viscosity at the end of treatment when, compared to the healthy subjects, hyperviscosity of the blood had been determined; no measurements of further rheologic parameters.

Clinical findings: At the end of treatment the authors reported that the pains at rest disappeared and that paresthesias no longer occurred in the patients with peripheral vascular disorders in stages III and IV according to Fontaine. A total of 65% of the patients said that the pains caused by exposure to cold did not reappear. In addition, there was an improvement in the ergometrically verified performance of patients with vascular disorders. Only in three out of 15 patients with angina pectoris and peripheral arteriopathies could a favorable clinical result be found.

In 1978 **Mashiah** and co-workers [620] studied the blood viscosity of 13 patients with intermittent claudication who had had these clinical symptoms for more than 1 year. The patients were treated for a period of 6 weeks with oral administrations of pentoxifylline (no data on dosage). The viscosity measurements were carried out at a temperature of 21°–25 °C (Contraves Low Shear viscosimeter, 1.24 and 124 s^{-1}). The hematocrit was mathematically corrected to a value of 45%. The results of the viscosity measurements were not uniform, and a significant blood viscosity decrease as a result of pentoxifylline could not be proven.

Note: The evaluation of this study is difficult because of the inconsistent temperatures for the viscosity measurements, the hematocrit values, which were only corrected mathematically, and the missing data on the dosage.

In 1978 **Pecker** and co-workers [689] studied erythrocyte deformability according to the method of Reid and co-workers in 12 patients with cerebrovascular disorders. The erythrocyte deformability was strongly increased in the group of patients compared to a healthy control group. Pentoxifylline was administered intravenously in daily doses of 600–800 mg. After 8 days the initial flow quantities of blood through the filters were doubled in 6 patients.

Note: See the summary and corresponding commentary in Sect. 3.3.4 for the validity of the filtration method used. The same applies to the following studies by Angelkort and co-workers, Gaillard and co-workers, Keller and Leonhart, Manrique, Martin and Vives, and Le Devehat and co-workers.

Satewachin and co-workers [751] carried out blood viscosity measurements using a Zachartschenko rotational viscosimeter at shear rates of 0.1–25 s^{-1} in patients with arterial thrombosis. After surgical removal of the thrombus, a combined therapy (infusion) with heparin, dextran, nicotinic acid, and dicumarol was started in order to prevent rethrombosis. For comparison, a control group received 200 mg pentoxifylline intravenously every 8 h. In the pentoxifylline group a stronger reduction of the yield stress was found, which was attributed to a decrease in erythrocyte aggregation.

Note: The results are difficult to interpret owing to the combination of various substances. In addition, there are no data on the temperature of the measurements and changes in the hematocrit.

In 1979 and 1980 **Angelkort** and co-workers [15, 16] studied erythrocyte deformability according to the method of Reid and co-workers in 25 patients with chronic, peripheral, occlusive diseases in stages IIa–III. The patients received 4 × 400 mg pentoxifylline orally per day for a period of 6 weeks. Using this method a clear difference was found in the erythrocyte deformability of healthy subjects and an untreated control group of patients. At the end of treatment, the flow rate rose significantly up to 50% higher than the initial value (filter flow rate from 1.7 ± 3 ml/min to 2.5 ± 0.3 ml/min.).

Note: As for Pecker and co-workers, this page.

Gaillard and co-workers [361] tested the intensity and the duration of action of different galenic preparations of pentoxifylline in 12 patients (no diagnostic details). For this the patients were given one tablet of 400 mg pentoxifylline and at the examination on the following day 4 times 100 mg pentoxifylline. There was a significant increase in the filtration rate, which was stronger and longer lasting with the single dose of 400 mg. Erythrocyte deformability was determined according to the method of Reid and co-workers.

Note: As for Pecker and co-workers, this page.

Metoki and co-workers [643] studied the blood viscosity (using a Brookfield-cone-plate viscosimeter) of patients with acute cerebral thrombosis and hemorrhages before and after pentoxifylline (no data on the dosage). In those patients in whom the total blood viscosity was raised, there was a decrease in the blood viscosity values after medication with pentoxifylline.

Note: An evaluation of this work is difficult owing to the lack of data on dosage and technical measurement details; no measurements of further rheologic parameters.

Keller and **Leonhardt** [486, 487] treated one female patient suffering from a painful sickle cell crisis with daily administration of 2.4 g pentoxifylline. Blood viscosity was measured using a cone-plate viscosimeter at a shear rate of 46 s^{-1}. Erythrocyte filtrability was tested according to the method of Reid and co-workers using Nucleopore filters (diameter 5-μm). There was a decrease in the blood viscosity and an increase in the erythrocyte deformability.

Note: As for Pecker and co-workers, p. 158. No temperature data.

Clinical findings: Although the hemolytic crises were not prevented by pentoxifylline, they were nonetheless no longer accompanied by pains. The use of analgesics could be avoided.

In 1979 **Manrique** [609] studied the effect of a 14-day treatment with 3 tablets of 400 mg pentoxifylline per day on the erythrocyte filtration (according to the method of Reid and co-workers) in patients with arterial occlusive disease. After a 7-day treatment the filtration rate rose significantly from 0.63 ± 0.21 ml/min to 0.98 ± 0.41 ml/min. It was shown, moreover, that the percentage increase in the filtrability varied considerably within the group of patients.

Note: As for Pecker and co-workers, p. 158.

In 1979 **Martin** and **Vives** [617, 618] studied erythrocyte rigidity according to the method of Reid and co-workers in 49 patients (average age of 68 years) which had had cerebrovascular attacks. In comparison to a control group of healthy subjects (average age of 46 years), there was a clear decrease in the flow rate of the erythrocytes (1.3 ± 0.2 ml/min to 0.61 ± 0.05 ml/min) in the patient collective. At the end of treatment, the filtration rate was 0.98 ± 0.06 ml/min in the 28 patients treated with pentoxifylline, which corresponds to a significant increase.

Note: As for Pecker and co-workers, p. 158.

Martin [619] studied the deformability of the red blood cells according to the filtration method of Reid and co-workers in 49 patients with cerebrovascular disorders compared to 65 healthy subjects. After an oral administration of 600 mg pentoxifylline per day, the authors found a statistically significant improvement in the filtration values.

Note: As for Pecker and co-workers, p. 158.

In 1981 **Angelkort** and co-workers [18] studied the effect of an oral treament with 1600 mg pentoxifylline per day over a period of 6–8 weeks in patients with chronic arterial occlusive diseases in stages IIa to IV according to Fontaine. Erythrocyte filtrability was measured through Nucleopore filters with a pore diameter of 5 μm and a driving force of 20 cm H_2O. There was a statistical increase in the erythrocyte filtration values after 6 weeks of therapy with pentoxifylline.

Similar results were also described by Angelkort and Kiesewetter elsewhere [17].

Note: No measurements of further rheologic parameters.
At the same time the authors studied the muscle blood flow using xenon clearance and found a statistically significant increase in the values during pentoxifylline treatment, especially in the case of initially impaired filtration rates.

In 1980 **Johnsson** and co-workers [477] reported on the effect of 1200 mg pentoxifylline per day on erythrocyte flexibility in 40 patients with a heart valve replacement. Nucleopore filters with a pore diameter of 3 μm were used. After the administration of pentoxifylline, an improved erythrocyte filtrability was observed in the group of patients with arteriosclerotic changes in the vessels, while no significant changes were found in the patients with artificial heart valves.

Note: An assessment of this work is difficult since the only technical details as to the method of measurement was the pore diameter of the filter were given.

In 1980 **Le Devehat** and co-workers [554] reported on the effect of pentoxifylline on erythrocyte deformability in diabetics with chronic occlusive arterial disease. The measurement of erythrocyte deformability was carried out according to the method of Reid and co-workers. The patients received 600 mg pentoxifylline per day. There was an improvement in erythrocyte deformability (no details as to significance).

Similar results were also reported by **Le Devehat** and co-workers in 1981 [557].

Note: As for Pecker and co-workers, p. 158.

In 1980 **Artale** and co-workers [3] described the effect of an intravenous infusion of 200 mg pentoxifylline in 250 ml saline solution on the whole blood viscosity (Wells-Brookfield viscosimeter at 37 °C) and on erythrocyte deformability (whole blood method according to Reid and co-workers). The authors found a reduction of the blood viscosity at various shear rates, while there was no statistical change in the hematocrit. There was an improvement in erythrocyte deformability.

Note: Measurements of erythrocyte deformability at 25 °C; inadequate method for erythrocyte filtration. The carrier solution may also have had an effect.

In 1981 **Chmiel** and co-workers [132] studied the effect of an intravenous administration of 3 mg pentoxifylline per kilogram body weight (short infusion with 50 ml saline solution over 15 min) The blood samples were taken 1, 15, 30, 60, and 120 min after the end of infusion. The viscoelasticity of the blood was measured using an oscillating monocapillary viscosimeter at a shear rate of 10^{10} s^{-1}at a temperature of 22 °C. The authors found a decrease in the viscoelasticity of the blood with pentoxifylline, which could be shown up to 1 h after the end of the infusion.

Note: The measurements were performed at 22 °C. It should be noted that six of the nine patients examined were treated with a diuretic at the same time.

In 1981 **Ott** and **Lechner** [677] studied the effect of a 25-day treatment with 1200 mg pentoxifylline orally on the whole blood viscosity (at various shear rates) and on the plasma fibrinogen concentration in 15 patients with subacute cerebrovascular disorders. During the almost 4-week long therapy, there was a statistically significant reduction of the viscosity values, with a maximum at the shear rate of 11 s^{-1}, as well as a reduction of the fibrinogen concentration. The hematocrit values did not change significantly.

Note: No exact details on the behavior of the hematocrit values; no measurements of further rheologic parameters.
In 1983 similar results were published elsewhere by the same team [678].

In 1981 **Sergio** and co-workers [791] reported on the effect of a 90-day treatment with 1200 mg pentoxifylline per day on the blood viscosity, plasma viscosity, and erythrocyte deformability in patients with chronic arterial occlusive disease. The blood viscosity was statistically significantly decreased at a shear rate of 450 s^{-1}, and erythrocyte filtrability was increased. Plasma viscosity, fibrinogen concentration, and erythrocyte aggregation were statistically significantly decreased (abstract, no details).

In 1981 **Donaldson** and co-workers [194] studied the hemorheologic effect of an oral dose of 200 mg pentoxifylline in 40 patients with intermittent claudication in a randomized double-blind study. There was a significant improvement in erythrocyte deformability while the placebo group did not show any such behavior. On the other hand, blood viscosity did not change statistically significantly. Both medication and placebo showed no statistical differences with regard to the pain-free walking distance and the blood flow in the region of the lower leg as measured by means of plethysmography (abstract, no details).

In 1981 **Guerrini** and co-workers [403] studied the effect of an acute intravenous administration as well as a long-term administration of pentoxifylline on the flow properties of the blood in patients with chronic arterial occlusive disease. Blood viscosity was measured using a Wells-Brookfield cone-plate viscosimeter at 37 °C and at various shear rates. Plasma viscosity was determined with the same instrument at 375 s^{-1}. Erythrocyte filtrability was measured using the method of Reid and co-workers at 37 °C. Ten patients were given 166 mg pentoxifylline per minute intravenously over a period of 30 min. Blood samples were taken up to 60 min after the end of infusion. Within the framework of the long-term study, 400 mg pentoxifylline were given to 25 patients three times daily for a period of 30 days. There was a decrease in the blood viscosity at all shear rates, both with the intravenous administration and with the long-term oral application of pentoxifylline. The hematocrit values, plasma viscosity, and the fibrinogen concentration remained constant. The filtrability of the blood was improved.

Note: No measurements of erythrocyte aggregation. The measurements of the erythrocyte filtrability according to the method of Reid is not adequate evidence. The authors attributed the reduction of the whole blood viscosity at unchanged hematocrit values to a decrease in erythrocyte deformability.

Similar results were reported by Guerrini and co-workers elsewhere [407].

In 1981 **Strano** and co-workers [834] described the effect of pentoxifylline on the flow properties of blood in patients with chronic arterial occlusive diseases (2 × 440 mg per day orally). The blood and plasma viscosities were measured using the Brookfield microviscosimeter and showed a statistically significant reduction during a 90-day therapy period.

Note: No comparative measurement of the hematocrit values.

In 1981, 1982 and 1983 **Perego** and co-workers [691, 692, 694] studied the effect of oral doses of pentoxifylline (3 × 400 mg per day) on the flow properties of blood in patients with peripheral arterial occlusive disease. Blood viscosity was measured using a Brookfield micro-cone-plate viscosimeter at 37 °C and at shear rates of 11.25, 90, and 450 s^{-1}. The hematocrit was mathematically corrected; the plasma viscosity was measured at 450 s^{-1}. Erythrocyte aggregation was measured indirectly from the erythrocyte sedimentation rate, which was corrected according to hematocrit and plasma viscosity. The erythrocyte deformability of the whole blood was determined according to the method of Reid and co-workers. After the 30-, 60-, and 90-day treatment with pentoxifylline, the investigators found a significant decrease in the whole blood viscosity at all shear rates, particularly at low shear rates. In addition, they also found a significant decrease in plasma viscosity, the erythrocyte sedimentation rate, and erythrocyte deformability.

Note: No quantitative measurements of erythrocyte aggregation.

In 1982 **Heilmann** and co-workers [428] studied the effect of an oral therapy with 1200 mg pentoxifylline daily over several weeks in patients with polycythemia vera. Blood viscosity was studied before and after this therapy using a falling ball viscosimeter at a standardized hematocrit of 40%. Eryhtrocyte deformability was determined according to the filtration method of Reid and co-workers. The authors found a statistically significant reduction of the blood viscosity both in the patients and in the treated controls, the erythrocyte filtrability was statistically significantly reduced only in the patients.

Note: No measurements of plasma viscosity and erythrocyte aggregation; both of the rheologic measurement techniques used are no longer considered to be relevant nowadays (see Sect. 3.3.4).

In 1982 **Heilmann** and co-workers [429] treated a 29-year-old female patient from Turkey with homozygotic sickle cell anemia and hemolytic crises which occurred several times a year. In this case, the pains were in the region of the major joints and the lower right calf. The authors tested erythrocyte filtration through Nucleopore filters with an average pore size of 3 μm using a standardized hematocrit of 20% before and after an 8-month therapy with pentoxifylline at a daily dosage of 800 mg orally. The result was that after pentoxifylline the time it took for the erythrocytes to pass through the filter was reduced from 20 to 14 min, which was traced back to an improvement in the flexibility.
These results are similar to those found by Keller and Leonhardt [486, 487] in sickle cell anemia.

Note: Single case report; no details on the temperature.

In 1983 **Strano** and co-workers [838] described the effect of an oral administration of 400 mg pentoxifylline per day for a period of several months in patients with chronic arterial occlusive diseases. The blood and plasma viscosities were determined with a Brookfield micro-viscosimeter and the erythrocyte filtrability with the whole blood method according to Reid. The authors found a statistically significant decrease in the blood and plasma viscosities as well as an improvement in erythrocyte filtrability.

In 1983 **Heilmann** [430] studied ten Patients with polycythemia vera, who were given 1200 mg pentoxifylline orally per day over the course of 4 weeks. The blood and plasma viscosities were measured using a Höppler falling ball viscosimeter; erythrocyte filtrability was determined with Nucleopore filters. There was a statistically significant decrease in the whole blood viscosity as well as an improvement in the erythrocyte deformability.

Note: The validity of the viscosimeter used is rather doubtful.

In 1983 **Pennarola** and co-workers [690] reported on a study with 25 patients with Raynaud's disease following vibratory trauma before and after a 60-day treatment with 3 × 400 mg pentoxifylline daily. The found an improvement in erythrocyte deformability (abstract, no details).

In 1983 **Strano** and co-workers [837] studied the effect of an oral administration of 800 mg pentoxifylline per day for a period of 90 days in patients with intermittent claudication and pains at rest. They determined the viscosity, according to Reid (1964); erythrocyte filtrability, according to Reid and co-workers [721]; and the plasma fibrinogen concentration. At the end of therapy, all factors were statistically significantly improved (abstract, no details).

In 1983 **Cabannes** and co-workers [106] studied the effect of pentoxifylline on the viscosity and the filtrability of blood in patients with sickle cell anemia (abstract, no results and no further details).

Urai and co-workers [869] studied the filtrability of blood in 20 patients with chronic arterial occlusive disease before and after a daily dose of 400 mg pentoxifylline for a period of 3 months. In 27% of the patients, in whom erythrocyte deformability was primarily changed, there was an improvement in the filtrability values (abstract, no details).

In 1983 **Seiffge** and co-workers [788] reported on one patient with thalassemia and with considerable pains due to capillary occlusions resulting from highly rigid erythrocytes. After the administration of 2 g pentoxifylline per day a clinical improvement was seen. Rheologically there was an improvment in the erythrocyte filtrability through Nucleopore filters (with a pore diameter of 5 μm) and an improvement in the results gained with an erythrocyte rigidometer.

Note: Single case study; measurements of erythrocyte deformability in a buffer containing albumin.

In 1983 **Metoki** and co-workers [644] reported on the effect of a 4-week therapy with an intravenous administration of pentoxifylline in patients with cerebrovascular diseases. They found a reduction of both the whole blood viscosity and the serum viscosity (original in Japanese, English abstract).

Manrique [611] studied the effect of pentoxifylline versus placebo in patients with acute myocardial infarction and found an improvement in the filtrability in the pentoxifylline group (abstract, no details).

In 1983 **Muggeo** and co-workers [665] reported on the effect of both a 1-week intravenous therapy with 600–800 mg and a 30- to 60-day oral therapy with 1.2–1.6 g pentoxifylline daily in patients with acute cerebrovascular diseases compared to healthy controls. The intravenous administration of pentoxifylline was infused in 250 ml saline solution. The blood and plasma viscosities were measured using a Wells-Brookfield LVT viscosimeter at 230 s^{-1}. There was a tendential decrease in the blood and plasma viscosities in both groups.

Note: No statistical evaluation of the results.

In 1983 **Moratti** and co-workers [655] described the effect of pentoxifylline on patients with diabetic vasculopathy. Twelve patients received 1200 mg of the substance per day orally, which led to an improvement in the erythrocyte filtration rate (whole blood method of Reid and co-workers).

In 1983 **Solerte** and co-workers [805] reported on the effect of pentoxifylline (1.2 g per day for 8 months) on erythrocyte filtrability and found a statistically significant improvement (abstract, no details).

In 1984 **Koul** and co-workers [527] gave 1200 mg pentoxifylline 3 times a day orally to patients due for open heart surgery and in addition administered 300 mg intravenously on the day of the operation. Erythrocyte filtrability was determined according to Reid and co-workers before and after the extracorporal circulation of the blood. On the first 2 days following the operation, the deformability was distinctly deteriorated both in the pentoxifylline group and in the placebo group, but then improved clearly in the pentoxifylline group.

Note: No data on plasma viscosity; the filtration method used is of limited relevance.

In 1984 **Strano** and co-workers [839] studied the effect of pentoxifylline (2 × 400 mg orally) over a period of 90 days in a double-blind cross-over study compared to placebo in patients with chronic occlusive arterial disease in stages II and III according to Fontaine. In addition to subparameters as pain-free walking distance, behavior of platelet function etc., the whole blood viscosity was measured at a shear rate of 230 s^{-1} and the plasma viscosity, erythrocyte filtrability, and plasma fibrinogen concentration were studied. During the treatment with pentoxifylline or placebo, blood samples were removed before and after 30 and 90 days. In the pentoxifylline group there was a reduction of the whole blood viscosity as well as of the plasma viscosity, and there was an increase in the erythrocyte filtration rate. In the placebo group there were no significant changes.

Note: No data on the temperature of the measurements and on the method used for the determination of the rheologic parameters; no data on the hematocrit.

Clinical findings: The authors found a statistically significant improvement of the pain-free walking distance with pentoxifylline compared to placebo.

In 1984 **Angelkort** and co-workers [20] reported on the effect of an oral dose of 1600 mg pentoxifylline per day over a period of 6 weeks in patients with severe chronic arterial occlusive disease. Erythrocyte filtrability was measured according to the method of Reid and co-workers. There was an improvement in the filtrability of the blood, which correlated with an improvement in the pain-free walking distance.

Note: For the validity of the method of Reid and co-workers see Sect. 3.3.4.

James and co-workers [473] treated ten patients with intermittent claudication for 2 and 4 weeks with pentoxifylline at a dosage of 1200 mg per day and determined the whole blood viscosity before and after therapy with a Wells-Brookfield viscosimeter at 37 °C and erythrocyte filtration according to the method of Reid and co-workers. There was an improvement in the erythrocyte filtrability, while no statistically significant differences could be found in the whole blood viscosity.

In 1984 **Donaldson** and co-workers [195] carried out a double-blind study to test the effect of 3 × 400 mg pentoxifylline on the whole blood viscosity and the erythrocyte deformability of 40 patients with intermittent claudication. The erythrocyte filtrability was determined according to the method of Reid and co-workers, the whole blood viscosity with several rotational viscosimeters at various shear rates. There was a significant improvement in the erythrocyte deformability, while the whole blood viscosity did not show any significant changes.

Note: Missing relevance of the erythrocyte filtration method; no data on the temperature of the measurements.

In 1985 **Sturani** and co-workers [849] infused 100 mg pentoxifylline over a period of 5 min into patients with chronic cor pulmonale. This resulted in a decrease in the whole blood viscosity at 94 s^{-1} and in an improvement in the erythrocyte filtrability (washed erythrocytes after elimination of the leukocytes through a filter) (abstract, no details).

In 1985 **Spürk** and co-workers [811] studied the effect of a therapy with 2 × 600 mg pentoxifylline in patients with severe arterial occlusive disease and found an improvement in erythrocyte filtrability, in erythrocyte aggregation, and in plasma viscosity (abstract, no details).

In 1985 **Ernst** and co-workers [330, 331] investigated the effects of an oral dose of 1200 mg pentoxifylline in 30 patients with intermittent claudication. In the course of the 4-week therapy, there was a significant improvement in the whole blood viscosity and in erythrocyte filtrability, as well as a an increase in the pain-free walking distance. The authors also suspected an improvement of the leukocyte deformability (abstract, no details).

In 1985 **Petralito** and co-workers [699] carried out an open study to test the effect of pentoxifylline on erythrocyte aggregation, as measured using a light transmission aggregometer after a 1–2 month therapy with 400 mg pentoxifylline daily. The authors described a significantly reduced erythrocyte aggregation as a result of pentoxifylline (abstract, no details).

In 1985 **Vaya** [874] described the effect of pentoxifylline on erythrocyte flexibility in patients with cerebrovascular diseases. An improvement in the measured values was found using the method of Reid and co-workers (abstract, no details).

In 1985 **Clivati** and co-workers [138] studied the effect of a 3-day therapy with 1200 mg pentoxifylline per day orally on the whole blood viscosity, erythrocyte aggregation, and erythrocyte filtrability in 36 children with β-thalassemia. The authors found an improvement in all the parameters studied (abstract, no details).

In 1985 **Antignani** and co-workers [24] reported on the effect of a combined oral and intravenous therapy with pentoxifylline on the flow properties of blood in patients with peripheral occlusive disease and found an improvement in the filtrability (abstract, no details).

In 1985 **Doletskii** and co-workers [193] studied blood viscosity in children after surgery and found an improvement in the whole blood viscosity at various shear rates as the result of an infusion of pentoxifylline (article in Russian).

In 1985, within the framework of a randomized double-blind study, **Gallus** and co-workers [365] studied the effect of a retarded version of pentoxifylline in patients with intermittent claudication. The authors studied the whole blood viscosity at 37 °C with a Wells-Brookfield viscosimeter, and erythrocyte flexibility both with a whole blood filtration test and according to the packing density after specific centrifugation. They found no changes at all in the hemorheologic parameters. The hematocrit values and the plasma fibrinogen concentration also remained constant.

Clinical findings: The authors found no statistically significant improvement in the pain-free walking distance.

Boisseau [79] studied the effect of pentoxifylline in patients with acute cerebrovascular insufficiency and found an improvement in the filtration values (abstract, no details).

In 1985 **Canovi** and co-workers [114] found an improvement in the erythrocyte filtration (method of Cordaro and Carandente) after a 2-week treatment with 800 mg pentoxifylline daily in 54 patients with various diseases.

Note: No measurement of further hemorheologic parameters.

In the year 1985 **Sternitzky** and **Seige** [816] treated 90 patients with peripheral chronic arterial occlusive disease by means of pentoxifylline. In addition to other parameters, they also studied the whole blood viscosity at intermediate and high shear rates (no data as to methododology) and found only minor and nonsignificant changes.

Note: No measurements of further hemorheologic parameters, but considerable clinical improvements.

Solerte and co-workers [806] administered 1.2 g pentoxifylline daily to diabetics for a period of 6 months. They found a reduction of the whole blood viscosity at 4.5 and 2.25 s^{-1} and an improvement in the whole blood filtrability (abstract, no details).

Solerte and **Ferrari** [807] determined erythrocyte filtrability in 70 patients with diabetes mellitus with and without vascular complications before and after a long-term therapy with oral doses of pentoxifylline (1200 mg daily). Erythrocyte filtrability was determined with a modification of the whole blood filtration method of Reid and co-workers. The authors found a statistically significant improvement in erythrocyte filtrability as well as a decrease in the plasma fibrinogen concentration.

Note: Further hemorheologic parameters were not studied.

In 1985 **Angelkort** and co-workers [21] described the improvement in the filtrability of the blood after oral administration of pentoxifylline (Nucleopore filters with 5-μm pore diameter, pressure 20 cm H_2O). Erythrocyte aggregation was measured using an instrument by Myrenne and plasma viscosity with a capillary viscosimeter. The patients with chronic arterial occlusive disease received daily infusions of 600 mg pentoxifylline over a period of several weeks. There was a statistically significant improvement in all the hemorheologic parameters studied.

Note: The improvement in the hemorheologic data correlated, to a great extent, with a clinical improvement.

Strano and co-workers [841] treated 16 patients with myocardial infarction by means of pentoxifylline at a dosage of 1200 mg orally each day over a period of 120 days in a cross-over experiment versus placebo. The blood and plasma viscosities were studied with a Brookfield micro-viscosimeter at 230 s^{-1}, and erythrocyte filtrability was tested according to the method of Reid and co-workers. There was a statistically significant reduction of the whole blood and the plasma viscosities in the pentoxifylline group, but not in the placebo group. The erythrocyte filtrability also improved only in the pentoxifylline group.

Parnetti and co-workers [685] studied four randomized groups of patients with cerebrovascular insufficiency. They administered 3 × 400 mg pentoxifylline per day orally. The measurement of erythrocyte filtrability was carried out with a modification of the whole blood filtration method of Reid and co-workers. There was a statistically significant improvement in erythrocyte filtrability.

Note: Simultaneous neurologic tests did not show any significant improvement.

Using a St. George filtrometer, **Matrai** and **Ernst** [627] determined erythrocyte filtrability in 20 male patients with chronic arterial occlusive disease before and after the oral administration of 1200 mg pentoxifylline over a period of 4 weeks. The authors described a small but statistically significant improvement in erythrocyte deformability and a marked decrease in filter pore blockage, which was explained by an improvement in the rheologic properties of the leukocytes.

Perego and co-workers [696] administered 400 mg pentoxifylline orally 3 times daily for a period of 90 days to 25 patients with intermittent claudication. They studied the whole blood viscosity with a Wells-Brookfield microviscosimeter at 37 °C. Erythrocyte filtrability was determined with the whole blood method according to Reid and co-workers. Changes in the erythrocyte sedimention rate served as a measure of erythrocyte aggregation. The authors described statistically significant improvements of all the hemorheologic parameters studied.

Note: See Sect. 3.3.4 for the methodologic limitations.

Schneider and co-workers [770] treated 40 patients with lacunar strokes with 800–1600 mg pentoxifylline orally over a period of months. Erythrocyte aggregation was studied with a mini-erythrocyte aggregometer, plasma viscosity with a capillary viscosimeter, erythrocyte deformability with a single-pore erythrocyte rigidometer, and the yield shear stress with a corresponding instrument. With hematocrit values that were minimally corrected there was a statistically significant decrease in erythrocyte aggregation and plasma viscosity, as well as an improvement in erythrocyte deformability. The yield shear stress decreased statistically significantly.

Strano and co-workers [842] studied the whole blood viscosity (at 230 s^{-1}), plasma viscosity, and erythrocyte filtrability in patients with arterial occlusive disease within the framework of a double-blind cross-over study, in which 400 mg pentoxifylline and placebo were administered orally twice daily over a period of 90 days. An improvement in the rheologic values was reported.

Note: No details as to the methodology and no numerical data on the results.

4.1.27.3 Discussion

After the teams of *Hess* and *Heidrich* described a viscosity-reducing property of the "vasoactive" substance pentoxifylline in the year 1973, there subsequently appeared a large number of rheologic studies on the improvement of the flow properties of blood with this drug. The varying results at the beginning may be primarily attributed to methodologic difficulties. The initial viscosimetric studies of *Jauch, Ott,* and *Heidrich* were mainly performed with inappropriate falling ball viscosimeters, the results of which are diffficult to interpret. The subsequent studies of various research teams using rotational viscosimeters were only carried out with a corrected hematocrit in isolated cases [800]; in many cases, on the other hand, this parameter was not considered at all or only calculated mathematically [620, 691, 692, 818, 819]. Variations in the blood viscosity caused by the circadian rhythm and changes in the measurement values due to hemodilution resulting from added fluid were also only discussed sporadically. Thus it is difficult to see that the viscosity reduction resulting from pentoxifylline could reach a similar extent as that from an infusion of 500 ml low molecular weight dextran (421, 444, 676). Based on the comparative evaluation of the studies available, the maintenance of a constant temperature of 37 °C seems to be particularly important; however, many investigators carried out measurements below body temperature.

The blood viscosity reduction observed by *Völker* [880] in 1976 after the oral administration of pentoxifylline corresponds to the results of *Störmer* and co-workers [818] after intravenous application of this substance. *Völker* set the hematocrit to a constant value during the viscosity measurements, and *Störmer* tried to eliminate mathematically the effect of a variable hematocrit value on the blood viscosity, *Perego* and co-workers [694] also reported on a blood viscosity reduction with a mathematically corrected hematocrit, while *Knarr* [514] and *Mashiah* [620] did not measure a decrease in the whole blood viscosity with pentoxifylline at room temperature.

Chmiel and co-workers [132] found a decrease in the viscoelasticity of the blood up to 1 h after the end of an infusion with pentoxifylline. The studies were, however, performed at 22 °C. Further findings on a reduction of the whole blood viscosity were made by *Sergio* and co-workers [791], while *Donaldson* and co-workers [195] could not find any effect on blood viscosity at the low dose of only 200 mg pentoxifylline.

Guerrini and co-workers [403] found a decrease in blood viscosity following the daily administration of 3 × 400 mg pentoxifylline for 30 days, while the plasma viscosity remained constant.

The decrease in the whole blood viscosity measured by *Ott* and *Lechner* [677] may possibly be traced back to a slight decrease in the hematocrit values, even though this reduction of the hematocrit is not statistically significant. The authors attribute the reduction of the whole blood viscosity to a decrease in the fibrinogen concentration; precise statements would only be possible if the total protein concentration in the plasma had been measured at the same time.

Heilmann and co-workers [428] also found a reduction of the whole blood viscosity using a falling ball viscosimeter, as did *Strano* and co-workers [841], who used a Brookfield device to measure the viscosity; *Artale* and co-workers [30], *Strano* and co-workers [834] and *Muggeo* and co-workers [665] used a cone-plate viscosimeter for the measurements. *James* and co-workers [473], on the other hand, could not find a decrease in the whole blood viscosity.

As explained in Sect. 2.1, a complete rheologic analysis, including the measurement of plasma viscosity, erythrocyte aggregation, and erythrocyte filtrability (rheologic status) is necessary in order to be able to make a definite statement as to the effect of a drug on the flow properties of blood.

Since *Ehrly* [266] discovered in 1975 that pentoxifylline improves erythrocyte filtrability, this special rheologic parameter has been increasingly tested by different research groups. While *Ehrly* [262, 266] described an improvement in erythrocyte filtrability after intravenous administration of pentoxifylline, *Völker* [880] did not find any effect on erythrocyte filtrability after the oral administration of pentoxifylline. *Knarr* [514] from Ehrly's team also found an improvement in the erythrocyte filtration rate after an intravenous infusion with pentoxifylline; however, his measurements were performed at 23 °C.

The majority of studies on erythrocyte filtration that were subsequently discussed were performed according to the method of Reid and co-workers which, from the today's standpoint is considered to be outdated so that these results must be relativized (see Sect. 3.3.4).

Schubotz and *Mühlfellner* [775], *Pecker* and co-workers [689], *Angelkort* and co-workers [15], *Manrique* [609], *Martin* and *Vives* [619], *Le Devehat* and co-workers [557], *Guerrini* and co-workers [403], *Heilmann* and co-workers [428], *Strano* and co-workers [837], *Moratti* and co-workers [655], *Pennarola* and co-workers [690], *Solerte* and co-workers [805], *Koul* and co-workers [527], *Strano* and co-workers [839], *James* and co-workers [473], *Donaldson* and co-workers [195], *Vaya* [874], *Solerte* and *Ferrari* [807], and *Perego* and co-workers [696] tested erythrocyte flexibility according to the above-mentioned method. Except for the reports of *Martin* and *Vives* [619], who treated patients with cerebrovascular incidents, the other subjects were all patients with chronic occlusive arterial disease.

After oral and intravenous administration of pentoxifylline, all of the authors found a partly statistically significant increase in the number of erythrocytes flowing through the Nucleopore filters. *Manrique* [609], on the other hand, observed that the increase in the erythrocyte filtrability in his group of patients was quite variable. *Johnsson* and co-workers [477] treated both patients with chronic occlusive arterial disease and with artificial heart valves. An improvement in erythrocyte filtrability could only be found in the patients with chronic occlusive arterial disease.

In 1981 *Donaldson* and co-workers [194] studied blood viscosity in patients with intermittent claudication and found a decrease in erythrocyte deformability after an oral administration of just 200 mg pentoxifylline per day. *Gallus* and co-workers [365] found no hemorheologic changes at all in their study.

The description of a single case by *Keller* and *Leonhardt* [486] is interesting as well. In one female patient with sickle cell disease there was an increase in erythrocyte filtrability after the oral administration of pentoxifylline, which correlated with the disappearance of the paroxysmal pains.

Heilmann and co-workers [429] reported on an improvement in erythrocyte filtrability and an improvement in the symptoms in one patient with homocygotic sickle cell anemia. *Seiffge* and co-workers [788] were able to report on a patient in a similar situation with thalassemia.

Seen as a whole, most authors determined an improvement in erythrocyte filtration (erythrocyte deformability) after the administration of pentoxifylline. However, methodologic difficulties must also be taken into consideration. In recent years an improvement in erythrocyte deformability or filtration has been described using more modern methods [849, 114, 627, 770]. A series of in vitro results also speaks for an improvement in erythrocyte filtration as a result of pentoxifylline [262, 564].

As regards the behavior of the plasma viscosity, the conclusions are contradictory. While some authors speak of a decrease, others find no significant change in this parameter. The statements of *Störmer* and co-workers [818], that the reduction of blood viscosity becomes evident especially at low shear rates, should theoretically be explained via a disaggregation of erythrocyte aggregates.

Measurements of the erythrocyte sedimentation rate as well as direct measurements of the erythrocyte aggregates (light transmission) resulted in a decrease in this parameter [691, 811, 699, 138, 21, 696, 770]. This could possibly be explained by a decrease in the plasma fibrinogen concentration described by many authors.

In summary, pentoxifylline can be described as a rheologically active drug with which the majority of the investigators found a decrease in the whole blood viscosity and almost all authors described an improvement in erythrocyte filtrability. The conclusions on the behavior of plasma viscosity are contradictory. The small number of studies on the behavior of erythrocyte aggregation speaks for a decrease in the aggregation tendency of the erythrocytes as a result of pentoxifylline. After dextran solutions pentoxifylline is the rheologicotherapeutic substance which has been studied most intensively; with no other drug have so many studies been carried out on erythrocyte deformability (erythrocyte filtrability).

4.1.28 Carbocromen

4.1.28.1 Introduction

Carbocromen is a coronary dilator. The primary indication for carbocromen is coronary heart disease.

4.1.28.2 Publications

In 1974 and 1975 **Chmiel** and co-workers [129, 130] reported on the hemorheologic effects of carbocromen in patients with recent myocardial infarction and increased viscosity values. Within 30 min the patients were intravenously infused with 200 mg carbocromen in 50 ml Ringer solution. The blood viscosity was determined with a Couette rotational viscosimeter which was equipped with a "guard-ring" to avoid surface effects. The plasma viscosity was measured using a capillary viscosimeter. The hematocrit was corrected mathematically. Compared to a placebo group the authors found a significant decrease in blood viscosity particularly at a shear rate of 0.488 s^{-1}. Basically the same blood viscosity reductions were found with heavy smokers. The plasma viscosity and the blood viscosity did not change at high shear rates (the measurement of which was considered by the authors to be an indirect method for the determination of changes in the erythrocyte deformability).

Note: The temperature was not given for the blood viscosity measurements; the correction for hematocrit was carried out mathematically; the erythrocyte deformability was not determined using a filtration test but rather indirectly with a capillary viscosimeter.

4.1.28.3 Discussion

There are only two publications by one research team available on the rheologic effects of carbocromen. In these cases the authors found a decrease in blood viscosity values at low shear rates in patients with recent myocardial infarction and raised viscosity values. The authors argued that the reduction of the blood viscosity at low shear rates may be traced back to a decrease in the erythrocyte aggregation. Quantitative measurements of the erythrocyte aggregation were not performed. Since the erythrocyte deformability was also only determined indirectly, an estimation of the rheologic properties of the drug is limited.

4.1.29 Clofibrate

4.1.29.1 Introduction

Clofibrate is known as a lipid-reducing substance. An effect of this drug, which is not so well known, is the reduction of the plasma fibrinogen level.
Since 1974 etofibrate, a combination of clofibrate and another lipid-reducing drug, nicotinic acid, has been used for the treatment of hyperlipoproteinemia.

4.1.29.2 Publications

In 1974 **Dormandy** and co-workers [201] brought out the first report on the improvement of the flow properties of blood with clofibrate. The 62 patients studied had been suffering for over 32 months from intermittent claudication and in part from pains at rest. The ankle pulses could not be felt in most of the patients. The patients received 2 g clofibrate per day orally for at least 6 months and were compared to a nontreated control group. The blood viscosity was measured at shear rates of 23–230 s^{-1} with a Brookfield micro-cone-plate viscosimeter and at shear rates of 0.77–2.62 s^{-1} using a Contraves viscosimeter. A mathematical correction of the hematocrit to a standardized value of 45% was performed.

At the end of treatment there was a significant reduction of the cholesterol and the triglyceride level from 265–205 mg per 100 ml and 185–135 mg per 100 ml respectively, and of the fibrinogen level from 426–330 mg per ml. There was a decrease in blood viscosity at all the shear rates measured.

Note: No data on the temperature for the viscosity measurements; no measurements of further rheologic parameters.

Clinical findings: In ten of the 62 patients there was an improvement of the clinical symptoms (a reduction of pains at rest and an increase in the pain-free walking distance).

In 1976 **Djurdjevic** and co-workers [190] presented a randomized study on 40 patients with increased cholesterol levels of the blood. The patients received 1.5 g clofibrate daily compared to a placebo group (n = 40). The blood viscosity was measured at 20 °C with a capillary viscosimeter (0.3 mm diameter). After 12 weeks of treatment, there was a reduction in the cholesterol and triglyceride levels and a significant decrease in blood viscosity.

Note: No correction for hematocrit; no measurements of further rheologic parameters; no blood viscosity measurements at body temperature.

In 1978 **Pfeiffer** and **Tilsner** [700] reported on 33 patients with hyperlipoproteinemia some of whom had an increased plasma viscosity compared to healthy subjects (plasma viscosity 1.4 ± 0.1 cP). The patients received 900 mg etofibrate (clofibrinic acid nicotinic acid esther) per day orally over a treatment period of 6 weeks. The plasma viscosity was measured using a Brookfield micro-cone-plate viscosimeter at 20 °C and at shear rates of 1–10 s^{-1}.
At the same time as the reduction of the triglycerides by 45% and of the cholesterol level by 21% there was a reduction in plasma viscosity which was most prominent in those patients with increased plasma viscosity. A reduction of the fibrinogen level was hypothesized.

Note: No correction for hematocrit; no measurement of further rheologic parameters; no measurements of the blood viscosity at body temperature.

In 1978 **Schwartzkopff** and **Roetz** [782] studied 53 patients with hyperlipoproteinemia of types IIa, IIb and IV in a cross-over double-blind study. During the 12-week period of study 41 patients still received a nonlipid-reducing accompanying medication (like e.g., pentoxifylline, and β-blockers). Within the framework of this experiment 1.5 g clofibrate or 1.5 g clofibrate and 60 mg standardized mucopolysaccharide complex were given to the patients each day. The blood viscosity measurements were performed using a micro-KPG-Ubbelohde capillary viscosimeter at 37 °C. A significant reduction of the blood viscosity compared to the placebo group was only found in the group treated with mucopolysaccharide. On the other hand, a significant decrease in the serum cholesterol level by 10%–20% and of the serum triglyceride level by 30%–70% could be varified both for the combination and clofibrate.

Note: No correction for hematocrit; no measurements of further rheologic parameters.

In 1979 **Arntz** and co-workers [26] studied the effect of a 7- to 9-week clofibrate therapy of 2 g per day on the flow properties of blood and on the lipoprotein level of blood in 36 patients with primary hyperlipoproteinemia of type II and IV and with increased serum and plasma viscosity values. The serum and the plasma viscosities were measured using a Schott capillary viscosimeter and the blood viscosity was measured using a Brookfield micro-cone-plate viscosimeter at shear rates of 46–115 s^{-1}. The hematocrit was set to a value of 40% for the blood viscosity measurements. At the end of treatment, a reduction of the triglyceride level of the patients by 30–147 mg% was observed. The serum cholesterol level decreased by 45 mg% only in the patients of the type II group. At the same time as the change in cholesterol level in the blood, there was only a significant decrease in the serum, plasma, and blood viscosities in type IV at a shear rate of 46 s^{-1}. A slight decrease could be observed in the fibrinogen level.

Note: No data on the temperature for the viscosity measurements; no measurements of further rheologic parameters.

In 1980 **Calvert** and co-workers [113] published a randomized double-blind study on the effect of a 12-day therapy with clofibrate in 22 patients with manifest diabetes. The blood viscosity was measured using a Brookfield micro- cone-plate viscosimeter at a hematocrit previously set to 45% at shear rates of 5.75–230 s^{-1}. The deformability of the erythrocytes (erythrocyte filtrability) was determined according to Reid and co-workers (Nucleopore filter with a pore diameter of 5 μm).
A significant reduction of 10% in the cholesterol level, of 33– 44% in the triglyceride level, and of 23% in the nonesterеd fatty acids could be observed. The plasma fibrinogen level also decreased by 23%, the reduction being strongest in those patients with a high plasma fibrinogen level. The reduction of blood viscosity at a corrected hematocrit was not clearly significant. The authors could not determine any change at all in the erythrocyte deformability under treatment with clofibrate.

Note: No temperature data; no measurments of further rheologic parameters.

In 1984 **Landgraf** and **Ehrly** [536] studied the effect of 2 g clofibrate per day orally for a period of 6 weeks on patients with chronic arterial occlusive disease and determined the whole blood viscosity and the plasma viscosity with an Ostwald capillary viscosimeter and the erythrocyte aggregation with an optical transmission method. There was a decrease in erythrocyte aggregation and plasma viscosity.

4.1.29.3 Discussion

A significant reduction of the blood and plasma viscosities under clofibrate or etofibrate were found by all the authors with the exception of *Schwartzkopff* and *Roetz*. While *Dormandy* found a decrease in the blood viscosity at all shear rates, this could either not be proven by other authors or, if so, only for certain shear rates.
In this connection it must be pointed out, as already mentioned in Sect. 3.3.4, that measurements with rotational viscosimeters produce results at low shear rates which are difficult to

evaluate. In addition, only two authors [26, 113] actually made a setting to a particular hematocrit.
The concurring results leave no doubt as to a reduction of the fibrinogen concentration, in particular with primarily increased fibrinogen levels [201, 26]. Therefore an effect of clofibrate or etofibrate on the flow properties of blood is also to be assumed. The quantitative measurement of erythrocyte aggregation resulted in a decrease in the values which correlated well with the behavior of the fibrinogen concentration. Whether the filtrability of blood is changed by the action of clofibrate has to be tested again with an improved filtration technique. The extent of the changes in the flow properties of blood as a result of fibrinogen reduction should not be set too high quantitatively, because this reduction is relatively small compared to that with streptokinase and ancrod. This might also explain the relatively small therapeutic response of patients with intermittent claudication. To what extent the reduction of the cholesterol level in the blood plays a direct role in the improvement of the flow properties of blood could not be clarified by the present studies. It seems plausible that due to an increased lipid content in the blood measurement artifacts could occur especially with cone-plate viscosimeters.

4.1.30 Pyridinol Carbamate

4.1.30.1 Introduction

Pyridinol carbamate is a drug, developed in Japan, which is supposed to improve the peripheral blood flow.

4.1.30.2 Publications

In 1975 **Blendin** [70] studied the effect of a 10-min intravenous infusion with 100 ml (150 mg) pyridinol carbamate on the flow properties of blood in patients with chronic occlusive arterial disease. The relative blood and plasma viscosities (H_2O = 1) of the heparinized blood samples were measured using an Ostwald capillary viscosimeter and the erythrocyte deformability (erythrocyte flexibility) was measured accordinng to the method of Ehrly and Rossbach. All studies were carried out at 23 °C. Fifteen and 45 min after the infusion the author found no changes in the hemorheologic parameters measured.
Note: The measurements of the rheologic parameters were made at 23 °C. No correction for hematocrit.

4.1.30.3 Discussion

There is only one publication available on the hemorheologic effects of pyrinidol carbamate. Under the experimental conditions chosen the author could determine neither a change in the blood and plasma viscosities nor in the erythrocyte deformability. The temperature of the measurements was 23 °C; a correction for hematocrit was not performed. No definite statement can therefore be made on whether pyrinidol carbamate affects the flow properties of blood or not.

4.1.31 Agonists and Antagonists at the Adrenoceptors

4.1.31.1 Introduction

Adrenoceptors are responsible for the regulation of the lumen diameter of vessels, especially of arteriols. One differentiates between blockers and stimulators of such adrenoceptors, various kinds of which have been described up to now. The best known are the α- and β-receptors; more details will not be explained here. Agonists and antagonists (stimulators or blockers) are, moreover, responsible for various functions in the heart and other organs. α-blockers like prazosin lead to a clear reduction of the systemic blood pressure via vasodilation. Agonists and antagonists at the adrenoceptors are used especially with coronary insufficiency and hypertension. There are only a few publications on the effect of α-blockers on the flow properties of blood; there are more publications on the hemorheologic effects of so-called β-blockers.
Since a comparably large number of publications are available on the β-mimetic drug isoxsuprin, this substance will be treated separately at the end of this section.

4.1.31.2 Publications

In 1970 **Aronson** and co-workers [2] reported on the effect of droperidol, a sympaticolytic drug. Patients about to have minor surgery received intravenous injections of droperidol (1 mg per kilogram body weight). The blood viscosity was studied with a Brookfield LVT-cone-plate viscosimeter at 37 °C and at various shear rates; the hematocrit was determined with a microhematocrit centrifuge. One hour after intravenous administration a reduction of the whole blood viscosity was found at all shear rates, as well as a reduction of the hematocrit values.

Note: The authors attribute the effect to endogenous hemodilution. No measurement of further hemorheologic parameters.

In 1976 **Dintenfass** and **Lake** [177] reported on the treatment of patients with ischemic cardiovascular diseases. The patients received 340 mg alprenolol per day orally. Blood viscosity was measured with blood samples anticoagulated with EDTA using a rotational viscosimeter (no details given) at a shear rate of 180 s^{-1} at 37 °C. Compared to a nontreated control group, the authors found a significant decrease in the blood viscosity. They assumed an improvement in the erythrocyte deformability.

Note: No correction for hematocrit; no direct measurements of the erythrocyte deformability; no measurements of further rheologic parameters.

In 1979 **Letcher** and co-workers [569] reported on the treatment of 11 patients with esential hypertension compared to a nontreated control group. The patients received 3 to 20 mg prazosin orally in several single doses each day during a treatment period of approximately 8 weeks. The blood and plasma viscosities were measured with a Couette viscosimeter at 37 °C and at high shear rates (208–52 s^{-1}) and at low shear rates (5.2–0.1 s^{-1}). At the same time as the decrease in the hematocrit value and the change in the plasma protein composition (reciprocal change in the fibrinogen and globulin fraction) there was a significant reduction of the blood viscosity, which was strongest at high shear rates. There was no significant change in the plasma viscosity.

Note: No direct measurement of the erythrocyte aggregation nor of the erythrocyte filtrability.

In 1981 **Palareti** and co-workers [682] studied the effect of metoprolol at a dosage of 10 mg per day in patients with uncomplicated essential hypertension. After a 7-day therapy there was a significant decrease in the hematocrit value and a tendency to a decrease in the whole blood viscosity, plasma viscosity and erythrocyte filtrability (abstract, no details).

In 1982 **Kirigaya** and co-workers [507] studied the effect of a daily dose of 30 mg propranolol on the blood viscosity in patients with ischemic cardiac diseases. The authors used a blood viscosimeter of their own and found no effect on the blood viscosity (abstract, no details).

In 1982 **Sarno** and co-workers [748] studied the effect of the β-blocker timolol on the flow properties of blood in 14 patients with essential hypertension. The whole blood viscosity was measured at 37 °C and at shear rates of 450–90 s^{-1} using a Wells-Brookfield rotational viscosimeter. Plasma viscosity was determined with the same apparatus at 450 s^{-1}. Erythrocyte filtrability was determined with the whole blood method of Reid. Following the administration of 10 and 20 mg timolol per day, there was a statistically significant reduction of the whole blood viscosity with a slight but statistically nonsignificant decrease in the hematocrit values. Plasma viscosity remained unchanged as a whole. There was no statistically significant change in erythrocyte filtrability.

In 1982 **Caimi** and co-workers [108] reported on the effect of metoprolol and oxprenolol on the flow properties of blood of patients with hypertension. The whole blood viscosity was measured with a Wells-Brookfield viscosimeter at various shear rates; the plasma viscosity was also measured with a Wells-Brookfield viscosimeter, but at high shear rates. Erythrocyte filtrability was determined using the whole blood method of Reid and co-workers. Eighteen patients with hypertension received 200 mg metoprolol, and 23 patients received 160 mg oxprenolol orally per day. The authors did not find any statistically significant changes in the hemorheologic parameters measured.

In 1982 **Waller** [895] reported on the treatment of patients with essential hypertension with atenolol (100 mg per day) and metoprolol (100/200 mg per day). The treatment period was 1 month. Blood viscosity was measured with a modified Deer rheometer at seven defined shear rates between 0.04 Nm^{-2} and 1.73 Nm^{-2}; plasma viscosity was measured with a capillary viscosimeter. At the end of treatment the author found a significant decrease in the systolic and the diastolic pressures, while, on the other hand, no significant changes from the initial values could be determined with regard to the hematocrit and the blood and plasma viscosities.

Note: A correction for hematocrit was not performed; no measurements of further rheologic parameters.

In 1983 **Pau** and co-workers [687] reported on the effect of atenolol on erythrocyte deformability in patients with Raynaud's syndrome. Erythrocyte filtrability was measured according to the method of Reid. The authors found no effect on erythrocyte filtrability (abstract, no details).

In 1983 **Heilmann** [436, 437] reported on the effects of an intravenous administration of hexoprenalin (0.3 μg/min) on the flow properties of blood in female patients in premature labor. In ten of these patients there was a decrease in the plasma viscosity, a decrease in the erythrocyte aggregation, and an improvement in erythrocyte deformability (abstract, no details).

Chien and co-workers [127] studied the effect of prazosin in patients with essential hypertension and found a reduction of the blood viscosity, which could be traced back to a reduction of the hematocrit value and of the plasma viscosity (abstract, no details).

Lang and co-workers [542] determined the blood viscosity using a Contraves LS 30 viscosimeter in 23 female patients with pre-eclampsia and in ten female patients with intrauterine growth inhibition. Compared to a control group, these patients had an increased blood viscosity, an increased hematocrit value, and an increased fibrinogen concentration. The administration of labetalol (a combined α- and β-blocker) for 1–2 weeks in nine female patients produced no change in the hemorheologic parameters (abstract, no details).

In 1983 **Dintenfass** [182] studied the effect of alprenolol on the flow properties of blood in patients. The whole blood viscosity, erythrocyte deformability, erythrocyte aggregation, and plasma viscosity were measured (no results given in the abstract, no further details).

In 1983 **Albanese** [7] reported on the effect of the α-blocker timoxamin on the microhematocrit, the whole blood viscosity (measured with a Brookfield cone-plate viscosimeter), plasma viscosity (Brookfield viscosimeter), and erythrocyte filtrability (according to Reid with the modification by Cordaro and Caradante). The drug (68 mg) was dissolved in 200 ml physiologic saline solution and infused into patients with arterial occlusive disease over a period of 30 min. The authors found a statistically significant decrease in the whole blood viscosity and in plasma viscosity. In vitro studies that were carried out in parallel showed no effect on the flow properties of blood. From this the authors concluded that the decrease in the hematocrit and the whole blood viscosity must be attributed to the α-blocking property of the drug.

Note: No measurement of further hemorheologic parameters.

In 1983 **Caimi** and co-workers [109] studied a group of 52 hypertensive patients and determined the blood viscosity with a Wells-Brookfield microviscosimeter at different shear rates and at 37 °C. Furthermore, the plasma and the serum viscosities were measured with the same instrument at 450 s^{-1}. Erythrocyte deformability was studied both mathematically according to Dintenfass and also according to the whole blood method of Reid and co-workers. Three medications (β-blockers) were compared: timolol (20 mg per day), metoprolol (200 mg per day), and oxprenolol (160 mg per day). While metoprolol and oxoprenolol had no effect at all on the hemorheologic parameters, with timolol there was a reduction of the whole blood viscosity at various shear rates in the course of the 30-day treatment period. The serum and the plasma viscosities increased during the 30-day therapy rather than decreased, which, however, was not statistically significant. There was also no change in the filtrability of the blood.

In 1984 **Blume** and co-workers [72] reported on the effect of the β-blocker metoprolol on the flow properties of blood in patients with peripheral vascular disorders. Patients with intermittent claudication and hypertension were given 100 mg metoprolol daily for a period of 4 weeks. Various hemorheologic parameters were studied: erythrocyte flexibility with a selecting erythrocyte rigidometer, erythrocyte aggregation using a photo-optic method, plasma viscosity with a capillary viscosimeter, and the flow yield stress. There was a statistically significant decrease in the erythrocyte aggregation, a decrease in the erythrocyte rigidity, as well as an improvement in the flow yield stress.

In 1984 **Siekmann** and **Heilmann** [798] reported on the tokolytic treatment with hexoprenalin and metoprolol in women in labor. They determined plasma viscosity and erythrocyte aggregation, as well as the hematocrit before and 120 min after isolated β-stimulation with hexoprenalin and after the additional administration of metoprolol. Plasma viscosity was measured with a capillary viscosimeter, erythrocyte aggregation by means of a photooptic method. In addition, the serum osmolarity was determined. While erythrocyte aggregation was affected only to a minimal extent, there was a statistically significant decrease in the hematocrit values as well as a reduction of the plasma viscosity. The authors attribute the change in the hematocrit values to inner hemodilution.

In 1985 **Turczynski** and **Znamirowska** [866] studied 52 patients with peripheral vascular disorders and found a statistically significant decrease in the whole blood viscosity (measured with a Gabe rotational viscosimeter) after the administration of 160 mg propranolol daily (original in Polish).

In 1985 **Furkalo** and co-workers [358] reported on the effect of antianginal drugs on the whole blood viscosity and plasma viscosity in patients with chronic coronary insufficiency. They found a decrease in these hemorheologic parameters using nitrates, β-blockers and molsidomin (original in Russian).

In 1985 **Toikawa** and co-workers [861] studied the effect of prazosin on plasma viscosity in patients with essential hypertension and found a reduction of the values.

In 1986 **Kiesewetter** and co-workers [499] studied the effect of metoprolol in 15 patients with essential hypertension in an open study. The patients were treated with a daily dose of 100 mg metoprolol for a period of 4 weeks. Plasma viscosity was studied with a capillary viscosimeter, erythrocyte aggregation with a light-transmission instrument, and erythrocyte rigidity with a single cell filtrometer. In addition, the flow yield stress was measured. There was a decrease in erythrocyte aggregation as well as an improvement in erythrocyte deformability, while plasma viscosity remained unchanged. There was also a marked change in the flow yield stress.

4.1.31.2.1 Isoxsuprin

4.1.31.2.2 Isoxsuprin is a β-sympathicomimetic drug used in cerebral and peripheral vascular disorders.

4.1.31.2.3 Publications

In 1976 **Schlichting** and **Heidrich** [755] were the first to report on an improvement in the flow properties of blood via an intravenous administration of 30 mg isoxsuprin in patients with peripheral arterial occlusive disease. The infusion period was 30 min. Blood viscosity measurements of the heparinized blood samples were carried out with a Brookfield micro-cone-plate viscosimeter at shear rates of $1.14-229.3\ s^{-1}$. The infusion with isoxsuprin led to a significant, shear rate-independent decrease in the blood viscosity immediately after infusion. The hematocrit did not change significantly. The reduction of the blood viscosity was more distinct at low shear rates than at high ones and was still unchanged 30 min after the infusion. The mode of action of the viscosity reduction could not be explained by the authors.

Note: No correction of the hematocrit; no measurements of further rheologic parameters; no data on the temperature of the measurements.

In 1976 **De Quiros** and **Hess** [166] published a study on the rheologic effects of intravenous infusions of isoxsuprin in patients with peripheral arterial occlusive diseases (no data on the stages). The blood and plasma viscosities of the blood samples, anticoagulated with EDTA, were measured using a Brookfield micro-cone-plate viscosimeter with a Mooney adapter at shear rates of 7.8–78 s^{-1}. In two experimental series the patients were infused with 30 mg isoxsuprin over a period of 30 or 60 min. Isoxsuprin led to a significant decrease in the blood viscosity at the end of infusion at all the shear rates measured, but particularly at low shear rates. With an infusion period of 30 min, the maximal viscosity reduction was immediately after the end of infusion, while, on the other hand, with a longer infusion period a significant decrease was found in the whole blood viscosity even 90 min after the end of infusion. Changes in plasma viscosity could not be determined. The authors interpreted the viscosity reduction as the result of a decrease in the tendency of the erythrocytes to aggregate and as an improvement in erythrocyte flexibility.

Note: No correction for the hematocrit; no details of the temperature of the measurements; no measurements of further rheologic parameters.

In 1978 **Di Perri** and co-workers [185] reported on the effect of an intravenous administration of isoxsuprin on the flow properties of the blood in patients with peripheral arterial occlusive diseases (no details on the Fontaine stage). The subjects received a dose of 20 mg kg^{-1} min^{-1} isoxsuprin for 30 min in a dilution of 1 mg isoxsuprin per 2.5 ml isotonic saline solution. The blood viscosity of the blood samples anticoagulated with EDTA was measured with a Brookfield micro-cone-plate viscosimeter, model 1/4 RVT, at shear rates of 3.75–750 s^{-1} and at a temperature of 37 °C. Plasma and serum viscosity measurements were also made with the above mentioned viscosimeter. Zero value measurements with a single administration of the carrier solution (isotonic saline solution) led to no changes in the rheologic parameters. At the end of the infusion there was a significant decrease in the blood viscosity at all the shear rates measured, the viscosity reduction reached a maximum 30 min after the end of infusion. The blood viscosity reduction was most marked at low shear rates. The plasma and serum viscosities, the fibrinogen level, and the hematocrit decreased simultaneously. Except for an increase in the free fatty acids in the blood, there were no changes in the lipid values of the blood. The blood viscosity-reducing effect of isoxsuprin was attributed to a reduction of the hematocrit and to a decrease in the erythrocyte aggregation and erythrocyte deformability. The reduction of the plasma viscosity was accounted for as the result of the decreased fibrinogen concentration and that of the serum viscosity as a result of changes in the protein and cholesterol levels.

Note: No correction of the hematocrit; direct measurements of erythrocyte aggregation and erythrocyte deformability were not performed.

In 1978 **Weber** [905] reported on results in the therapy of peripheral arterial occlusive diseases using isoxsuprin; he found a viscosity reduction.

In further study in the year 1980, **Weber** and co-workers [906] investigated the effect of intravenous isoxsuprin doses on the blood viscosity in patients with chronic peripheral circulatory disturbances in stages III and IV in a placebo-controlled double-blind study. A total of 30 mg isoxsuprin was administered intravenously over a period of 3 days. Blood viscosity was measured with a capillary viscosimeter constructed specifically for this purpose. The authors found a reduction of the whole blood viscosity in the isoxsuprin group. The hematocrit values were also significantly decreased. The authors suspected the decrease in viscosity was not the result of hemodilution, despite the fact that the hematocrit values during the administration of isoxsuprin were lower than during the administration of placebo.

Note: No correction for hematocrit; no details on the temperature of the viscosity measurements; no measurements of further rheologic parameters.

In 1980 **Guerrini** and co-workers [405] reported on the effect of isoxsuprin on the blood viscosity in patients with chronic arterial occlusive disease. The drug was either given over a period of 30 min or twice a day for 1 week. The authors described an effect on the hyperviscosity of the blood (abstract, no details).

In 1981 **Di Perri** and co-workers [187] described the effect of isoxsuprin on the blood viscosity with a simultaneous therapy using the nonselective β-blocker propranolol and with the selective β-blocker metoprolol in patients with chronic arterial occlusive disease. The blood and the plasma viscosities were measured with a Wells-Brookfield viscosimeter at 37 °C; erythrocyte filtrability was determined according to the method of Reid and co-workers. In 8 patients with chronic arterial occlusive disease there was a reduction of the blood and plasma viscosities with isoxsuprin alone. After pretreatment with propranolol and the continuation of this therapy, there was no change in the blood and plasma viscosities after an infusion of isoxsuprin. If, on the other hand, the selective β-blocker metoprolol was given before and at the same time as isoxsuprin, then the known viscosity-reducing effect of isoxsuprin ensued.

In 1982 **Kirigay** and co-workers [507] reported on a reduction of blood viscosity, using a viscosimeter developed by the research group, in patients with ischemic heart disease (abstract, no details).

In 1982 **Bode** and co-workers [73] reported on the effect of oral administerations of isoxsuprin (80 mg per day) on the flow properties of blood in patients with adult-onset diabetes. The blood viscosity (mathematically corrected hematocrit) and the plasma viscosity were measured at shear rates of $0.08-94.5\ s^{-1}$ using a Contraves Low Shear 30 rheometer. Following a 3-month treatment period, the researchers found no significant changes in the blood and plasma viscosities.

Note: No data on temperature; no measurements of further rheologic parameters.

In 1985 and 1986 **Aarts** and co-workers [1, 2] carried out a double-blind cross-over study to study the effect of isoxsuprin in patients with intermittent claudication. Either 40 mg isoxsuprin or placebo was given three times daily. The viscosity of the blood and the plasma was measured with a Contraves LS 30 viscosimeter at $130\ s^{-1}$, and erythrocyte deformability was determined indirectly according to Dintenfass. The authors found an improvement in the erythrocyte deformability. Significant changes in the whole blood viscosity and the plasma viscosity could not be determined.

4.1.31.2.4 Discussion

The results of the hemorheologic studies following the administration of agonists and antagonists at the adrenoceptors are not uniform. This is not surprising since they are substances with various modes of action – even though they are combined under one generic term. In other words, all β-blockers are not the same in action, and α-blockers cannot be set on the same level as β-stimulators either clinicaly or pharmacologically. In addition, the individual authors also used various dosages, and it is known that one and the same substance can produce contrary effects on the vasoactivity of the peripheral vessels at different dosages. With the great variety of drugs used, it seems to be difficult to give an evaluation; isoxsuprin is discussed separately.

According to the studies of *Aronson* and co-workers, *Dintenfass, Letcher* and co-workers, as well as *Toikowa* and *Albanese,* the α-blockers droperidol and prazosin reduce the blood viscosity with a simultaneous decrease in the hematocrit values. *Aronson* and co-workers and *Letcher* and co-workers attribute the reduction of the blood viscosity to an increased inflow of interstitial fluid into the intravasal system.

Waller as well as *Paul* and co-workers found no changes in the blood and plasma viscosities with the use of atenolol and isoproterenol as far as erythrocyte filtrability was concerned. *Waller* and *Caimi* and co-workers also found no change in the blood viscosity with the use of metoprolol, while, on the other hand, *Palareti* and co-workers found a tendential decrease in the whole blood viscosity and the plasma viscosity.

Kiesewetter as well as *Blume* and co-workers found a decrease in erythrocyte aggregation and an improvement in erythrocyte deformability with a daily dose of 100 mg metoprolol. *Siekmann* and *Heilmann* also found an effect on hemorheologic parameters with metoprolol.

According to *Kirigaya* and co-workers, the administration of 30 mg propranolol per day has no effect on blood viscosity, while *Turczynski* and *Znamirowska* did find a reduction of the whole blood viscosity with 160 mg propranolol per day.

With timolol *Sarno* and *Caimi* found a statistically significant reduction of the whole blood viscosity with a nonsignificant decrease in the hematocrit value. According to the studies of *Heilmann,* hexoprenaline caused a reduction of the plasma viscosity as well as a decrease in erythrocyte aggregation.

Lang and co-workers could not find any changes in the hemorheologic parameters after the administration of labetalol. According to the studies of *Caimi,* oxprenolol also had no hemorheologic effect.

With the exception of *Bode* and co-workers and *Aarts* and co-workers, all the other authors found an improvement in the flow properties of blood after the intravenous administration of isoxsuprin. In the studies of *Schlichting* and *Heidrich, De Quiros* and *Hess, Di Perri* and co-workers, and *Weber* and co-workers, the changes in the viscosity were not corrected for hematocrit, i.e., a substance-specific effect on the viscosity other than a possible hemodiluting effect cannot be proven.

The viscosity measurements were carried out with the rotational viscosimeters which are now common. A reduction of the blood viscosity was found particularly at low shear rates. This could indicate a decrease in erythrocyte aggregation; however, measurements of this kind were not carried out by any of the research teams.

While *De Quiros* and *Hess* did not find any changes in plasma viscosity, *Di Perri* and co-workers determined a reduction of this parameter in addition to a reduction of the whole blood viscosity.

Since a hematocrit correction was not carried out for the viscosity measurements, and since two authors found a hematocrit reduction, a conclusion on the blood viscosity-reducing effect of isoxsuprin and on its rheologic mode of action is not possible as yet.

A compilation of various publications on the subject of β-blockers and hemorheology appeared in 1983 [434].

4.1.32 Choline Esters ("Essential Phospholipids"), Unsaturated Fatty Acids, Fish Oil Preparations

4.1.32.1 Introduction

"Essential" phospholipids are choline phosphoric acid diglyceride esthers of natural origin with a predominant proportion of unsaturated fatty acids like linoleic acid (approximately 70%), linolenic acid, and oleic acid. The essential phospholipids are used for the treatment of peripheral and cerebrovascular disorders with increased cholesterol values as well as with angina pectoris and diabetic angiopathies. In addition, other choline esthers, unsaturated fatty acids like eicosapentaenic acid, and fish oil preparations were studied with regard to their hemorheologic effect.

4.1.32.2 Publications

In 1976 **Ehrly** and **Blendin** [272] reported on the improvement of the flow properties of blood as a result of essential phospholipids in patients with peripheral vascular disorders in stage II according to Fontaine. The patients were infused with 750 mg of the substance intravenously over a period of 15 min. The blood and the plasma viscosities (heparin as anticoagulant) were measured with an Ostwald capillary viscosimeter. Additional blood viscosity measurements at shear rates of $5.6-46\ s^{-1}$ were performed with a Brookfield micro-cone-plate viscosimeter. All viscosity measurements were carried out at 23 °C. Erythrocyte deformability was tested according to the method of Ehrly and Rossbach. The investigators found no change in the blood and plasma viscosities. There was a tendential but nonsignificant improvement in erythrocyte deformability.

Note: No correction for hematocrit; no viscosity measurements at body temperature; no measurement of erythrocyte aggregation.

In 1976 **Klemm** [509] studied patients with peripheral arterial occlusive disease. The patients received 600 mg essential phospholipids orally three times a day for a treatment period of 18 days. At the end of treament the author found a 2.7% decrease in the blood viscosity compared to the initial values.

Note: No detailed data on the method for the blood viscosity measurements; no measurements of further rheologic parameters.

Auteri and co-workers [33] treated patients with chronic ischemic diseases of the lower extremities with a single intravenous dose of 1 mg cytidine diphosphate choline and studied the whole blood, the plasma, and the serum viscosities with the help of a Wells-Brookfield microviscosimeter 10, 30, and 60 min after infusion. The authors found that cytidine diphosphate choline caused a statistically significant reduction of the whole blood viscosity, reaching a maximum after 30 min, and an improvement in erythrocyte filtrability, also reaching a maximum 30 min after infusion. The plasma viscosity and the fibrinogen concentration showed no significant changes.

In 1984 **Falchi Delitala** and co-workers [345] described the effect of 1 g cyticholine in comparison to isotonic saline solution with regard to a series of clinical and laboratory-clinical parameters, including the whole blood viscosity which was measured with a Contraves Low Shear 2 viscosimeter at 37 °C, and found a decrease as a result of this therapy.

Note: No measurement of further hemorheologic parameters.

Crolle and co-workers [152] treated ten patients with vascular diseases with 2 × 1000 mg cyticholine for 6 days and found an improvement in the erythrocyte filtration index. This was determined according to the original method of Reid and co-workers (whole blood method).

In 1984 **Merchan** and **Dona** [633] studied 25 patients with cerebrovascular or cardiovascular insufficiency, who were given highly unsaturated phosphatidyl choline intravenously in single doses of 500 mg or as a long-term therapy of 30 days (200 mg daily). The blood and plasma viscosities were determined with a Wells-Brookfield microviscosimeter at 37 °C; erythrocyte deformability was studied with the whole blood filtration test according to Reid and co-workers. There was a reduction of the whole blood viscosity and an improvement in the filtration values.

Note: No measurement of erythrocyte aggregation; minimal validity of the applied filtration method.

In 1984 **Woodcock** and co-workers [924] studied the effect of a diet that had either high or low eicosapentaenic acid within the framework of a double-blind experiment. The flow properties of blood were determined before and after 7 weeks of therapy. Blood viscosity was measured with a Wells-Brookfield microviscosimeter at two shear rates and at 37 °C. Plasma viscosity was determined with a capillary viscosimeter at 25 °C. The hematocrit was determined using the microhematocrit method. There was a statistically significant decrease in the blood viscosity in the group on a diet rich in eicosapentaenic acid. Plasma viscosity and the hematocrit did not decrease statistically significantly.

Note: No measurement of other hemorheologic parameters.

In 1985 **Simons** and co-workers [799] studied the effect of an encapsulated fish oil preparation in comparison to placebo on the blood viscosity of patients with hyperlipidemia. Blood viscosity was measured with a Wells-Brookfield microviscosimeter. After a therapy of several months, no decrease in the viscosimetric data could be found as a result of lipidic fish oil, even though the plasma glyceride concentration decreased significantly.

Note: No measurement of other hemorheologic parameters.

In 1985 **Kobayashi** and co-workers [515] found a decrease in whole blood viscosity and an improvement in erythrocyte filtrability in 40 patients with hyperlipidemias and ischemic diseases after a daily dose of 2.7 g eicosapentaenic acid over a period of 16 weeks. The whole blood viscosity was measured with a cone-plate viscosimeter at various shear rates and at a temperature of 37 °C. The plasma viscosity was determined with the same instrument at 375 s^{-1}. Erythrocyte filtrability was determined according to the method of Reid and co-workers with minor modifications.

4.1.32.3 Discussion

Ehrly and *Blendin* found that the intravenous administration of essential phospholipids does not influence the macrorheologic parameters such as blood, plasma, and structural viscosities while, on the other hand, erythrocyte deformability does tend to improve. Changes in the lipid pattern of the erythrocyte membranes or another blood lipid composition were given as a possible mode of action for the improved erythrocyte deformability.
Klemm found that essential phospholipids lower blood viscosity. Since no correction of the hematocrit was performed and since there were no details on the viscosimeter used the results must be relativized. *Merchan* and *Donu* found a reduction of the blood viscosity as a result of phosphatidylcholine as well as an improvement in the erythrocyte filtrability within the framework of a long-term therapy. These results must also be relativized owing to methodologic reasons.
With another substance cytidine (diphosphate holine) that was studied by *Auteri* and co-workers, there was an improvement in the whole blood viscosity and the erythrocyte filtrability while the plasma viscosity remained unchanged. Cyticholine led to a reduction of the whole blood viscosity and to an improvement in erythrocyte filtrability [152, 345]; however, the method used limits the relevance of these findings.
The hemorheologic results on the effect of highly unsaturated fatty acids, specifically eicosapentaenic acid, on the flow properties of blood were investigated in various studies. In one experiment encapsulated fish oil was administered, but no viscosimetric changes were found [799]. With a diet that was either rich or low in eicosapentaenic acid, there was, on the other hand, a statistically significant decrease in the blood viscosity with a diet rich in fatty acids. The studies of *Kobayashi* and co-workers also showed changes in the flow properties of blood as a result of the direct administration of eicosapentaenic acid.
On the whole, the results on the effect of phospholipids, i.e., lipid acids or fish oil are not uniform, so that a definite hemorheologic action of the individual specific drugs or of their mode of application cannot be postulated.

4.1.33 Calcium Dobesilate

4.1.33.1 Introduction

Calcium dobesilate is used for the treatment of the varicose symptom complex, chronic venous insufficiency, diabetic retinopathies, and post-thrombotic syndrome. Calcium dobesilate has an angioprotective action in the sense of a reduction in the capillary permeability, which is supposed to inhibit the pathologically raised outflow of fluid from the intravasal into the interstitial space.

4.1.33.2 Publications

In 1977 **Hudomel** and co-workers [457, 458] reported on an improvement in the flow properties of blood as a result of calcium dobesilate. Within the framework of a double-blind study, 17 patients with diabetic retinopathies and raised blood viscosity received 750 mg calcium dobesilate or placebo daily for a treatment period of some months. The blood viscosity of the blood samples, anticoagulated with sodium citrate, was measured with a capillary viscosimeter at temperatures of 20°, 25° and 37 °C. At the end of treatment, the authors found a significant decrease in the blood viscosity at 20 and 25 °C. At body temperature no reduction could be determined in the blood viscosity. There were no changes in the total lipids, cholesterol, total proteins, albumins, and globulins. The authors were of the opinion that calcium dobesilate prevented an increase in the blood viscosity by decreasing the capillary permeability and the erythrocyte aggregation (determined via the blood cell sedimentation rate).

Note: No correction for hematocrit; no measurements of further rheologic parameters.

In 1978 **Watzlaw** [904] from the Ehrly team studied the hemorheologic effects of calcium dobesilate in eight patients with diabetes mellitus and peripheral vascular disorders in stage II according to Fontaine. The patients received a single intravenous application of 500 mg calcium dobesilate. The measurement of the blood and plasma viscosities (heparin as anticoagulant) was carried out with an Ostwald capillary viscosimeter ($H_2O = 1$) at a temperature of 37 °C. In addition, erythrocyte filtration was measured according to the method of Ehrly and Rossbach. Erythrocyte aggregation was quantitatively photometrically. After a single, intravenous injection of 4 ml calcium dobesilate, the investigators found no changes in the rheologic parameters.

Note: No correction for hematocrit; no measurements of the blood viscosity at low shear rates.

In 1980 **Barras** and **Graf** [48, 49] studied the flow properties of blood in 21 patients with diabetic retinopathies, 11 of whom had an increased plasma viscosity compared to a healthy control group. The patients received three capsules of calcium dobesilate daily (3 × 500 mg). The blood viscosity of the heparinized blood samples was determined with a Contraves Low Shear Couette rheometer at a shear rate of 128 s^{-1} and the plasma viscosity was measured with a capillary viscosimeter, a Couette, and a Brookfield micro-cone-plate viscosimeter (20 °C). After 3 months, the authors were able to record a significant decrease in the blood and plasma viscosities in the patients with increased blood viscosity. The results of the plasma viscosity obtained with the three viscosimeters were comparable, although the difference from to the initial values was smaller when using a Brookfield micro-cone-plate viscosimeter. A reduction in the relative blood viscosity (blood viscosity divided by plasma viscosity) was determined. The authors explained the decrease in the blood viscosity by a reduction of the plasma viscosity and not by a change in the hematocrit value which remained unchanged during therapy. They found a significant decrease in the plasma fibrinogen level and in the α_1- and α_2-globulin levels with a simultaneous increase in the plasma albumin concentration.

Note: No viscosity measurements at body temperature; no measurement of the blood viscosity at low shear rates; no measurement of further rheologic parameters.

In 1984 **Ernst** and **Marshall** [323] studied the effect of calcium dobesilate on the hemorheologic parameters in patients with intermittent claudication. Eighteen patients received oral daily doses of 2 × 500 mg calcium dobesilate for a period of 6 weeks. Erythrocyte filtrability was determined using a filtration method according to Dodds and Dormandy. The authors found an improvement in the erythrocyte filtrability, especially in those patients with an initially deteriorated erythrocyte deformability.

In 1984, in a double-blind study, **Vojnikovic** [885] investigated the effect of calcium dobesilate on the blood and the plasma viscosities of patients with diabetic retinopathies. The blood and plasma viscosities were measured with a Determann capillary viscosimeter at 37 °C. The authors found a statistically significant decrease in both the blood viscosity and the plasma viscosity in the calcium dobesilate group, but not in the placebo group.

In 1985 in a double-blind study with patients with diabetic retinopathy **Benarroch** and co-workers [59] studied the effect of 500 mg calcium dobesilate three times daily for a period of 3 months on the blood viscosity in comparison to placebo. Blood viscosity was measured with a Brookfield viscosimeter with a special adapter for low shear rates. Erythrocyte deformability was determined according to the method of Sirs. In the calcium dobesilate group the authors described a statistically significant decrease in the blood viscosity, while there was no statistically significant change in the plasma viscosity.

In 1985 **Galigani** and co-workers [362] reported on changes in the blood and plasma viscosities and in the whole blood filtrability following a daily oral dose of 1500 mg calcium dobesilate over several months in 20 patients with diabetic microangiopathy. They found a significant improvement in the plasma viscosity values and an improvment in the filtrability of the blood (abstract, no details).

4.1.33.3 Discussion

The first publication on the improvement of the flow properties of blood with calcium dobesilate in patients with diabetic retinopathy came from *Hudomel* and co-workers in 1977. Their results regarding the blood viscosity-reducing effect of oral calcium dobesilate administration must be relativized since the viscosity measurements were performed at 20 °C and 25 °C; at 37 °C no blood viscosity-reducing effect could be found. The results of this team are not contrary to the dissertation of *Watzlaw,* who also could not determine a reduction of the blood viscosity at 37 °C after short-term intravenous administration of calcium dobesilate in patients with peripheral arterial occlusive disease.

The results of *Barras* and *Graf,* who found a reduction of blood viscosity, measured at 20 °C, after oral administration of calcium dobesilate in patients with diabetic retinopathies, are rather difficult to evaluate, especially since *Hudomel* and co-workers did see a temperature-dependent effect. *Vojnikovic* and *Benarroch* and co-workers, on the other hand, found a reduction of the blood viscosity in their studies. Measurements of the plasma viscosity were performed by *Watzlaw, Barras and Graf, Vojnikovic, Benarroch* and co-workers and *Galigani* and co-workers. While *Watzlaw* and *Benarroch* did not find any significant reduction of the plasma viscosity, *Barras* and *Graf* as well as *Vojnikovic* and *Galigani* determined a significant reduction of the plasma viscosity. Measurements of erythrocyte aggregation (photometric technique) were only performed by *Watzlaw;* no changes were found in this subparameter. While *Watzlaw* could not show any significant changes in erythrocyte filtrability, *Ernst* and *Marshal, Vojnikovic, Benarroch* and co-workers, and *Galigani* and co-workers did find an improvement in this parameter.

Seen as whole, after oral as well as the intravenous administration of calcium dobesilate, there are indications but no definite proofs of a rheologic effect on the flow properties of blood in the short-term experiment. The rheologic results are not uniform enough even after the administration of calcium dobesilate over a longer period.

4.1.34 Cinnarizine/Flunarizine

4.1.34.1 Introduction

Cinnarizine and its difluorized derivative flunarizine have been used for the treatment of peripheral and cerebral vascular disorders since 1968 and 1974, respectively. The same mode of action is claimed for both drugs: the inhibition of the inflow of calcium into the unstriated muscles. Both substances have vasodilating properties.

4.1.34.2 Publications

In 1977 **Di Perri** and co-workers [184] investigated the effect of cinnarizine on the flow properties of the blood of 25 patients with peripheral arterial occlusive diseases (intermittent claudication) and an increased blood viscosity compared to a healthy control group. The blood viscosity of the blood samples anticoagulated with EDTA was measured with a Brookfield micro-cone-plate viscosimeter at shear rates of 3.75–750 s^{-1} and at a temperature of 37 °C. The shear rate for the plasma and serum viscosity measurements was 750 s^{-1}. Fourteen patients received a single dose of 300 mg cinnarizine orally, and 11 patients received 75 mg cinnarizine orally every day for a period of 1 week. After a single administration of cinnarizine, there was a significant reduction of the blood viscosity at all shear rates, the decrease in the whole blood viscosity being strongest at low shear rates. In addition there was a tendential but nonsignificant decrease in the fibrinogen level and in the plasma viscosity. Changes in the serum viscosity and osmolarity were not found. Similar results were obtained in long-term experiments. In 1979 *Di Perri* and co-workers published an article in which these results were discussed once again [186].

Note: No correction for hematocrit; no measurement of further rheologic parameters.

In 1978 **Mashiah** and co-workers [620] reported on rheologic studies in 13 patients with intermittent claudication, who had been suffering from these symptoms for more than 1 year and who showed increased blood viscosity compared to a control group. The patients were treated with an oral administration of cinnarizine over a period of 6 weeks (no details on the dose). The viscosity measurements were carried out at a temperature of 21°–25 °C (Contraves Low Shear, 1.24 and 124 s^{-1}). The hematocrit was corrected mathematically to a value of 45%. Owing to the nonuniform results of the viscosity measurements, the authors were not able to prove a blood viscosity reduction as a result of cinnarizine.

Note: No viscosity measurements at body temperature; no measurements of further rheologic parameters; no details on the dosage.

In 1979, within the framework of a double-blind study **De Cree** and co-workers [160], treated ten patients with intermittent claudication, who had increased blood viscosity values compared to a healthy control group. The viscosity measurements of the heparinized blood samples were carried out with a capillary viscosimeter at 37 °C, and the hematocrit was previously set at a value of 45%. In this experiment, the subjects received 20 mg flunarizine or placebo tablets daily. A significant reduction of the blood viscosity was found. In addition, the patients felt a local, painful ischemia after interruption of the blood circulation of the upper arm (blood pressure manchette) and simultaneous muscular exercise of the hand and lower arm muscles, which led to an increase in the blood viscosity. This blood viscosity increase was reduced significantly by previous flunarizine medication. The plasma viscosity was influenced neither by the duration of the ischemia nor by the medication. In a further double-blind experiment in this series of studies, 22 patients with myocardial infarction, diabetes mellitus, and arteriosclerosis received oral flunarizine applications (20 mg per day). The erythrocyte deformability measurements were performed according to the method of Reid and co-workers. Compared to a control group of the same age, the zero values of the patient group were greatly increased. After the administration of flunarizine, there was a significant increase in erythrocyte flexibility in the patient group compared to the placebo group.

Note: No measurement of erythrocyte aggregation.

Regarding the question of whether a long-term therapy with flunarizine has an effect on the disturbed microcirculation,in 1979 **Rudofsky** and co-workers [744] studied the blood viscosity with a falling ball viscosimeter in 14 patients with arterial vascular disorders in stage IIb according to Fontaine. In the 1st month the patients received 2 × 2 capsules of 10 mg flunarizine orally every day. Already after 4 weeks there was a reduction of the blood viscosity, which could only be verified after physical exercise (no detailed data).

Note: The viscosity measurements were performed with an inadequate falling ball viscosimeter; no temperature data; no measurements of further rheologic parameters.

Clinical findings: The pain-free walking distance could be increased during the course of treatment.

In 1980 **Heilmann** [431] examined 12 pregnant women with placenta insufficiency. Blood viscosity was measured at a hematocrit of 40% at shear rates of 0.39–78.6 s^{-1} using a Brookfield micro-cone-plate viscosimeter with a Mooney adapter and plasma viscosity was measured with a Schott falling ball viscosimeter. The aggregation behavior was quantitatively measured with a photometer, and erythocyte deformability was determined using a Nucleopore filter system with a pore diameter of 5 μm at a hematocrit of 10% and a pressure of 5 cm H_2O. The patients received 30 mg flunarizine per day for up to 14 days. A significant reduction of the blood viscosity at low shear rates and a reduction of the plasma viscosity were found. There were no changes in the high molecular weight proteins. The aggregation behavior of the erythrocytes was not affected. On the other hand, the author observed a significant increase in erythrocyte filtrability.

Note: No temperature given for the measurements.

In 1981 **Guerrini** and co-workers [406] reported on the effect of flunarizine and cinnarizine on the rheologic parameters of patients with chronic arterial occlusive disease and hyperviscosity. One group of patients received flunarizine or cinnarizine at a single dose of 50 or 300 mg orally; the measurements were carried out 3 and 6 h or 1.5 and 3 h after intake, respectively. In the second group, cinnarizine was given at a dose of 75 mg three times per day for 1 week and flunarizine at a dose of 5 mg three times per day for 30 days. Both drugs led to a decrease in the viscosity of the blood and to an improvement in erythrocyte filtrability (abstract, no details).

Perhoniemi and co-workers [697] studied the effect of flunarizine (5 mg per day) compared to pentoxifylline (400 mg per day) on erythrocyte rigidity in 31 patients with intermittent claudication. The treatment lasted 3 months. Erythrocyte deformability was measured with a rotational viscosimeter for which the erythrocytes were suspended in physiologic saline solution. The whole blood viscosity and the plasma viscosity were measured with a Wells-Brookfield viscosimeter. After therapy the authors found an improvement in the filtration values with both flunarizine and pentoxifylline but without a statistically significant difference between the two. The plasma viscosity and the whole blood viscosity did not change statistically significantly.

Note: The significance of the filtration method must be relativized owing to the suspension of the erythrocytes in saline solution. No measurement of the erythrocyte aggregation.

Clinical findings: The pain-free walking distance was longer after therapy.

In 1985 **De Clerck** [159] found a decrease in the blood viscosity and an improvement in erythrocyte deformability as a result of flunarizine in patients with vascular disorders (abstract, no details).

In a randomized double-blind study, **Catalano** and co-workers [118] studied the effect of flunarizine (10 mg per day) on the blood viscosity and erythrocyte filtrability. They found a significant improvement in both parameters compared to placebo. Erythrocyte filtration was carried out according to the suggestions of the Italian Hemorheologic Society (no exact details). Blood viscosity was measured with a Wells-Brookfield viscosimeter.

Note: No data on the plasma viscosity and no hematocrit correction.

4.1.34.3 Discussion

The results on the reduction of the blood viscosity as a result of the oral administration of cinnarizine are not uniform. While *Di Perri* and co-workers found a significant decrease in the blood viscosity especially at low shear rates, *Mashiah* and co-workers found no reduction of the whole blood viscosity at the shear rates measured and with a mathematically corrected hematocrit. The patients in both research groups were suffering from intermittent claudication and had an increased blood viscosity compared to a healthy control group. Changes in the plasma and serum viscosities were not found. Measurements of further rheologic parameters were not carried out. *Di Perri* and co-workers postulated that the reduction of the blood viscosity found at high shear rates could be explained by improved erythrocyte deformability. The decrease in the blood viscosity at low shear rates found by the research team of *Di Perri* should theoretically be explained by a disaggregation of the erythrocyte aggregates and not by an improved deformability of the erythrocytes. But since the fibrinogen level did not change significantly, hematocrit changes could also have caused the reduction of the blood viscosity.

Mashiah and co-workers were not able to determine a blood viscosity reduction after oral administration of cinnarizine and postulated that a drug-induced improvement of the flow properties of the blood (except in defibrinogenation with ancrod and in hemodilution) was very unlikely. *Guerrini* and co-workers, on the other hand, reported on a reduction of the viscosity values and an improvement in the erythrocyte filtrability after administration of cinnarizine. Further details were not given in the abstract.

In summary, it may be said, that based on the literature available, it is controversial whether cinnarizine can have a favorable effect on the flow properties of blood or not.

The results available on the reduction of the blood viscosity with the oral administration of flunarizine are nonuniform. While *De Cree* and co-workers and *Heilmann* found a reduction of the blood viscosity with corrected hematocrit, *Rudofsky* and co-workers did not observe a reduction of the blood viscosity after giving flunarizine orally. However, when the patients were subjected to physical stress, Rudofsky also observed a decrease in the blood viscosity.

While, on the one hand, *De Cree* and co-workers did not find any change in the plasma viscosity, *Heilmann* determined a reduction of the plasma viscosity without being able to find simultaneous changes in the high molecular weight protein concentration. *De Cree* and co-workers, *Heilmann, De Clerck,* and *Catalano* found an improvement in erythrocyte deformability. This must be qualified by pointing out that the relevance of the measurements of erythrocyte flexibility by *De Cree* was limited since the filtration of the whole blood according to the method of Reid and Dormandy is no longer relevant today; there was no precise information on the methods of *De Clerck* and *Catalano*.

Measurements of the tendency of the erythrocytes to aggregate were only performed by one investigator; there were no changes in this parameter. *Rudofsky* found that a reduction of the blood viscosity after oral administration of flunarizine could be found particularly after physical exercise.

Heilmann pointed out that, while the viscosity studies showed a statistically significant reduction of the blood viscosity, these must be interpreted with circumspection owing to the special apparatus conditions of low shear viscosimeters and owing to the measurements in the prestasis region. *Heilmann* believed that the improved flow properties of the blood were caused by an improved erythrocyte deformability and not by a reduction of the plasma viscosity. He presumed that, in the stagnation zones of the placental flow unit, a decreased metabolite removal led to a decrease in the flexibility of the erythrocytes resulting from a hindrance of the outflow. Thus this pathologic change in erythrocyte deformability may be prevented by the oral administration of flunarizine.

Guerrini and co-workers found a decrease in the whole blood viscosity and an improvement in the erythrocyte filtrability after oral administration of flunarizine; there were no further methodologic details in the abstract. *De Clerck* and *Catalano* and co-workers also found a decrease in the blood viscosity values and an improvement in erythrocyte filtrability.

On the whole, the results of the hemorheologic studies on cinnarizine and flunarizine are partly contradictory, even though an improvement in the hemorheologic parameters can be inferred from the majority of the publications.

4.1.35 Hydroxyethyl Starch Solutions

4.1.35.1 Introduction

Hydroxyethyl starch is a polysaccharide which is obtained by the hydroxyethylization of amylopectin hydrolysates. The bonds of this polysaccharide are mainly 1–4 glycosidic, the ethynol group is at position 2 to glucose. Hydroxyethyl starch solutions are used, among other things, after loss of fluid due to shock, i.e., to control hypovolemia, and in vascular disorders. The hydroxyethyl starch solutions are available as 6%–10% solutions with various molecular weights (MW 40 000, MW 200 000, MW 450 000). In addition to the average molecular weight, the degree of substitution is also given, with hydroxyethyl starch, e.g., MW 200 000/0.5.

4.1.35.2 Publications

In 1977 **Keppeler** [489] from Ehrly's team reported on the effect of hydroxyethyl starch solutions on the flow properties of the blood of patients with various diseases and an increased blood cell sedimentation rate. The patients were infused intravenously with 500 ml 6% hydroxyethyl starch solution (MW 450 000) over a period 30–40 min. The blood and the plasma viscosities of the heparinized blood samples were measured with an Ostwald capillary viscosimeter at 37 °C; erythrocyte deformability was tested according to the method of Ehrly and Rossbach (hematocrit 10%, Millipore filter diameter 8 μm, 37 °C, driving force 12.9 mm H_2O). Erythrocyte aggregation was determined photometrically (light transmission) according to the method of Ehrly and Schmitt. At the same time as the significant reduction of the hematocrit, the investigator found a slight, nonsignificant reduction of the blood viscosity. In addition, he found a significant increase in plasma viscosity and erythrocyte aggregation and a significant decrease in erythrocyte deformability.

Note: No viscosity measurements at low shear rates.

In 1980 **Harke** and co-workers [417] infused 500 ml 10% mid-molecular weight hydroxyethyl starch solution (MW 200 000/0.5) as well as 1000 ml 10% low molecular weight dextran solution in hysterectomy patients over an infusion period of 2 h in order to avoid a drop in the colloidosmotic pressure during the 30-min operation. Blood viscosity was determined with an Ubbelohde capillary viscosimeter at 37 °C. The investigators found a significant decrease in the hematocrit and in blood viscosity after infusions with hydroxyethyl starch, while, on the other hand, there was no significant drop in the blood viscosity after dextran infusions.

Note: No measurements of further rheologic parameters; no correction for hematocrit; patients with different amounts of blood loss.

In 1982 **Landgraf** and co-workers [534] reported on the improvement in the flow properties of blood in patients with various diseases (pneumonia, ulcus cruris, arterial occlusive diseases, chronic lymphadenosis) and an increased blood cell sedimentation rate as a result of the infusion of 500 ml 6% low molecular weight hydroxyethyl starch solution (MW 40 000/0.5). The blood and plasma viscosities were measured with an Ostwald capillary viscosimeter at 37 °C; erythrocyte deformability was determined according to the method of Ehrly and Rossbach (hematocrit 10%, Millipore filter diameter 8 μm, 37 °C, driving force 12.9 mm H_2O. Erythrocyte aggregation was determined photometrically according to the method of Ehrly and Schmitt (light transmission). The investigators found a significant decrease in the hematocrit, the blood and plasma viscosities as well as in erythrocyte aggregation. Changes in erythrocyte deformability were not found.

In 1982 **Müller-Bühl** and co-workers [664] reported on a pilot study investigating the effect of 500 ml 10% mid-molecular weight hydroxyethyl starch solution (MW 200 000/0.5) on the flow properties of blood in patients with peripheral arterial occlusive diseases in stage II according to Fontaine. The patients received one infusion per day for 3 days. At the end of treatment, the authors found a significant decrease in the blood viscosity and a non-significant decrease in the plasma viscosity.

Note: No measurements of further rheologic parameters; no data on the temperature.

Clinical findings: The researchers found that the pain-free walking distance was slightly longer.

In 1983 **Landgraf** and **Ehrly** [535] reported on the hemorheologic changes in hypervolemic hemodilution as a result of dextran 40 and HAES (hydroxyethyl starch solution) 200/0.5 in patients with intermittent claudication. At an interval of 8 days 500 ml 9% solution of both plasma substitutes was infused intraindividually over a period of 30 min. The whole blood viscosity and the plasma viscosity were determined with an Ostwald capillary viscosimeter at 37 °C. Measurements of the erythrocyte aggregation (light transmission method) and of the hematocrit were also performed. After the infusion of the hydroxyethyl starch solution, there was a reduction of the hematocrit and the blood viscosity, which was statistically significant for the entire study period of 120 min after the end of infusion. Plasma viscosity, however, increased statistically significantly only up to the 60th min past infusion. Erythrocyte aggregation was clearly decreased for a long time.

Note: No measurement of erythrocyte filtration.

In 1983 **Müller** and co-workers [662] studied the effect of hydroxyethyl starch solution (MW 200 000) on the flow properties of the blood of patients with cerebrovascular disorders and found a reduction of the whole blood viscosity and erythrocyte aggregation as well as an improvement in erythrocyte filtrability.

In 1984 **Krepp** and co-workers [528] studied the effect of a 5-day hypervolemic hemodilution therapy with 10% hydroxyethyl starch solution, MW 200 000/0.5, in a controlled prospective study on 50 patients with cerebrovascular disorders. A total of 500 ml infusion solution was given inravenously over a period of 6 h. Plasma viscosity was determined with a capillary viscosimeter. There was a decrease in plasma viscosity during the treatment with HES 200, while the plasma viscosity values rose after an analogous infusion with dextran 40 (10%). Urine viscosity, also measured with a capillary viscosimeter, rose statistically significantly with both plasma substitutes, however, to a lesser degree with HES 200 than with dextran 40.

In 1985 **Landgraf** and co-workers [539, 540, 541, 316] reported on a randomized cross-over study of patients with intermittent claudication who were infused with various plasma substitutes hypervolemically, i.e., without previous venesection. The infusion lasted 2 h; 10% dextran 40, 6% hydroxyethyl starch MW 40 000/0.5 and 10% hydroxyethyl starch MW 200 000/0.5 were infused intraindividually. Between the individual infusions there was a wash-out period of 8 days. An infusion of 500 ml physiologic saline solution served as a control. There was a reduction of the whole blood viscosity with all infusion solutions used, however, to a slighter degree with the saline solution. Plasma viscosity was only decreased with the use of saline solution and low molecular hydroxyethyl starch solution. After the administration of hydroxyethyl starch of MW 200 000, the plasma viscosity increased slightly, and after the infusion of low molecular weight dextran it increased markedly. Erythrocyte aggregation was decreased immediately after the infusion of dextran 40. With hydroxyethyl starch MW 200 000 and with saline solution the decrease occurred more towards the end of the experiment.

In 1986 in a prospective randomized cross-over study, **Staedt** and co-workers [812–814] studied the effect of two different colloidal plasma substitutes (10% low molecular weight dextran 40 and 10% hydroxyethyl starch 200 000/0.5) in patients with ischemic apoplexy. The infusion lasted 6–8 h; blood samples were taken before, during and after the end of infusion. The whole blood viscosity was measured with a Contraves Low Shear viscosimeter and erythrocyte aggregation with a light transmission method. There was a statistically significant reduction of the whole blood viscosity with the infusion of both plasma substitutes. The plasma viscosity decreased more clearly with the infusion of hydroxyethyl starch than with dextran 40. Erythrocyte aggregation was clearly reduced with mid-molecular hydroxyethyl starch, but not with low molecular dextran.

Note: No measurement of erythrocyte filtrability.

Within the framework of a long-term therapy, **Haass** and co-workers [411] infused patients with acute cerebrovascular disorders with both 10% low molecular dextran and 10% mid-molecular hydroxyethyl starch (200 000/0.5). Each infusion was carried out over a period of 4–6 h, at first 2 × 500 ml daily alternately with an electrolyte solution and later each one in a single dose of 500 ml daily. The whole blood viscosity was measured with a Wells-Brookfield viscosimeter at various shear rates, plasma viscosity with a capillary viscosimeter and erythrocyte aggregation with an optical transmission method. The whole blood viscosity decreased more clearly with a single administration of hydroxyethyl starch solution than with the administration of dextran, but with dextran the effect was longer-lasting than with the starch solution. Plasma viscosity increased after each individual infusion with dextran solution, but decreased slightly with the hydroxyethyl starch solution. Erythrocyte aggregation showed a statistically significant decrease in the values both during the individual infusions and during the total period of treatment with hydroxyethyl starch solution. On the other hand, a short-term administration of dextran resulted in a decrease in the erythrocyte aggregation values, while a long-term administration effected an increase in the values. This was explained as being the result of an accumulation of the higher molecular fractions of dextran 40 in the body.

Note: No measurement of erythrocyte filtrability.

In 1986 **Kiesewetter** and co-workers [503] treated patients with cerebrovascular microangiopathies with low molecular hydroxyethyl starch (hypervolemic hemodilution, no blood sampling) and found a decrease in the flow shear stress, the erythrocyte aggregation, and the erythrocyte deformability. In addition, plasma viscosity was decreased. With the simultaneous administration of pentoxifylline, an additional improvement was achieved in the erythrocyte deformability.

4.1.35.3 Discussion

In the studies discussed above, 6%–10% hydroxyethyl starch solutions of various molecular weights (MW 40 000 = low molecular, MW 200 000 = mid-molecular, MW 450 000 = high molecular) were infused hypervolemically, i.e., without venesection. It was to be expected that the small number of studies on the effect of hydroxyethyl starch solutions on the flow properties of blood in patients are difficult to reduce to a common denominator since they were carried out with various molecular weights and various concentrations. With high molecular weight hydroxyethyl starch solution [489] only a slight decrease in blood viscosity was found despite the significant reduction of the hematocrit; plasma viscosity was increased, and erythrocyte deformability was decreased.

The reduction of the blood viscosity found by *Harke* and co-workers after the infusion of 1000 ml 10% mid-molecular weight hydroxyethyl starch solution (MW 200 000) must also be considered with regard to the effect of the blood loss during the operations performed on the patients. In addition, the patients were not given food or fluids for 12 h before the operation, which of course also affected the viscosity values after the infusions with the plasma substitutes and which must have an influence on their comparability. *Harke* and co-workers found a reduction of the viscosity after infusions with hydroxyethyl starch, but not, however, after infusions with low molecular weight dextran. This could have been caused, among other things, by the previous fluid deficiency, which can result in dehydration after the infusion of the highly oncotic dextran solution.

Müller-Bühl and co-workers also carried out viscosity studies with 10% mid-molecular weight hydroxyethyl starch of MW 200 000 in patients with peripheral occlusive disease in stage II. Following a 10-day treatment with daily infusions of 500 ml, they found a significant decrease in blood viscosity, but no significant decrease in plasma viscosity. Measurements of the erythrocyte aggregation and the erythrocyte deformability or filtrability were not performed. The authors reported on an extension of the pain-free walking distance, which, however, was not marked according to the corresponding dia-

gramm. The treatment period of 10 days may also possibly be too short in the therapy of chronic vascular disease.
The studies of *Landgraf* and co-workers and *Landgraf* and *Ehrly* showed that, with the use of hydroxyethyl starch solution of MW 40 000 as well as a MW 200 000/0.5, erythrocyte aggregation was decreased much more strongly than could be expected on the basis of a reduction of the fibrinogen concentrarion as a result of hemodilution. Therefore, a specific disaggregating effect on the erythrocyte aggregates was postulated for hydroxyethyl starch. The studies of *Staedt* and co-workers, *Haass* and co-workers, and *Müller* and co-workers corroborate these observations.
While the whole blood viscosity in all these studies was decreased after the administration of hydroxyethyl starch, both in a relatively fast infusion period of 30 min or 1 h and with an infusion time of 6–8 h, the behavior of plasma viscosity was not uniform. With the relatively short-term infusion 500 ml hydroxyethyl starch of MW 200 000, *Landgraf* and *Ehrly* found a slight increase in the viscosity; *Müller-Bühl* and co-workers reported similar results. If, however, the infusion was extended to several hours (*Krepp* and co-workers, *Staedt* and co-workers, *Haass* and co-workers), then there was a more or less marked decrease in plasma viscosity. In the comparative studies investigating dextran 40 and hydroxyethyl starch solution MW 200 000/0.5 (*Landgraf* and co-workers, *Staedt* and co-workers, and *Haass* and co-workers), hydroxyethyl starch showed an improved rheologic behavior compared to low molecular weight dextran with MW 40 000. In addition, it could be shown that a longer infusion period with hydroxyethyl starch could also influence the flow properties of blood favorably. With regard to erythrocyte filtrability, most of the authors speak of an improvement in this parameter, which can be explained as being the result of a decrease in the plasma viscosity.
In summary, it may be said that hydroxyethyl starch solutions are of particular interest from the rheologic point of view, if one disregards the high molecular substances with a molecular weight of 450 000. Hydroxyethyl starch solutions of MW 200 000/0.5 or 40 000/0.5 effect a multifactorial improvement in the flow properties of blood and have a specific disaggregating effect on erythrocyte aggregates. Despite the fact that there are still very few publications on hydroxyethyl starch solutions and their influence on the flow properties of the blood, it is to be expected that in future this plasma substitute will become increasingly important as a rheologic therapeutic.

4.1.36 Suloctidil

4.1.36.1 Introduction

Suloctidil is a vasodilator used for the therapy of peripheral vascular disorders.

4.1.36.2 Publications

The initial results of a pilot study on the hemorheologic effects of suloctidil on the flow properties of the blood of geriatric patients were published by **Roba** and co-workers in 1977 [734]. At the same time as the reduction of the blood viscosity, a decrease in the fibrinogen level was observed.

Note: Details of the viscosity measurements were not given; no measurements of further rheologic parameters.

In 1979 **Roncucci** and co-workers [737] reported on the effect of suloctidil on the blood viscosity in 40 diabetics suffering from retinopathies, nephropathies, neuropathies, and microangiopathies. Within the framework of a double-blind study, the subjects received 200 mg suloctidil per day in addition to insulin or oral diabetic therapy. The treatment period was 5 months. The blood viscosity of the samples, anticoagulated with heparin or with EDTA, was measured with a Brookfield micro-cone-plate viscosimeter at 230 s^{-1} and with a Contraves Low Shear 100 viscosimeter at 0.77 s^{-1}. Erythrocyte deformability (erythrocyte flexibility) was investigated using a polycarbonate filter (pore size 5-μm, no further details). The authors found a significant reduction of the blood viscosity only at a shear rate of 230 s^{-1}, whereas at a shear rate of 0.77 s^{-1} no reduction of the values could be found. An explanation could not be given for the significant increase in the hematocrit value and in the fibrinogen level in both the placebo and the suloctidil groups. Erythrocyte deformability was not changed by the therapy with suloctidil.

Note: No correction for hematocrit; no temperature data, no measurements of further rheologic parameters.

In 1981 **van Stalle** and **Lambelin** [873] reported on a clinical study in which 20 diabetic patients with vascular diseases received 200 mg suloctidil daily for 5 months. This group was compared to a second group of patients who received placebo. The whole blood viscosity was studied at shear rates of 230 and 0.77 s^{-1} the plasma fibrinogen concentration and erythrocyte deformability were studied as well. While no changes in the viscosity values could be proven in the placebo group, in the suloctidil group there was a statistically significant reduction of the whole blood viscosity (at 230 s^{-1}). There were no data on the behavior of erythrocyte deformability (abstract, no details).

In 1983 **Caimi** and co-workers [110] studied the effect of a long-term therapy with suloctidil (600 mg per day) on the whole blood viscosity (Wells-Brookfield viscosimeter at three different shear rates), the plasma viscosity (measured with the same viscosimeter), and erythrocyte filtrability (determined according to the whole blood method of Reid and co-workers) in 17 patients with peripheral arterial occlusive diseases. After several months of therapy, the authors found an improvement in erythrocyte filtrability, while the other hemorheologic parameters were not statistically significantly changed.

In 1985 **Merchan** and co-workers [634] studied the effect of 200 mg suloctidil three times daily on the flow properties of blood in 15 patients with cerebrovascular insufficiency. The blood and the plasma viscosities, measured with a Wells-Brookfield microviscosimeter at 37 °C, and erythrocyte deformability, measured using the whole blood method according to Reid and co-workers, were each determined before and after the 180-day therapy. The authors found an improvement in the hemorheologic parameters.

4.1.36.3 Discussion

Roncucci and co-workers described a viscosity reduction as a result of suloctidil, however, only at a particular shear rate and despite the fact that the hematocrit values were raised significantly. *Caimi* and co-workers were not able to find a decrease in the blood viscosity and the plasma viscosity during long-term therapy, while *Merchan* and co-workers did find a reduction of the blood and the plasma viscosities after an 180-day therapy. The conclusions on erythrocyte deformability are just as contradictory: while *Roncucci* and co-workers found no effect, *Caimi* and co-workers as well as *Merchan* and co-workers could determine a favorable effect. However, the last two groups of authors used the whole blood method according to Reid and co-workers, which is now no longer adequate. Seen as a whole, the publications at hand do not give a clear indication of a possible hemorheologic effect of suloctidil.

4.1.37 Butalamine

4.1.37.1 Introduction

Butalamine is a musculotropic vasodilator which is used for the treatment of peripheral and cerebrovascular disorders.

4.1.37.2 Publications

In 1978 **Stödter** [817] from the Ehrly team studied the effect of oral butalamine administration on the flow properties of blood in patients with peripheral arterial vascular diseases. The patients received 120 mg butalamine orally three times per day. The blood and plasma viscosities of the heparinized blood samples were determined with an Ostwald capillary viscosimeter and changes in erythrocyte deformability (erythrocyte flexibility) with a capillary filter system (Millipore filter, diameter 8 μm; hematocrit 10%; 12.9 mm H_2O) according to the method of Ehrly and Rossbach. The temperature of the measurements was 37 °C. The investigator was not able to determine any changes in the flow properties of blood under the experimental conditions chosen.

Note: Measurements of erythrocyte aggregation were not carried out. A hematocrit correction was not made for the measurement of the blood viscosity.

4.1.38 Naftidrofuryl

4.1.38.1 Introduction

Naftidrofuryl is a vasodilator which is widely used for the treatment of peripheral and central vascular cisorders.

4.1.38.2 Publications

The first publication on the hemorheologic effects of the vasoactive substance naftidrofuryl were published by **Ott** [676] from the Heidrich team in 1973. The subjects included patients with vascular disorders and other diseases. The Haake falling ball viscosimeter was used; the blood viscosity measurements were performed at 37 °C. Five patients received a single infusion of 200 mg naftidrofuryl in 250 ml 5% levulose solution intravenously over a period of 1 h. Viscosity measurements were performed within 4–8 h after the end of infusion. In the meantime, the heparinized blood samples were stored at 4 °C. At the end of infusion there was a 4.24% drop in the blood viscosity with a decrease in the hematocrit of 0.64%. The administration of the carrier solution alone in a control group of five subjects led to a decrease of 6.61% in the blood viscosity with a drop of 1.12%in the hematocrit. Following the administration of levulose with naftidrofuryl, there was a blood viscosity reduction which was clearly less than that following the infusion of the carrier solution levulose alone. In comparison, the infusion of 500 ml low molecular weight dextran led to a 15.26% reduction of the blood viscosity with a hematocrit decrease of 13%.

In 1974 and 1975 **Heidrich** and **Ott** [421, 424] published major parts of this study.

Note: The viscosimetric results found with a falling ball viscosimeter are not relevant because of methodologic reasons; no correction for hematocrit; no measurement of further rheologic parameters.

In 1979 **Savi** and co-workers [752] studied the effect of naftdrofuryl on the flow properties of blood. Thirty patients with peripheral vascular disorders received two intravenous infusions daily of 40 mg naftidrofuryl in 250 ml isotonic saline solution (= 80 mg naftidrofuryl in 500 ml isotonic saline solution). The blood and plasma viscosities of the heparinized blood samples were measured with a Wells-Brookfield LVT viscosimeter at 37 °C and at shear rates of 11.5, 23, 46, 115, and 230 s^{-1}. At the same time as a significant decrease in the blood and plasma viscosities the investigators found a reduction of the fibrinogen level.

Note: No measurements of further rheologic parameters; no correction for hematocrit.

In 1984 **Kiesewetter** and co-workers [494] studied the effect of naftidrofuryl on the hemorheologic parameters in 15 patients with chronic arterial occlusive disease in the severe stages. In a double-blind study 400 mg naftidrofuryl in physiologic saline solution was administered intravenously for 2 weeks, followed by 600 mg naftidrofuryl orally for 22 weeks. In a smaller group of patients placebo was given as a comparison. The flow shear stress, the erythrocyte aggregation (light transmission method), the erythrocyte deformability (selecting erythrocyte rigidometer), and the plasma viscosity (a capillary viscosimeter method) were studied. There was a significant decrease in the so-called rigidity index and a statistical reduction of the erythrocyte aggregation, while the reduction of the plasma viscosity could not be proven statistically. The flow shear stress was statistically significantly improved compared to the placebo group.

In 1984 **Jung** and co-workers [482, 483] studied the effect of naftidrofuryl on the flow properties of the blood of 12 patients with chronic artérial occlusive disease in stage II according to Fontaine, who had a hematocrit of 45% or higher. First, these patients took part in a 10-week walking therapy, during which they received a daily dose of 250 ml 5% levulose solution as "comparative therapy". Then, while continuing the physiotherapy, 600 mg naftidrofuryl was administered daily via a short-term infusion with 250 ml 5% levulose. Erythrocyte deformability was studied with an optical transmission method, erythrocyte rigidity with a selecting erythrocyte rigidometer, and plasma viscosity with a capillary viscosimetric method. There was a statistically significant decrease in the erythrocyte rigidity as a result of the pharmacologic therapy, but not as a result of the physical monotherapy. The hematocrit, the erythrocyte aggregation and the plasma viscosity remained practically unchanged following both the physical therapy alone and the combination therapy.

Clinical findings: The authors found there was an increase in the pain-free walking distance after comparative therapy and an additional improvement in this parameter after administration of the drug.

Note: No measurement of erythrocyte filtrability.

In 1986 **Heilmann** and **Hirche** [439] studied the effect of naftidrofuryl on the flow properties of blood in 20 pregnant women with gestosis. Infusions of 400 mg naftidrofuryl were given in 500 ml glucose solution for over 3 weeks. Erythrocyte aggregeation was studied with a cone-plate aggregometer, erythrocyte deformability with a Teitel polymicroviscosimeter, and plasma viscosity with a capillary viscosimeter. Following naftidofuryl there was an improvement of erythrocyte deformability, while erythrocyte aggregation and plasma viscosity remained unchanged.

In 1986 **Kiesewetter** and co-workers [502] studied the effect of naftidrofuryl on the flow properties of blood of patients with basilar insufficiency. Fifteen patients received 600 mg naftidrofuryl per day orally for 2 months. The flow shear stress was determined, as were erythrocyte aggregation (light transmission aggregometer), erythrocyte rigidity (selecting erythrocyte rigidometer), and plasma viscosity (capillary viscosimeter). There was a statistically significant improvement in the above-mentioned hemorheologic parameters with almost unchanged hematocrit values.

4.1.38.3 Discussion

The early studies of *Ott* and co-workers are not very relevant owing to their methodologic insufficiencies. In the studies of *Sari* and co-workers, in addition to the substance naftidrofuryl larger amounts of isotonic saline solution were infused as a carrier, and there was no hematocrit correction, so that the viscosity decrease found must be relativized. This reticence in interpretation is supported by the studies of *Kiesewetter* and co-workers, *Jung* and co-workers and, *Heilmann* and *Hirche* who could not show any influence of naftidrofuryl on plasma viscosity. On the other hand, the authors did find a significant improvement in erythrocyte rigidity, measured with a selecting erythrocyte rigidometer. While *Heilmann* and *Hirche* and *Jung* and co-workers could not find any change in erythrocyte aggregation, this factor was significantly decreased in the studies of *Kiesewetter* and co-workers.

In conclusion, it may be said that partially contradictory statements were made on the effect of a therapy with naftidrofuryl on hemorheologic parameters. On the other hand, in the relatively small number of publications available, there are reports that agree on an improvement in erythrocyte deformability.

4.1.39 Vincamine

4.1.39.1 Introduction

Vincamine is an alkaloid obtained from Vinca minor *and it is used for the treatment of cerebrovascular disorders. Further areas of application for vincamine are vascular disorders of the inner ear and of the retina. Vinpocetine is a vincamine derivative with areas of application similar to those of vincamine.*

4.1.39.2 Publications

In 1979 **Klein** and co-workers [508] studied the effect of vincamine on the flow properties of blood in patients with cerebrovascular disorders, some of whom also had neurologic deficiency symptoms. The patients received 2 × 30 mg vincamine per day for a treatment period of 84 days. Blood viscosity was measured with a capillary viscosimeter at 20 °C. At the end of treatment there was a significant reduction of the blood viscosity; a decrease in the hematocrit value was also noted.

Note: No correction for hematocrit; the blood viscosity measurements were made at 20 °C. No measurements of further rheologic parameters.

Clinical findings: The symptoms accompanying cerebrovascular insufficiency, such as headaches, dizziness, tinnitus, sleep disturbances, and impaired memory, were favorably influenced by the vincamine therapy.

In 1981 **Ehrly** and **Landgraf** [293] studied the effect of a single intravenous infusion of 15 mg vincamine in patients with intermittent claudication. They found no changes in the blood and plasma viscosities nor in erythrocyte filtrability. In in vitro studies with hyperosmolar blood there was, however, a dose-dependent improvement in the filtrability values.

In 1985 **Osawa** and **Maruyama** [673] reported on the effect of vinpocetine on the flow properties of blood. Patients with cerebrovascular diseases received 3 × 1 tablet of 5 mg vinpocetine a day for 2 weeks. Blood viscosity was measurd with a cone-plate viscosimeter at 37 °C and at shear rates of 18, 75, and 375 s^{-1}; plasma viscosity was determined with the same instrument at 375 s^{-1}. The hematocrit and the fibrinogen levels only changed slightly during treatment, but the whole blood viscosity sank statistically significantly at all the shear rates studied. The plasma viscosity was also statistically significantly decreased.

Note: No measurements of erythrocyte aggregation and erythrocyte deformability.

In 1985 **Kuzuya** and co-workers [532] studied the effect of vinpocetine, a derivative of vincamine, on blood viscosity in ten male patients with chronic cerebrovascular disease and increased blood viscosity. Vinpocetine was given at a dose of 5 mg per day for 2 weeks. Blood viscosity was determined with a cone-plate viscosimeter at various shear rates; plasma viscosity was determined with the same instrument at 375 s^{-1}. The studies were carried out at 37 °C. While there was no statistically significant reduction of the hematocrit value and the plasma fibrinogen concentration, the blood viscosity decreased statistically significantly especially at low shear rates. The plasma viscosity also decreased slightly but statistically significantly.

Note: No studies of erythrocyte aggregation and erythrocyte deformability.

4.1.39.3 Discussion

While *Klein* and co-workers found a decrease in the whole blood viscosity and in the hematocrit after the administration of vincamine, these findings could not be corroborated by *Ehrly* and co-workers. With vinpocetine *Osawa* and *Maruyama* as well as *Kuzuya* and co-workers found a reduction of the blood viscosity at low shear rates and a slight decrease in the plasma viscosity, while the hematocrit and plasma fibrinogen concentration remained practically unchanged. The currently available studies on the effect of vincamine and similar substances on the flow properties of blood are not sufficient to support a definite conclusion on a hemorheologic effect.

4.1.40 Ticlopidine

4.1.40.1 Introduction

Ticlopidine is an platelet function inhibitor. For this reason it is recommended, among other things, as an antithrombotic drug.

4.1.40.2 Publications

In 1979 **Shigehiro** and co-workers [795] reported on the effect of ticlopidine in patients with angina pectoris. The blood and plasma viscosities of these patients were studied before, 2, and 6 weeks after the administration of ticlopidine. The authors found a reduction of the blood viscosity values as well as a decrease in the anginal complaints (abstract, no details).

In 1982 **Aukland** and co-workers [32] studied the effect of ticlopidine (2 × 250 mg orally per day) for a period of 12 months in patients with intermittent claudication. Blood viscosity was measured at 25 °C and at shear rates of 128 and 0.2 s^{-1} using a Contraves LS 30 viscosimeter at a standardized hematocrit of 45%. Plasma viscosity was measured at 25 °C with a capillary viscosimeter. There was no statistically significant reduction of the viscosity values compared to a double-blind control group.

Note: No measurements of further rheologic parameters; viscosity measurements were carried out at 25 °C.

Clinical findings: There was no extension of the pain-free walking distance in the patients with intermittent claudication.

In 1982 **Tekeres** and co-workers [854] reported on the effect of a 6–12-month treatment with 500 mg ticlopidine orally per day in patients after myocardial infarction. This group of patients was compared to another group who did not receive a platelet function inhibitor. Even after the 1 week of therapy, the blood viscosity, measured with a Contraves LS 100 viscosimeter, showed a clear reduction. In addition, a decrease was observed in the fibrinogen concentration. The frequency of reinfarction was reduced (abstract, no details).

Neumann and co-workers [669] studied the effect of ticlopidine (250 mg twice a day orally) in 12 patients with diabetes mellitus. Blood viscosity was measured with a Contraves LS 30 viscosimeter at 25 °C and plasma viscosity with a capillary viscosimeter at 25 °C. Erythrocytes deformability was measured at room temperature according to the method of Kenny and co-workers using washed erythrocytes. The treatment period was 12 weeks. There was a reduction of the plasma viscosity and the whole blood viscosity which correlated with an 11% reduction of the serum globulin concentration. The improvement in the filtration values measured could no longer be found if the blood samples were set to a constant number of leukocytes. The authors attributed the decrease in the viscosity values to a decrease in the total protein concentration.

Note: The measurements were performed at 25 °C; no determination of erythrocyte aggregation.

Clinical findings: There was no improvement in the diabetic symptoms during the 12-week therapy.

In 1983 **Strano** and co-workers [836] studied the effect of a 60-day therapy with 500 mg ticlopidine orally per day in a cross-over study compared with placebo in 16 patients with obliterative arteriosclerosis. Erythrocyte filtrability was determined according to the whole blood method according to Reid. The authors found a clear improvement in erythrocyte filtrability during therapy.

Note: The filtration method used is not very relevant; further rheologic parameters were not studied.

In 1985 **Randi** and co-workers [718] studied the effect of ticlopidine on the flow properties of blood in ten patients with chronic arterial occlusive disease. The patients were treated with a daily dose of 750 mg orally for a period of 3 months; a conventional nicitonic acid drug was given for comparison. The blood and plasma viscosities were measured with a Brookfield LTV viscosimeter at 37 °C and at 230 s^{-1}. Following ticlopidine therapy there was a reduction of the blood viscosity which, however, was not statistically significant. In the nicotinate group there was no change in the viscosity values.

Clinical findings: There was an extension of the pain-free walking distance following ticlopidine therapy.

4.1.40.3 Discussion

While *Shigehiro* and co-workers and *Tekeres* and co-workers found a reduction of the blood viscosity values in the course of an oral therapy with ticlopidine, *Aukland* and co-workers as well as *Randi* and co-workers could not find an influence on this parameter. However, *Neumann* and co-workers did find a reduction of the blood and plasma viscosities in patients with diabetes, which they attributed to a decrease in the serum globulin concentration.

While *Strano* and co-workers determined an increase in blood filtrability following ticlopidine, *Neumann* and co-workers were able to show that, while this effect could also be seen with the filtration model chosen by this team, it was, on the other hand, no longer present in the blood samples after a correction of the leukocyte number. The filtration method used by *Strano* is limited in its meaningfulness owing to methodologic reasons.

Seen as a whole, the indications for a hemorheologic effect of ticlopidine are not sufficient to prove a corresponding mode of action beyond any doubt.

4.1.41 Antiepileptics

4.1.41.1 Introduction

Antiepileptics are drugs which are suitable for the symptomatic therapy of the various forms of epilepsy without otherwise reducing the motor excitability. Diphenylhydantoin and carbamazepine belong to this group of drugs.

4.1.41.2 Publications

In 1981 **Liewendahl** and co-workers [574] reported on patients who were treated with two different antiepileptic drugs. The patients who received diphenylhydantoin showed an improved erythrocyte flexibility compared to a healthy control group. This was not the case in patients who received carbamazepine.

Note: It was not explained how the erythrocyte flexibility was determined. There were no measurements of further rheologic parameters.

4.1.41.3 Discussion

A clear evaluation of the effect of these drugs on the flow properties of the blood is not possible on the basis of the one study.

4.1.42 Buflomedil

4.1.42.1 Introduction

Buflomedil is a vasoactive substance which is widely used for the treatment of peripheral and cerebrovascular diseases.

4.1.42.2 Publications

In 1980 **Hildebrandt** [448], from the Ehrly team, tested the effect of oral and intravenous administrations of buflomedil on the flow properties of blood in patients with peripheral arterial vascular disorders in stage IIb according to Fontaine. In an short-term experiment the patients received 200 mg buflomedil in 250 ml isotonic saline solution intravenously, and in a long-term experiment of 6 weeks the patients received 450 mg buflomedil orally per day. The blood and plasma viscosities of the heparinized blood samples were measured with an Ostwald capillary viscosimeter at 37 °C, and erythrocyte deformability was tested through a Millipore filter with a pore diameter of 8 μm. The temperature was 37 °C. Erythrocyte aggregation was measured photometrically (light transmission) at a hematocrit of 40%. Simultaneously with a slight decrease in the hematocrit, Hildebrandt found a reduction in the blood viscosity after oral and intravenous administrations of buflomedil. There were no changes in other rheologic parameters.

Note: No correction for hematocrit.

In 1981 **Dormandy** and co-workers [209] studied the effect of buflomedil on the flow properties of blood in ten patients with vascular disorders of the lower extremities (no precise details on the severity). The patients received 200 mg buflomedil intravenously at an infusion interval of 5 min. The measurement of erythrocyte deformability was performed using Nucleopore filters (diameter 5 μm, 20 °C). Thirty minutes after the end of the infusion, the investigators found a significant improvement in erythrocyte deformability. Four hours after the end of the infusion the erythrocyte deformability was still significantly increased.

Note: No measurement of further rheologic parameters.

In 1981 **Sergio** and co-workers [791] reported on the effect of a long-term therapy with buflomedil in patients with chronic peripheral arterial vascular disorders and found a decrease in the whole blood viscosity and an improvement in erythrocyte deformability (abstract, no details).

In 1981 **Materazzi** and co-workers [621] reported on the effect of the intravenous administrations of various dosages of buflomedil on the whole blood viscosity and on blood filtrability. Changes in erythrocyte filtrability, measured according to the whole blood method of Reid and co-workers at a temperature of 37 °C, were determined in the femoral venous blood both before and after physical exercise. This procedure was repeated after the intravenous administration of 100 mg buflomedil. Metabolic parameters were tested in addition to the rheologic ones. After the administration of buflomedil ten of the patients showed a statistically significant improvement of 5% in erythrocyte filtrability which had been worsened by muscular exercise. From this the authors concluded that buflomedil counteracts exercise-induced deterioration of erythrocyte deformability.

Note: Further rheologic parameters were not determined.

In 1982 **Coccheri** and co-workers [140] studied the effect of a 20-day oral administration of 150 mg buflomedil on the flow properties of blood in patients with diabetes mellitus. The results were compared to another group of diabetics who did not receive buflomedil. The blood samples were taken from the antecubital vein before therapy and on the day after the oral therapy: they were anticoagulated with EDTA. Blood viscosity was measured with a Wells-Brookfield micro viscosimeter at 23 s^{-1} and 230 s^{-1} and at 37 °C. Erythrocyte deformability was determined according to a modification of the whole blood method of Reid and co-workers whereby the leukocytes and the platelets were mostly eliminated and the samples set to a hematocrit of 40%. The temperature was 37 °C. The investigators found a statistically significant decrease in the hematocrit value in the buflomedil group as well as a reduction of the blood viscosity at both the shear rates measured. The erythrocyte filtration values also improved, while there were no changes in the rheologic parameters of the control group.

Note: No measurement of erythrocyte aggregation.

Similar results were presented by **Coccheri** and co-workers elsewhere [139].

In 1982 **Perego** and co-workers [692, 693] studied the effect of buflomedil on the flow properties of blood in patients with chronic occlusive arterial disease in stage II according to Fontaine. For 20 days the patients received an intravenous infusion of 100 mg buflomedil in 50 ml isotonic saline solution over a period of 60 min twice a day. The blood and plasma viscosity measurements were carried out with a Brookfield micro-cone-plate viscosimeter (37 °C; 450, 90, and 11.25 s^{-1}). The hematocrit was mathematically corrected to a value of 45% for the blood viscosity measurements. The measurement of erythrocyte deformability was carried out according to the method of Reid and co-workers. The erythrocyte sedimentation rate was determined at a corrected hematocrit of 45%. After the administration of buflomedil, the investigators found a significant decrease in the blood viscosity, which was most prominent at intermediate and high shear rates. This result correlated with a significant improvement in erythrocyte deformability. There were no changes in the plasma viscosity, the erythrocyte sedimentation rate, and the plasma fibrinogen level. The infusion of the carrier solution alone (50 ml isotonic saline solution) did not lead to any significant changes in the flow properties of blood.

Note: No quantitative determination of erythrocyte aggregation; inadequate filtration method.

Le Devehat and co-workers [559] studied the effect of buflomedil on the whole blood filtrability of patients with diabetic angiopathy and found a normalization of the values.

In 1983 **Pieragalli** and co-workers [701] reported on the protective effect of buflomedil on changes in the flow properties of blood caused by stress in patients with peripheral vascular diseases. Buflomedil was given orally at a dosage of 2 × 600 mg per day for a period of 4 weeks. Before and after this treatment, the blood and plasma viscosities and the blood filtrability were controlled before and after ergometric exercise on a treadmill. The results showed that the deterioration in blood filtrability caused by physical exercise disappeared with buflomedil (abstract, no details).

In 1984 **Leonardo** and co-workers [563] reported on the effect of buflomedil on the flow properties of blood. Two groups of 12 patients each were given 200 mg buflomedil either intravenously or intra-arterially. The whole blood viscosity was measured with a Wells-Brookfield LVT cone-plate viscosimeter at 37 °C and at various shear rates; plasma viscosity was measured with the same instrument at 230 s^{-1}. Erythrocyte filtrability was determined according to the original method of Reid and co-workers. The authors found a slight, but statistically significant improvement in the whole blood filtrability and in the blood viscosity.

Note: No measurements of further hemorheologic parameters.

Forconi and co-workers [354] treated ten patients with chronic arterial occlusive diseases in stage II according to Fontaine using 2 × 300 mg buflomedil orally per day for 30 days. The blood and plasma viscosities were measured with a Wells cone-plate viscosimeter at a shear rate of 90 s^{-1} and at a temperature of 37 °C. The whole blood filtrability was measured with a modified whole blood method. The authors found no significant changes in the hemorheologic factors. If, however, these parameters were studied after physical exercise (treadmill), then there was a statistically significant improvement in erythrocyte filtrability.

In 1985 **Dorigo** and co-workers [196] treated ten patients with chronic arterial occlusive disease both by means of a single intravenous administration of buflomedil and a repeated application for 5 days. The viscosity was measured with a Brookfield LVT cone-plate viscosimeter at 23 and 230 s^{-1}, and the erythrocyte filtrability was determined with a modified whole blood method. The authors found a significant decrease in the whole blood viscosity, while erythrocyte filtrability and plasma viscosity showed no significant changes. The hematocrit values were, on average, slightly lower after therapy than before.

In 1985 **Mozzi** and co-workers [661] described the effect of 2 × 300 mg buflomedil orally per day for 2 weeks in patients with chronic arterial occlusive diseases of the legs. A Wells-Brookfield rotational viscosimeter was used at 37 °C for the measurement both the whole blood and the plasma viscosities. The shear rates for the whole blood were 23 and 115 s^{-1} and 230 s^{-1} for the plasma. The authors could not find any statistically significant changes in the blood and plasma viscosities.

Note: No measurements of other rheologic parameters.

In 1985 **Boisseau** and co-workers [80] treated patients with chronic arterial disease using buflomedil. They carried out measurements of the whole blood filtrability according to the method of Reid and co-workers as well as with a Hanss hemorheometer using ectacytometry. They found no effect on the parameters studied (abstract, no details).

In 1985 **Avellone** and co-workers [34] treated 20 patients with arteriosclerosis of the cerebrovascular system using 450 mg buflomedil per day orally for 90 days and compared the hemorheologic effects in a cross-over study with sulfinpyrazone. The whole blood viscosity, plasma viscosity, and erythrocyte deformability were studied according to the whole blood method of Reid. Positive changes were reported in the hemorheologic parameters (abstract, no details).

Rossi and co-workers [742] treated patients with chronic arterial occlusive disease with an intravenous administration of buflomedil followed by oral ones for 12 weeks. The authors found a slight decrease in blood filtrability (abstract, no details).

Saitta and co-workers [747] treated a group of patients with peripheral occlusive diseases using 600 mg buflomedil orally for 20 days and determined the erythrocyte filtrability with the Reid whole blood filtration method. They found a significant decrease in the erythrocyte filtration index.

4.1.43.3 Discussion

Hildebrandt, Coccheri and *Leonardo* as well as *Dorigo* found a slight but significant decrease in the whole blood viscosity simultaneously with a slightly decreased hematocrit. *Perego,* on the other hand, found a significant decrease in the blood viscosity after the hematocrit had been mathematically corrected, as did *Avellone* and co-workers, while *Mozzi* found no changes at all in the flow properties of the blood after the oral administration of buflomedil. With regard to erythrocyte filtrability, most authors found an improvement in the values, which can be interpreted in the sense of an increased deformability. However, most of these studies were carried out with the whole blood method according to Reid and co-workers. *Hildebrandt* and *Boisseau* and co-workers, on the other hand, could not find an effect and *Rossi* and co-workers even found a slight decrease in blood filtrability. Of particaular interest are the results of *Pieragalli* and co-workers on the one hand and *Forconi* and co-workers on the other, who both found that the oral administration of buflomedil improves the erythrocyte deformability after muscular excercise. In summary, the results are contradictory, which, to a great extent, may be attributed to methodologic reasons. Further studies are necessary in order to reach a definite conclusion regarding the hemorheologic effect of buflomedil.

4.1.43 Nitroglycerine and Nitrates

4.1.43.1 Introduction

Nitroglycerine and nitrates are vasodilators which are used for the treatment and prophylaxis of attacks of angina pectoris, for the follow-up treatment of myocardial infarction, in left heart insufficiency, and in chronic cor pulmonale. Most of the hemorheologic studies were carried out with the substance isosorbide dinitrate.

4.1.43.2 Publications

The first reports on hemorheologic studies investigating the effect of isosorbide dinitrate on the flow properties of blood were presented by **Caesar** and co-workers [107] in 1980. Twenty patients with peripheral arterial vascular diseases of the lower extremities in stages II and III according to Fontaine, a further group of patients suffering from hypertentsion, and a healthy control group received a single oral dose of 20 mg isosorbide dinitrate. The investigators found a significant decrease in the blood viscosity at low shear rates (0.19 s^{-1}) in the patients as well as in the healthy subjects. The hemorheologic changes were accompanied by a decrease in hematocrit.

In 1982 **Hossmann** and co-workers [453] studied the apparent whole blood viscosity and the plasma viscosity using a Contraves LS 2 low shear viscosimeter and a Brookfield LVT viscosimeter at shear rates of 0.03–229 s^{-1} and at a temperature of 37 °C. Thirty minutes after the oral administration of 20 mg isosorbide dinitrate, there was a reduction of the blood viscosity especially at low shear rates. The plasma viscosity also decreased slightly but statistically significantly. The decrease in the apparent blood viscosity after isosorbide dinitrate was more marked the more severe the stage of the disease was. The hematocrit values decreased as well.

Note: The rheologic studies were carried out with viscosimeters which are prone to error, especially at low shear rates (heterophase effect). No measurements of further rheologic parameters.

In 1982 **Kirigaya** and co-workers [507] reported that, following sublingual administration of 0.3 mg nitroglycerine in patients with coronary heart disease, there was no change in the blood viscosity (abstract, no details).

In 1983 **Ehrly** and **Landgraf** [299, 300] reported on the effect of isosorbide dinitrate on the flow properties of blood in patients with chronic arterial occlusive disease. After oral administration of 20 mg of this drug there was a statistically significant decrease in the whole blood and plasma viscosity, while erythrocyte aggregation and erythrocyte filtrability did not show any statistical changes.

In 1984 **Murashko** and co-workers [666] studied the effect of nitrites on flow properties of the blood in patients with vascular diseases and found a decrease in the whole blood viscosity after a single dose of the drug (article in Russian).

In 1985 **Furkalo** and co-workers [358] reported on the effect of antianginal medications on the whole blood viscosity and plasma viscosity in patients with chronic coronary insufficiency and found a decrease in these hemorheologic parameters using nitrates, β-blockers, and molsidomine (article in Russian).

In 1985 **Weinberger** and co-workers [911] studied the plasma viscosity and the fibrinogen concentration before and after an acute administration of 5 mg isosorbide dinitrate sublingually in patients with myocardial infarction. They found a statistically significant decrease in the plasma viscosity, which reached a maximum 30 min after administration of the substance, while the hematocrit and the fibrinogen concentration remained unchanged (abstract, no details).

4.1.43.3 Discussion

After oral administration of isosorbide dinitrate *Caesar* and co-workers, *Hossmann* and co-workers, and *Ehrly* and *Landgraf* found a decrease in the blood and plasma viscosities. This result correlated with a decrease in the hematocrit value. In addition, *Ehrly* and co-workers found no changes in erythrocyte filtrability and erythrocyte aggregation. The studies of *Furkalo* and co-workers and *Weinberger* and co-workers point in the same direction. In summary, it may be said that, according to the present findings, isosorbide dinitrate can probably decrease the blood and plasma viscosities by way of an inner hemodilution (see Sect. 3.4). There is only one report on the substance nitroglycerine.

4.1.44 Insulin and Oral Antidiabetic Drugs

4.1.44.1 Introduction

Insulin, an endocrine enzyme of the pancreas, is used for the treatment of diabetes mellitus. In diabetes mellitus, fatty acid synthesis is disturbed owing to the insufficient production of insulin. This is followed by metabolic acidosis because of increased ketone formation. Since large quantities of water are necessary to discharge the ketones and the glucose (glucosuria), there follows in early or manifest coma diabeticum, secondary exsiccosis as well as disturbances in the electrolyte milieu, which cause a deterioration of the flow properties of blood. Deterioration of the flow properties of blood has been found even with well stabilized diabetics. The question as to the extent that the administration of insulin affects the deteriorated flow properties of the blood of diabetics has been studied by a number of research teams. Single studies on the effect of oral antidiabetics (sulphonylurea) were also carried out.

4.1 44.2 Publications

The first publication on an improvement in the flow properties of blood as a result of the administration of insulin was published by *Skovborg* and co-workers [801] in 1969. In addition to the insulin, 21 precomatose diabetics received infusions of isotonic saline solution and sodium bicarbonate. The blood viscosity of the blood samples anticoagulated with EDTA was determined with a Brookfield micro-cone-plate viscosimeter at shear rates of 2.3–230 s^{-1}; the plasma viscosity was measured only at 230 s^{-1}. At the end of treatment the authors found a significant reduction of the blood and plasma viscosities and the total protein concentration, the albumin, and the β and γ globulins at the same time as a decrease in the hematocrit value. If the hematocrit was mathematically corrected, the reduction in the blood viscosity was less.

Note: The temperature for the viscosity measurements was not given, and an evaluation of the hemorheologic effect of the insulin is difficult owing to the simultaneous intravenous administration of fluid.

In 1979 **Stoltz** and co-workers [822] and **Drouin** and co-workers (218) reported on changes in the blood viscosity in insulin dependent diabetics who had been inadequately treated. The patients were connected to an artificial pancreas. The viscosity of the blood samples, anticoagulated with EDTA, was measured with a Couette rotational viscosimeter at shear rates of 0.232, 0.346, and 1.16 s^{-1}. The hematocrit was set to a value of 40%. Simultaneously with the normalization of the blood glucose level, there was a significant decrease in the blood viscosity.

Note: Measurements of further rheologic parameters were not carried out, and the temperature for the blood viscosity measurement was not given.

In 1980 **Drouin** and co-workers [219] reported on the changes in the flow properties of blood during controlled diabetes therapy with an artificial pancreas. The blood viscosity of the blood samples, anticoagulated with EDTA, was measured at a corrected hematocrit of 40% with a Couette rotational viscosimeter at shear rates of 0.204–128.5 s^{-1}; plasma viscosity was determined with a capillary viscosimeter. Erythrocyte flexibility was determined with polycarbonate filters (Nucleopore, pore diameter of 5 μm) according to the method of Reid and co-workers. At the beginning of treatment, the patients showed an increased glucose level, and the macrorheologic parameters of the blood (blood and plasma viscosities) were significantly deteriorated compared to those of a healthy control group. During the course of the insulin treatment, there was a reduction of the blood and plasma viscosities as well as an improvement in erythrocyte deformability following the normalzation of the increased blood glucose levels.

Note: Measurements of erythrocyte aggregation were not carried out. No data on the temperature of the measurements.

In 1980 **Paisy** and co-workers [681] reported on the effect of an optimal stabilization of the blood sugar values as a result of insulin and sulfonylurea therapy in patients with diabetes mellitus. They studied the flow properties of the blood and the plasma with a capillary viscosimeter as well as with a Contraves rotational viscosimeter. The hematocrit was set at 45%. The authors found a statistically significant decrease in the whole blood viscosity and the plasma viscosity after stabilization of the diabetes, which correlated well with a decrease in the plasma fibrinogen concentration.

Note: No measurement of erythrocyte filtrability and erythrocyte aggregation.

In 1980 **Juhan** and co-workers [478, 479] studied the effect of insulin on the flow properties of blood in diabetics. The insulin was administered via an artificial pancreas. Erythrocyte deformability was determined according to the filtration technique of Reid and co-workers. Following the use of the artificial pancreas, there was an improvement in erythrocyte deformability and an increase in the intraerythrocytic adenosine triphosphate in parallel to the normalization of the plasma glucose level.

Note: No data on the temperature; no measurements of further rheologic parameters.

In 1980 **Rouselle** and co-workers [743] reported on the hemorheologic changes in poorly stabilized insulin-dependent diabetics. The blood viscosity measurements were carried out at a constant hematocrit (40%) using a Couette rotational viscosimeter (0.2–128 s^{-1}); plasma viscosity was measured with a capillary viscosimeter. Erythrocyte deformability was measured according to the method of Reid and co-workers using polycarbonate filters (pore size: 5 μm). At the same time as the normalization of the blood glucose values as a result of the artificial pancreas, there was a significant decrease in the plasma viscosity and the fibrinogen level. The filtrability measurements showed a slightly significant improvement in the erythrocyte deformability. The authors argued that the blood viscosity at a shear rate of more than 50 s^{-1} was mainly determined by erythrocyte deformability. Since the investigators found only a significant decrease in the blood viscosity at very low shear rates, they believed that the improvement observed in the flow properties of blood was to be attributed more to a decrease in erythrocyte aggregation than to an improvement in erythrocyte deformability. The authors were of the opinion that the filtration tests carried out with whole blood depended on the erythrocyte aggregation to a very considerable extent.

Note: No temperature data; no direct measurements of the aggregation behavior of the erythrocytes.

In 1981 **Juhan** and co-workers [480] reported on the effect of insulin, using an artificial pancreas, on erythrocyte deformability in diabetics. The whole blood method of Reid and co-workers was used. After the administration of insulin, the authors found an improvement in blood filtrability, which could already be seen a few hours after starting therapy.

The authors believed that it was not the reduction of the glucose concentration in the blood, but rather a direct effect of the insulin which was responsible for the improvement in blood filtrability.

Similar results were reported elsewhere [481].

Note: No measurements of other rheologic parameters.

In 1981 **Diamantopoulos** and co-workers [169] reported on filtrability studies using Nucleopore filters in healthy subjects and diabetics undergoing various kinds of therapy (diet, glibenclamide, glibenclamide phenophormine, insulin). The authors found that there was no difference in erythrocyte filtrability as a result of both dietary stabilization of the diabetes and the administration of insulin or oral antidiabetic drugs. They concluded that the various types of therapy for diabetes did not have different effects on blood filtrability (abstract, no details).

In 1981 **Le Devehat** and co-workers [558] studied the effect of insulin on blood filtrability in diabetics. They assumed that the improvement in blood filtrability was not correlated with the reduction of the blood glucose level, but with an "insulinization" which influenced the metabolism of the erythrocytes favorably (abstract, no details).

In 1981 **Akiyama** and co-workers [6] reported on the effect of an artificial pancreas on the flow properties of blood in diabetics and found a statistically significant decrease in the whole blood viscosity and the plasma viscosity (abstract, no details).

Voisin and co-workers [884] found that the flow properties of blood blood could be improved after the blood sugar values had been stabilized with an artificial pancreas. The blood viscosity was measured with a Contraves LS 30 viscosimeter at various shear rates and at a constant hematocrit of 40%. Measurement of plasma viscosity was carried out with an automatic capillary viscosimeter. The authors found that there was a significant decrease in blood viscosity 24 h after the start of treatment and that after 48 h of treatment the viscosity values of the patients not longer differed from those of the controls.

In 1982 **Poon** and co-workers [706] described a prospective study with 74 insulin-dependent patients who were divided into two groups. The first group was stabilized especially well, the second group was monitored in the standard way. Venous blood, anticoagulatd with EDTA, was examined with a Contraves LS 30 viscosimeter at various shear rates and at 37 °C. In both groups the blood and plasma viscosities sank at similar rates; however, in the group treated in the standard way, the decrease was somewhat more marked than in the other group. The authors discussed the question of whether blood viscosity was a direct pathogenetic factor of diabetes mellitus.

In 1982 **Lowe** and co-workers [592] described hemorheologic measurments in 22 patients with diabetes mellitus, who were treated with animal insulin or human insulin. All rheologic parameters (blood viscosity, hematocrit, plasma viscosity, erythrocyte deformability) had deteriorated in the patients with diabetes mellitus compared to the healthy subjects, but they were not affected by either the human or the animal insulin (abstract, no details).

Tymms [867] studied blood viscosity in poorly stabilized diabetics before and after a 10-day subcutaneous administration of insulin. There was a reduction at low shear rates, while the whole blood viscosity remained unaffected at high shear rates, as did plasma viscosity and the blood filtration rate. Since the hematocrit decreased slightly as the result of therapy, the author believed in a certain hemodiluting effect (abstract, no details).

Uchimura [868] studied the behavior of blood viscosity in 42 diabetics following both the administration of insulin and a diet for a period of 5 years. The whole blood viscosity was studied with a Brookfield microviscosimeter at shear rates of $230-23\ s^{-1}$ and at 37 °C (EDTA-anticoagulated blood). Plasma viscosity was measured with a microcapillary viscosimeter. The author found no change in the blood and plasma viscosities during standard therapy over a period of 5 years.

Stufano and co-workers [848] studied erythrocyte filtrability with the whole blood filtration method according to Reid and co-workers in 23 diabetics and 18 controls. They compared the values found to those of the blood sugar, the cholesterol, and the glycosylized hemoglobin. The authors found no statistically significant difference in erythrocyte filtrability between healthy subjects and diabetics and no correlation to the diabetic stabilization, which was assessed according to the blood sugar level.

Louis and co-workers [583] studied the effect of an intravenous administration of porcine insulin on the whole blood viscosity, the plasma viscosity, and erythrocyte filtrability in ten insulin-dependent diabetics. During therapy the blood sugar values were maintained constantly high by the simultaneous infusion of glucose. The authors found that, compared to the healthy subjects, the diabetics had clearly raised blood and plasma viscosities as well as reduced filtrability values. The administration of insulin with the simultaneous maintenance of the raised blood sugar level did not effect any changes in the hemorheologic parameters. The authors concluded that insulin did not have any effect on the hemorheology of the blood in vivo if the blood sugar level was kept at a constant high (abstract, no details).

In 1983 **Oughten** and co-workers [680] studied the behavior of whole blood viscosity and the viscosity of almost plasma-free, compact erythrocytes at 23 and

230 s^{-1}. They treated 13 diabetics by applying insulin in two different ways and found significantly lower viscosity values after 8 months of therapy. Other rheologic parameters, on the other hand, were not affected. The authors believed that improved diabetic control and therapy did not affect the rheologic parameters significantly (abstract, no details).

Lakomek and co-workers [533] studied erythrocyte aggregation with a photo-optic technique in adult diabetics and in diabetic children compared to healthy controls. They found no differences in the whole blood viscosity and differences erythrocyte aggregation only in adult diabetics. The plasma viscosity was, however, raised in the poorly stabilized diabetics compared to the well stabilized ones (abstract, no details).

In 1984 **Leiper** and co-workers [562] reported on the effect of animal insulin compared to biosynthetically produced human insulin on the flow properties of blood in 22 diabetics. A Contraves LS 30 viscosimeter was used to measure blood viscosity at 37 °C and a capillary viscosimeter to measure the plasma viscosity. In addition, erythrocyte deformability was tested with a filter system. The authors found no difference in the effect of the two types of insulin on the flow properties of blood.

In 1985 **Le Devehat** and co-workers [561] reported on an improved erythrocyte filtrability in patients with diabetes mellitus after the diabetes was well stabilized with an insulin pump (abstract, no details).

In 1985 **Sarno** and co-workers [750] reported on the effect of an artificial pancreas in patients with diabetes mellitus and found a slight but significant decrease in the whole blood viscosity and no significant decrease in erythrocyte filtrability. Erythrocyte filtrability was tested according to the method of Dodds and co-workers; no data were given for the other hemorheologic methods.

4.1.44.3 Discussion

The investigators agreed in their reports that, in the treatment of precomatose or poorly stabilized diabetics, controlled doses of insulin (artificial pancreas) effect a decrease in the blood viscosity values even when a hematocrit correction was performed. With regard to patients with uncomplicated diabetes mellitus the conclusions concerning the influence of insulin on the whole blood viscosity are contradictory. Three research teams reported on decrease in plasma viscosity [219, 743, 801]. In the majority of the erythrocyte flexibility measurements performed with whole blood [219, 478], the reservations regarding this method must be taken into consideration (see Sect. 3.3.4). Measurements of the erythrocyte aggregation were only performed by *Lakomek*. In the study by *Skovborg* and co-workers, the specific rheologic effect of insulin cannot be isolated from the possible additional rheologic action of sodium carbonate.

Regarding all available studies, no final conclusion can be made on the hemorheologic effect of insulin on the flow properties of the blood. This is in part due to the inappropriate measurement methods used. It is likely that the hemorheologic effects of metabolic acidosis and of exsiccosis (hemoconcentration) in the clinical picture of diabetes mellitus could be normalized with the controlled administration of insulin if the intravasal loss of fluid is compensated for at the same time. In addition, the decrease in plasma viscosity in an adequate long-term therapy could be explained via a reduction of the plasma fibrinogen concentration [743].

4.1.45 Acetylsalicylic Acid

4.1.45.1 Introduction

Acetylsalicylic acid and its soluble salt (lysine acetylsalicylate) are known worldwide as analgesics, antirheumatic drugs, and thrombocyte aggregation inhibitors.

4.1.45.2 Publications

In 1980 **Bonniot** and co-workers [82] reported on the effect of lysine acetylsalicylate on the flow properties of blood in 86 patients with cerebral and peripheral vascular disorders. They were given 500 mg lysine acetylsalicylate orally per day. For the measurement of the viscosity of the blood samples, anticoagulated with EDTA, a coaxial cylindrical viscosimeter (LS 30) was used at 37 °C and at shear rates of 0.01–100 s^{-1}. The hematocrit correction was carried out mathematically. The investigators found a decrease in blood viscosity, especially at low shear rates.

Note: No measurement of further rheologic parameters.

In 1983 **Vittoria** and co-workers [876] studied the effect of an intravenously administered form of acetylsalicylic acid on the flow properties of blood in 14 patients with chronic arterial occlusive disease. The whole blood viscosity was measured with a Brookfield viscosimeter at 37 °C and at various shear rates; the erythrocyte filtrability was determined with Nucleopore filters at 37 °C. Either 500 or 2000 mg acetylsalicylic acid was administered intravenously. The measurements were carried out before and after muscular excercise. The authors found that neither dose of acetylsalicylic acid effected a change in the hemorheologic parameters measured at rest. On the other hand, the increase in viscosity after defined ergometric stress was reduced compared to the blank values. The same was true for the erythrocyte filtrability which did not deteriorate after exercise to the same extent as it did without the administration of acetylsalicylic acid.

In 1984 **Ernst** and **Marshall** [324] studied patients with various diseases in order to see if the oral administration of 1000 mg acetylsalicylic acid led to hemorheologic changes. The authors found that there was no effect on the flow properties of blood with acetylsalicylic acid.

In 1985 **Passero** and co-workers [686] studied the effect of acetylsalicylic acid compared to placebo on the flow properties of blood in patients with cerebrovascular diseases. The treatment lasted 6–12 months. The dose was 1 g per day. The whole blood viscosity was determined with a cone-plate viscosimeter (Wells-Brookfield ¼ LVT) at 37° and at various shear rates; plasma viscosity was measured with the same instrument at 375 s^{-1}. The hematocrit was studied according to the method of Wintrobe. The authors found a statistically significant decrease in the plasma viscosity as well as a decrease in the whole blood viscosity at different shear rates; the hematocrit remained unchanged.

4.1.45.3 Discussion

Vittorio and co-workers found no influence on the hemorheologic parameters following an intravenous administration of acetylsalicylic acid in resting patients with arterial occlusive disease.

However, after defined ergometric stress, the increase in the blood viscosity and the reduction of the erythrocyte filtrability was less marked. *Bonniot* and co-workers were able to determine a decrease in the whole blood viscosity following an oral administration of only 500 mg lysine acetylsalicylate; this was, however, carried out with a corrected hematocrit. *Passero* and co-workers also found a statistically significant decrease in the whole blood viscosity and the plasma viscosity with no hematocrit correction. *Ernst* and *Marshall* did not find a hemorheologic effect following the oral administration of acetylsalicylic acid. Based on the small number of studies at hand it is not possible to prove a hemorheologic effect of acetylsalicylic acid on the flow properties of blood under resting conditions.

4.1.46 Dipyridamole

4.1.46.1 Introduction

Dipyridamole is a vasodilator which is used in the therapy of coronary heart disease. It effects thrombocyte aggregation and function.

4.1.46.2 Publications

In 1980 **Molaro** and co-workers [651] described the effect of dipyridamole (400 mg per day for 1 week) on the whole blood viscosity (Wells-Brookfield viscosimeter) and erythrocyte filtrability (whole blood method according to Reid and co-workers) in patients with hereditary spherocytosis and polycythemia vera. The authors found no effect on the whole blood viscosity in spherocytosis; however, they did find a decrease in the values in polycythemia vera. The erythrocyte filtrability was improved slightly but statistically significantly in both groups.

In 1985 **Passero** and co-workers [686] studied the effect of dipyridamole in comparison to placebo on the flow properties of blood in patients with cerebrovascular diseases. The treatment period was 6–12 months; the dose was 225 mg dipyridamole orally per day. The whole blood viscosity was determined with a cone-plate viscosimeter (Wells-Brookfield ¼ LVT) at 37 °C and at various shear rates. Plasma viscosity was measured with the same instrument at 375 s^{-1}. The hematocrit was studied with the method of Wintrobe. Compared to the non-treated patients there was a decrease in the fibrinogen concentration and the plasma and blood viscosities, especially at low shear rates. A statistical signifcance was not given.

In 1985 **Vaya** [874] described the effect of dipyridamole on erythrocyte deformability in patients with cerebrovascular diseases. Erythrocyte filtrability was determined according to the whole blood method of Reid and co-workers. There was an improvement in the filtration values (abstract, no details).

4.1.46.3 Discussion

The results of the available studies on the effect of dipyridamole on the flow properties of blood are contradictory. *Molaro* and co-workers found no effect of the substance on the whole blood viscosity in patients with spherocytosis, but a decrease in the values in polycythemia vera. *Passero* and co-workers, on the other hand, described a decrease in the plasma viscosity and the blood viscosity, especially at low shear rates; there was no statement as to the statistical significance. *Molaro* and co-workers and *Vaya* studied the effect of dipyridamole on the erythrocyte deformability and found a positive influence. Since both study groups used the whole blood method according to Reid and co-workers, the subsequent results must be interpreted carefully.

4.1.47 Eburnamonine

4.1.47.1 Introduction

Eburnamonine is a semisynthetic alkaloid obtained from Voacanga africana *which is used for the treatment of peripheral and cerebrovascular disorders.*

4.1.47.2 Publications

In 1981 **Guerrini** and co-workers [402] studied the effect of eburnamonine on the flow properties of blood. Patients with generalized vascular disorders received 40 mg eburnamonine in 10 ml isotonic saline solution intravenously every 2 days. The blood, plasma, and serum viscosities of the blood samples, anticoagulated with EDTA, were measured with a Brookfield micro-cone-plate viscosimeter at 37 °C and at shear rates of 750 and 150 s^{-1}. The authors found a significant decrease in blood viscosity. The plasma and serum viscosities, the fibrinogen level, and the hematocrit did not change significantly.

Note: A correction of the hematocrit was not carried out; no measurements of further rheologic parameters.

In 1981 **Le Devehat** and co-workers [56] tested the effect of eburnamonine on erythrocyte deformability. Twenty-five diabetics with no sign of arterial occlusion received 45 mg eburnamonine intravenously for 9 days (no details of the carrier solution; infusion time, 2–3 h). Erythrocyte deformability was determined according to the whole blood method of Reid and co-workers. The investigators found a significant improvement in erythrocyte deformability.

Note: No measurements of further rheologic parameters.

In 1983 **Palareti** and co-workers [684] described the effect of an intravenous infusion with 40 mg eburnamonine in patients with acute cerebral ischemia. As well as this substance, all patients also received mannitol or glycerine solution as routine treatment. Blood viscosity was measured with a Brookfield LVT viscosimeter at 37 °C and at shear rates of 23 and 230 s^{-1} after the blood samples had been corrected to a standard hematocrit. Plasma viscosity was also measured with a

Wells-Brookfield instrument. Blood filtrability was determined according to the method of Reid and co-workers. The group of patients was compared to a group of healthy subjects. There was a reduction of the whole blood viscosity at the high shear rate, while the reduction of the viscosity at the low shear rate was not statistically significant. Plasma viscosity decreased statistically significantly. The filtrability increased statistically significantly.

Note: No measurement of erythrocyte aggregation. The substance eburnamonine was not studied by itself, but rather combined with an infusion of either mannitol or glycerine.

In 1985 **Boisseau** and co-workers [80] studied the effect of eburnamonine (vinburnine). The filtrability was measured according to the method of Reid and co-workers; the filtrability of isolated erythrocytes was measured with a Hanss hemorheometer; erythrocyte deformability was determined by means of ectacytometry. Vinburnine showed no effect on the parameters studied (abstract, no details).

4.1.47.3 Discussion

Owing to the few publications available and their contradictory results, a conclusive improvement in the flow properties of blood cannot be shown with eburnamonine.

4.1.48 Stanozolol

4.1.48.1 Introduction

Stanozolol is an anabolic steroid which increases fibrinolysis and reduces the fibrinogen concentration of plasma. It has been used clinically for patients with Raynaud's syndrome and for postoperative thrombosis prophylaxis.

4.1.48.2 Publications

In 1981 **Ayres** and co-workers [35] described the effect of 5 mg stanozolol twice a day in ten patients with secondary Raynaud's syndrome. Blood viscosity was studied at 37 °C and 25 °C with both a Wells-Brookfield and a Contraves LS viscosimeter at various shear rates. In addition, the viscosity values were also measured with a corrected hematocrit. After a 3-month therapy with this drug, the authors found no statistically significant changes either in the whole blood viscosity or in the corrected whole blood viscosity. The hematocrit increased and the fibrinogen concentration dropped statistically significantly. In the patients with Raynaud's syndrome, the authors found an increase in the flow rate in the hand despite there being no improvement in the blood viscosity. They assumed that the increase in the hematocrit value and the drop in the fibrinogen concentration compensated each other hemorheologically, and they deduced from this that the improvement in the circulation of the hand was not caused rheologically.

Note: No measurements of further rheologic parameters.

Burns and co-workers [99] tested the effect of an intravascular injection of 50 mg stanozolol in a double-blind study with 20 patients compared to 20 control patients 1 day before an abdominal surgery and found no difference in the expected post operative increase in the plasma fibrinogen concentration and in the blood viscosity compared to the control group (abstract, no details).

Blamey and co-workers [69] studied the effect of a single preoperative intramascular injection of 50 mg stanozolol in 14 patients with a high postoperative risk of deep venous thrombosis. There was no difference compared to the control group as far as the behavior of the blood viscosity was concerned, as measured with a Contraves LS 30 viscosimeter at 37 °C and at various shear rates. Plasma viscosity, measured at 25 °C, also showed no difference between the two groups.

In 1985 **Lowe** and co-workers [595] treated patients with venous occlusions of the retina with stanozolol (10 mg per day for 16 weeks). There was a statistically significant decrease in the blood viscosity and the plasma viscosity and a reduction of the fibrinogen concentration (abstract, no details).

4.1.48.3 Discussion

According to the studies of *Burns* and co-workers the acute administration of stanozolol had no effect on blood viscosity. *Ayres* and co-workers did not find a decrease in the whole blood viscosity: while the fibrinogen concentration was reduced as a result of stanozolol, the hematocrit was increased owing to the anabolic effect of the drug. *Lowe* and co-workers, on the other hand, found a reduction of both the blood and plasma viscosities. In summary, it may be said that, based on the present reports, the question of a rheologic effect as a result of stanozolol cannot be fully answered yet.

4.1.49 Nicergoline

4.1.49.1 Introduction

Nicergoline is a α-sympatholytic agent which increases the peripheral blood flow and which has an inhibiting effect on thrombocyte aggregation. The indications are cerebral and peripheral vascular disorders.

4.1.49.2 Publications

In 1981 **Guerrini** and co-workers [404] studied the effect of a single, intravenous dose of 8 mg nicergoline over a period of 30 min in patients with chronic peripheral arterial occlusive diseases. In a double-blind study versus placebo, the blood and the plasma viscosities, the hematocrit, and erythrocyte filtrability were studied. After the administration of this drug, there was a difference to placebo manifesting as a reduction of the whole blood viscosity and an increase in erythrocyte filtrability (abstract, no details).

4.1.49.3 Discussion

Since only one group has studied this substance rheologically, the question as to its rheologic efficacy must remain open.

4.1.50 Nifedipine

4.1.50.1 Introduction

Nifedipine is a calcium antagonist which is among other things used for the therapy of coronary heart disease.

4.1.50.2 Publications

In 1981 **Slonim** and co-workers [80]) studied the effect of the administration of 20 mg nifedipine sublingually in patients with ischemic heart disease and impaired blood filtrability. The authors used the whole blood method of Reid and co-workers and found an improvement in blood filitrability which was still measurable 45 min and 4 h after administration of the drug (abstract, no details).

In 1983 **Waller** and co-workers [896, 897] reported on the effect of a single dose of nifedipine sublingually on erythrocyte deformability at rest and following physical stress in patients with angina pectoris. At the time of the study all patients were also treated with β-blockers. The authors found that 1 h after the oral administration blood filtrability was improved in all patients, while after placebo there was no change (abstract, no details).

Montagnani and co-workers [652] treated 20 patients with essential arterial hypertension with 40 mg nifedipine daily for 2 days and found an improvement in erythrocyte deformability as measured with two filtration instruments (abstract, no details).

In 1985 **Korabelshchikowa** and co-workers [525] found a reduction of the whole blood viscosity (measured with a rotational viscosimeter) and an increase in erythrocyte aggregation (determined according to a photometric method of O'Brien) after treating 40 patients with stable angina pectoris by means of daily doses of 30–40 mg nifedipine for 8 weeks (original in Russian).

In 1985 **Weinberger** and co-workers [911] studied plasma viscosity, the hematocrit value, and the fibrinogen concentration before and after an acute administration of 10 mg nifedipine sublingually in patients with myocardial infarction. They found a statistically significant decrease in plasma viscosity, with a maximum 30 min after administration of the drug, while the hematocrit and the fibrinogen concentration remained unchanged (abstract, no details).

In 1986 **Schürmann** and co-workers [777] studied the effect of an oral administration of 3 × 10 mg nifedipine for a period of 2 weeks in ten patients with chronically stable angina pectoris. While the erythrocyte aggregation values improved in all the patients, an improvement in erythrocyte filtrability was found in only six out of ten cases (abstract, no details).

In 1986 **Waller** and **Roath** [89], in a double-blind study described the effect of a long-term oral administration of 10 mg nifedipine on the blood flow properties in patients with angina pectoris. The authors found no effect on erythrocyte filtrability after chronic administration of nifedipine.

Note: No measurements of other hemorheologic parameters.

4.1.50.3 Discussion

While *Slonim* and co-workers, *Waller* and co-workers, *Montagnani* and co-workers, as well as *Schürmann* and co-workers found an improvement in erythrocyte deformatbility, no improvement in erythrocyte filtrability could be shown by *Waller* and *Roath* after chronic administration of the drug in patients with angina pectoris. *Korabelshchikowa* and co-workers found a decrease in the whole blood viscosity. *Weinberger* and co-workers were the only ones who studied plasma viscosity and found an improvement in the values. While *Korabelshchikowa* and co-workers described an increase in erythrocyte aggregation as a result of nifedipine, *Schürmann* and co-workers found a statistically significant reduction of this parameter. Seen as a whole, the hemorheologic results following the use of nifedipine are relatively contradictory, so that not definite evaluation is possible with regard to a rheologic mode of action.

4.1.51 Prostaglandins/Prostacyclins

4.1.51.1 Introduction

Prostaglandins as well as prostacyclins and analogous substances are tissue hormones which have a range of different pharmacologic characteristics. Among other things, they have a vasodilatory effect and inhibit platelet aggregation. Clinically these substances are used for the therapy of chronic occlusive arterial disease.

4.1.51.2 Publications

In 1981 **Dowd** and co-workers [215] described the effect of infusions of prostaglandin I_2 and prostaglandin E_1 on blood filtrability. Twelve patients with Raynaud's syndrome and systemic scleroderma were studied. After the infusion of 5–7.5 ng prostaglandin I_2 per kilogram body weight per minute, as well as after 6–12 ng prostaglandin E_1 per kilogram per minute (administerd via a central vein catheter), there was, compared to the 12 control patients, a statistically significant increase in erythrocyte filtrability as determined by a filtration method using Nucleopore filters. For the filtration the erythrocytes were separated from the plasma and suspended in a phosphate buffer solution with the addition of albumin.

Note: No measurements of other rheologic parameters.

These results were also presented elsewhere [216].

In 1981 **Belch** and co-workers [58] studied the effect of prostaglandin on erythrocyte deformability in ten patients with peripheral vascular diseases. Erythrocyte deformability was measured using Nucleopore filters with a pore diameter of 5 μm. With a dosage of 5 ng per kilogram body weight and per minute, the authors found a decrease in erythrocyte deformability (letter to the editor, no further details).

In 1984 **Yamaguchi** [927] reported on the effect of a 5- to 14-day intravenous infusion of 80–120 μg prostaglandin E_1 per day on erythrocyte deformability in patients with peripheral vascular disorders. Erythrocyte filtrability was determined according to the whole blood method of Reid. The authors found a significant improvement in erythocyte filtrability after prostaglandin E_1 (abstract, no details).

In 1984 **Lucas** and co-workers [596] studied the effect of a 72-h intravenous infusion of prostaglandin E_1 on erythrocyte deformability (special method for the elimination of leukocytes and additional washing processes) and plasma viscosity, measured at 25 °C, in patients with Raynaud's disease.

In 1985 **Grasselli** and co-workers [391] reported on the effect of prostacyclin on the flow properties of blood in 13 patients with chronic arterial occlusive disease. A dose of 10 ng per kilogram body weight per minute were infused intravenously for 1–2 days. Erythrocyte filtrability was determined according to the whole blood method of Reid and co-workers. The authors did not find any change in erythrocyte deformability but an improvement in leukocyte deformability (abstract, no details).

In 1985 **Mikita** and co-workers [645] reported on a double-blind study in patients with Raynaud's disease who were given 5 ng prostacyclin intravenously per kilogram body weight per minute. The whole blood viscosity was measured at various shear rates using a Contraves viscosimeter. In addition, erythrocyte and leukocyte filtrability was determined with polycarbon filters. The preliminary results showed no significant differences in the hemorheologic parameters between prostacyclin and placebo (abstract, no details).

4.1.51.3 Discussion

The results available on the effect of prostaglandin E_1 on erythrocyte filtrability are contradictory. While *Belch* and co-workers found a deterioration of erythrocyte deformability, *Dowd* and co-workers as well as *Yamaguchi* described an improvement in the filtration values. The studies of *Belch* and co-workers as well as *Yamaguchi* were carried out with an inadequate filtration method; the measurements of *Dowd* were carried out with washed erythrocytes and their validity is therefore also limited.

Lucas and co-workers, *Grasseli* and co-workers, as well as *Mikita* and co-workers studied the effect of prostacyclin on the flow properties of blood. *Grasseli* and co-workers found no effect on the erythrocyte filtrability with the drug, and *Mikita* and co-workers also could not determine a significant change in either erythrocyte filtrability or in the whole blood viscosity as a result of prostacyclin.

4.1.52 Indomethacin and Other Antirheumatic Drugs

4.1.52.1 Publications

In 1981 **Mercke** and **Kjellen** [635] reported on measurements in patients with malignant tumors before and after the administration of the drug indomethacin and found an improvement in erythrocyte deformability (abstract, no details).

In 1984 **Bedi** and co-workers [56] described the effect of the antirheumatic drugs fenclofenac and diclofenac on the plasma viscosity in patients with rheumatic diseases. They found a statistically significant decrease after a 2- or 6-month therapy.

Note: No studies of further rheologic parameters.

In 1985 **Nenci** [668] studied the effect of indobufen at a dose of 200 mg orally per day in patients with cardiovascular diseases and found an improvement in erythrocyte filtrability measured before, during, and 4 weeks after treatment using a method with Nucleopore filters (abstract, no details).

4.1.52.2 Discussion

The individual reports available on three different anti inflammatory substances (antirheumatic drugs) do not allow a definite conclusion on the hemorheologic efficacy of the substances. A reduction of the plasma viscosity is feasible owing to the inhibition of fibrinogen synthesis.

4.1.53 Hydroxychloroquine

4.1.53.1 Introduction

Hydroxychloroquine is a drug used in the treatment of malaria.

4.1.53.2 Publications

In 1981 **Rose** and co-workers [738] reported on measurements of the whole blood viscosity at various shear rates in blood samples from patients undergoing surgery who had been prescribed 600–1400 mg hydroxychloroquine preoperatively. Compared to the control patients who had not received the drug, there was a reduction of the plasma viscosity as well as the whole blood viscosity during the operation, while the viscosity increased in the control group (abstract, no details).

In 1984 **Ernst** and co-workers [326] studied the effect of hydroxychloroquine sulphate (600–1400 mg preoperatively compared to placebo). The operations were carried out because of various diseases. The whole blood viscosity and the plasma viscosity were measured with a Contraves LS 30 viscosimeter. Erythrocyte filtrability was determined according to Dodds and co-workers. In addition, the hematocrit and the fibrinogen concentration were determined. In the hydroxychloroquine group there was a reduction of the blood viscosity at various shear rates, while the plasma viscosity remained unchanged. Erythrocyte filtrability decreased slightly. The viscosity measurements were carried out with a corrected hematocrit. The uncorrected hematocrit in the hydroxychloroquine group decreased more markedly than in the placebo group. The slight deterioration in blood filtrability was less markedly expressed in the hydroxychloroquine group than in the placebo group both before and after the operation.

4.1.53.3 Discussion

There are two publications available on the effect of hydroxychloroquine. *Rose* and co-workers found a difference in operated patients between hydroxychloroquine and placebo as far as the blood and the plasma viscosities were concerned. *Ernst* and co-workers also studied patients having undergone surgery and found a reduction of the whole blood viscosity, while the plasma viscosity remained unchanged. Erythrocyte filtrability deteriorated. Since both studies were carried out in patients who were subjected to a number of exogenous and medicinal influences within the framework of their operation, the results at hand cannot be taken as proof of a hemorheologic effect of hydroxychloroquine.

4.1.54 Fludrocortisone

4.1.54.1 Introduction

Fludrocortisone is a halogenized glucocorticoid.

4.1.54.2 Publications

Fludrocortisone was used by **Humphrey** and co-workers [463] in 1981 for the therapy of polycythemia (10 mg orally per day). The authors found no significant changes in the hematocrit value and the whole blood viscosity (abstract, no details).

4.1.55 Ketanserin

4.1.55.1 Introduction

Ketanserin is a 5-hydroxytryptamine antagonist with vasodilatory properties and is used for the treatment of essential hypertension and peripheral chronic occlusive arterial disease.

4.1.55.2 Publications

In 1981 **De Cree** and co-workers [161] studied the effect of the selective 5-hydroxytryptamine antagonist ketanserin on blood filtrability in patients with essential hypertension and acute myocardial infarction. Before the beginning of therapy erythrocyte filtrability was clearly reduced in both groups of patients. This filtrability decrease was attributed to changed plasma factors. After acute intravenous administration of 10 mg ketanserin, there was, in addition to a decrease in blood pressure, an improvement in the filtrability values. In a further double-blind study in 30 patients with acute myocardial infarction, erythrocyte filtrability was clearly improved after the oral administration of 60 mg ketanserin in comparison with placebo (abstract, no details).

Jacobs and co-workers [470] treated 15 patients with either a single dose of 140 mg ketanserin or 2 × 40 mg daily in a long-term therapy of 3 weeks. The following hemorheologic parameters were measured: hematocrit, plasma viscosity, and erythrocyte aggregation. There was no statistically significant difference between the hemorheologic parameters before and after treatment with ketanserin (abstract, no details).

In 1983 **De Cree** and co-workers [162] reported on studies in 30 patients with acute myocardial infarction who received either placebo or 60 mg ketanserin orally. It was shown in the filtration study that the filtrability was clearly improved in the blood with a high platelet count, while the effect was not as distinct in blood with a low platelet count. It was assumed that the aggregation-inhibiting effect of ketanserin on the thrombocytes also affected erythrocyte deformability (abstract, no details).

De Cree and co-workers [163] treated 20 patients with intermittent claudication for 3 months using 3 × 60 mg ketanserin orally per day. They found a statistically significant improvement in blood filtrability in the ketanserin group, while in the placebo group there were no changes (abstract, no details).

In 1985, in a double-blind study **Walker** and co-workers [894]) studied the effect of ketanserin (2 mg per h) and placebo in 14 patients with chronic arterial occlusive disease over a period of 7 days. The whole blood viscosity was measured with a Contraves LS 30 viscosimeter, the plasma viscosity with a capillary viscosimeter. Erythrocyte filtrability was measured with a St. George filtrometer. The authors found a statistically significant decrease in the whole blood viscosity at various shear rates and a slight but statistically significant decrease in the plasma viscosity and the hematocrit values. With regard to erythrocyte filtrability there was an increase in the transit time and a decrease in the clogging rate of white blood cells.

In 1985 **Bogar** and co-workers [77] described the effect of the 5-hydroxytryptamine antagonist ketanserin on the flow properties of blood in patients with ischemic diseases. The drug was infused continuously in 20 patients over 7 days (50 mg per day in 1000 ml dextrose/saline infusions). For comparison, ten patients received infusions without ketanserin. The hemorheologic measurements were performed with heparinized venous blood. The whole blood viscosity was measured with a Contraves Low Shear viscosimeter at various shear rates; a Harkness viscosimeter was used to measure plasma viscosity. The whole blood viscosity was mathematically set to a standard hematorcrit of 45%. Erythrocyte filtrability was measured with a Carrie-Med instrument. In the patients treated with 5-hydroxytryptamine there was a reduction of the hematocrit and the whole blood viscosity. There was also a reduction of the plasma viscosity values. There were slight but statistically significant changes in erythrocyte filtration manifesting as a decrease in the number of leukocytes which were clogging the pores. Hemodilution was considered to be the mechanism of action of ketanserin. The results were also presented in a shorter form elsewhere [78].

4.1.55.3 Discussion

While *De Cree* and co-workers in a number of studies described an effect of ketanserin on blood filtrability in patients with various diseases , *Walker* and co-workers as well as *Bogar* and co-workers attributed the slight but statistically significant change in the filtrability value to a medicinal effect of the leukocytes which normally clog the pores. Since *Bogar* and co-workers found a reduction of the hematocrit, whole blood viscosity, and plasma viscosity, they argued that hemodilution was the mechanism of action for the changes they found. *Walker* also found a slight but statistically significant decrease in the plasma viscosity values, while *Jacobs* and co-workers could not determine an influence on the hemorheologic parameters the studied. All in all the results available up to now on the influence of ketanserin on hemorheologic parameters are contradictory and must be investigated further.

4.1.56 Plant Extracts (Bioflavonoids)

4.1.56.1 Introduction

Flavonoids are found in horse chestnut extracts among other things and are used for the therapy of chronic venous diseases. A capillary-sealing and edema-protecting effect is presumed to be the mechanism of action.

4.1.56.2 Publications

In 1982 **Klemm** and **Enghofer** [510] treated 30 patients with varicose veins in the lower extremities in order to study the effect of a 12-day oral administration of 3 × 1800 mg of horse chestnut extract (with a standardized escin content) daily. The whole blood viscosity of the heparinized blood was determined with a microviscosimeter at a shear rate of $200\,s^{-1}$ and at a temperature of 37 °C. The viscosity of the whole blood was statistically significantly decreased by 5% on average as a result of the drug.

Note: No hematocrit correction; no measurements of other rheologic parameters.

These findings were also presented elsewhere [511].

In 1983 **Witte** and co-workers [919] reported on the effect of an administration of flavon glycosides obtained from *Ginkgo biloba* on the viscoelasticity of whole blood and on plasma viscosity (measured with an oscillating capillary rheometer). The blood samples were set at a standard hematocrit of 45% and the temperature was 22 °C. The values given were measured at a shear rate of 10 s^{-1}, and the oscillation frequency was 2 Hz. The group of patients treated with the flavon glycoside substance showed a decrease in the viscosity, especially with regard to the elasticity (abstract, no details).

In 1984 the **Shanghai Cooperative Group for the Study of Tanshinone IIA** [794] reported on the results of a double blind-study on the effect of the drug sodium tanshinone IIA sulphonate (an extract from *Radix salviae miltiorrhizae*) on the flow properties of blood in patients with coronary heart disease. The water-soluble drug was administered for 2 weeks at a dosage of 160 mg per day. There were no statistically significant changes in the blood and plasma viscosities.

Note: No data on the hemorheologic methods used.

In 1985 **Kiesewetter** and co-workers [498] reported on the effect of bencianol (a bioflavonoid) at a single dose of 2 mg in patients with abnormal hemorheologic parameters. There was an improvement in the hemorheologic parameters, especially in erythrocyte deformability and erythrocyte aggregation (abstract, no details).

4.1.56.3 Discussion

There are five studies available on four different flavonoids whose effect on the flow properties of blood was studied at different dosages and with various measurement methods. Therefore it is not surprising that conclusions on the influence on hemorheologic parameters in patients are not uniform. In individual cases further studies would be necessary in order to decide whether individual bioflavonoids or the drug group as a whole has an effect on the flow properties of blood or not. For publications on rutosides see Sect. 4.1.23.

4.1.57 Zilazep

4.1.57.1 Introduction

Zilazep is a caronary dilator and is used for the treatment of ischemic cardiac diseases.

4.1.57.2 Publications

In 1982 **Kirigaya** and co-workers [507] reported on measurements of the whole blood viscosity in patients with ischemic cardiac disease. They used a laboratory viscosimeter of their own design. After the oral administration of 300 mg zilazep, there was a reduction of the blood and plasma viscosities (abstract, no details).

4.1.58 Erythrocyte Transfusion in Sickle Cell Anemia

4.1.58.1 Publications

In 1982 **Jan** and co-workers [474] studied the effect of a transfusion of normal erythrocytes on the flow properties of blood in 15 patients with sickle cell anemia. The viscosity was measured with a rotational viscosimeter. After the infusion of the erythrocyte concentrate, there was an increase in the hematocrit and in the blood viscosity. The simultaneous measurement of oxygen pressure in the blood led the authors to the assumption that an increase of the hematocrit to more than 35% did not increase the oxygen capacity in these patients owing to the ensuing hyperviscosity.

Note: No measurement of further rheologic parameters.

4.1.59 Iron Therapy in Polycythemia

4.1.59.1 Introduction

Iron preparations (mostly containing bivalent iron) are usually prescribed for iron deficiency conditions and iron deficiency anemia.

4.1.59.2 Publications

In 1982 **Milligan** and **Davis** [649] reported on a study in polycythemia patients with iron deficiency. Viscosity measurements were carried out before and after treatment with an iron preparation. The whole blood viscosity was measured at high and low shear rates. Despite a slight increase in the hematocrit values, the whole blood viscosity showed a decrease both at high and low shear rates after the therapy with iron. However, the authors did mention that the Coulter instrument used did not give an exact hematocrit value.

4.1.60 Piracetam

4.1.60.1 Introduction

Piracetam is a substance that is used to improve cerebral capabilities in patients with cerebrovascular insufficiency.

4.1.60.2 Publications

In 1983 **Kiesewetter** and co-workers [496] treated patients with diabetic retinopathy by means of 12 mg piracetam orally per day for 4 weeks. They determined the plasma viscosity with a capillary viscosimeter, erythrocyte aggregation with an optical transmission instrument, and erythrocyte deformability with the single pore rigidometer. In addition, the yield shear stress was determined. The authors described a decrease in erythrocyte aggregation, an improvement in the erythrocyte deformability, and a decrease in the yield shear stress values.

Clinical findings: There was an improvement in the visus and in the visual field.

Ciuffetti and co-workers [134] described the effect of piracetam on the flow properties of blood in patients with cerebrovascular insufficiency. The whole blood viscosity and plasma viscosity were measured with a Wells-Brookfield viscosimeter at 37 °C; erythrocyte filtrability was determined according to a modification of the original whole blood method of Reid. The temperature of the measurements was 37 °C. Piracetam was compared to pentoxifylline, placebo, and a combination of piracetam plus pentoxifylline (piracetam dosage 3 × 800 mg daily, pentoxifylline dosage 400 mg daily). The authors found that an improved erythrocyte filtrability was only achieved with a combination of both drugs.

4.1.60.3 Discussion

While *Kiesewetter* and co-workers found an improvement in erythrocyte aggregation and deformability at a dosage of 12 g per day, *Ciuffetti* and co-workers could not determine an influence on erythrocyte filtrability at a dosage of 3 × 800 mg per day. Since there is possibly a dosage dependency, further hemorheologic studies are necessary with regard to the question of a hemorheologic effect of piracetam.

4.1.61 Lipid Solutions

4.1.61.1 Introduction

Fatty emulsions (lipid solutions) are used for parenteral nutrition, especially in infants and patients with a high caloric requirement.

4.1.61.2 Publications

In 1983 **Sengespeik** and co-workers [790] studied the effect of a lipid infusion on the flow properties of blood. Ten premature and ten mature babies received an infusion of 0.15 g per kilogram body weight per hour for 4 hours. The blood samples taken before and after the infusion period showed no changes in the blood and plasma viscosities and in erythrocyte aggregation, as measured with a rheoscope.

4.1.62 Exercise, Physiotherapy, Diet

4.1.62.1 Introduction

In this and the following sections (4.1.63, 4.1.64 and 4.1.65), nonmedicinal measures which have shown changes in the flow properties of blood are cited. In this section there are a number of publications on studies in which various physiotherapeutic or dietary measures were used. In 1904 ***Lommel*** *[582] had already reported on changes in blood viscosity with sweating procedures.*

4.1.62.2 Publications

In 1983 **Hall** and co-workers [414] reported on the effect of a low-fat, low-salt, and high-carbohydrate diet plus simultaneous training on the flow properties of blood in patients with arterial hypertension. With a therapy lasting a total of 26 days there was a statistically significant reduction of the whole blood viscosity and of the plasma viscosity with a simultaneous decrease of the systemic blood pressure values (abstract, no details).

In 1984 **Ernst** and co-workers [322] reported on the effect of carbon dioxide baths on the flow properties of blood in eight patients with angiographically proven chronic arterial occlusive disease. Heparin-anticoagulated blood was measured in a Carri-Med viscosimeter at a standardized hematocrit of 45% and at a temperature of 37 °C. Plasma viscosity was studied with a Harkness capillary viscosimeter, erythrocyte flexibility according to Dodds and Dormandy, and erythrocyte aggregation with an instrument by Myrenne and Co. In the short-term experiment there was an improvement in erythrocyte flexibility, a decrease in plasma viscosity, and a decrease in blood viscosity which, however, was significant only at one shear rate. In the long-term experiment there was only an improvement in erythrocyte filtrability.

In 1984 **Matrai** and co-workers [624] reported on an improvement in the flow properties of blood as a result of carbon dioxide baths in patients with arterial occlusive disease. A corresponding effect could not be shown in healthy subjects (abstract, no details).

In 1984 **Diehm** and co-workers [170] reported on hemorheologic changes after 8 weeks of endurance training in patients with peripheral arterial occlusive disease in stage II according to Fontaine. Blood viscosity was studied with a Wells-Brookfield viscosimeter with a Mooney adapter, blood viscosity with a Harkness capillary viscosimeter, erythrocyte aggregation photometrically, and erythrocyte deformability with a Teitel micropolyviscosimeter. While the plasma and the serum viscosities hardly changed at all, there was a reduction of the whole blood viscosity as well as a decrease in the tendency of the erythrocytes to aggregate. Erythrocyte filtrability improved. After a 12-week break in training the values returned to the initial values (abstract, no details).

In 1985 **Selvetella** and co-workers [789] described the effect of fever therapy (pyretotherapy) with a polyvalent vaccine in patients with inflammatory arteriopathies. The vaccine was applied intravenously and led to rises in temperature from 39 °C to 41 °C. During the course of this 1–2-week therapy, the hematocrit and erythrocyte filtrability were determined according to the method of Reid and co-workers. The authors described a statistically significant decrease in the hematocrit values and in the erythrocyte filtrability during the fever therapy.

Note: No further hemorheologic data.

In 1985 **Matassi** and co-workers [628] reported on the effect of intra-arterially administered ozone in patients with chronic arterial vascular diseases. The authors found an improvement in erythrocyte deformability (no data on methods; abstract, no details).

In 1985 **Gallasch** and co-workers [363] studied to what extent the hemorheologic parameters of patients with chronic arterial occlusive disease at the stage of intermittent claudication could be influenced or not by an 8-week period of endurance training. The whole blood viscosity was measured with a Wells-Brookfield viscosimeter with Mooney adapter. Erythrocyte aggregation was measured photometrically; erythrocyte deformability was determined with a Teitel polymicroviscosimeter. The serum viscosity and the plasma viscosity fell statistically significantly, while the fibrinogen concentration and the hematocrit remained unchanged. The apparent whole blood viscosity from native blood remained the same; when it was corrected to a standard hematocrit there was a slight decrease in the viscosity. Erythrocyte aggregation decreased, and erythrocyte deformability was also improved.

Note: There was a statistically significant extension of the pain-free walking distance.

In 1985 **Perego** and co-workers [695] reported on the effect of a 3-month training period on the blood filtrability in patients with intermittent claudication and found an increase in the parameters at rest. The decrease in erythrocyte filtrability immediately after acute physical stress was statistically less significant at the end of the training period than at the beginning of treatment (abstract, no details).

In 1985 **Ernst** and **Matrai** [328] studied the effect of controlled training on the flow properties of blood in patients with intermittent claudication and found an improvement (abstract, no details).

In 1985 **Barnard** and co-workers [43] studied the effect of a combined therapy of training and a low-fat and low-salt diet as well as a specific carbohydrate diet on the systemic blood pressure of patients with essential hypertension. After a 26-day treatment the whole blood viscosity (citrated blood) was measured with a Sienco-Sonoclot viscosimeter at 37 °C. The authors found a reduction of the blood viscosity which did not correlate with the hematocrit values.

Note: No measurements of further rheologic parameters.

In 1986 **Ernst** and co-workers [334] studied 20 patients who were undergoing special rehabilitation therapy (balneolic, physical, and other measures) after cerebral stroke. Before and after the 6-week therapy there was a decrease in the whole blood viscosity at various shear rates (Contraves rotational viscosimeter) and a slight, but statistically signifcant decrease in plasma viscosity. There were no significant differences in erythrocyte aggregation.

In 1986 **Kharkarov** and **Tirulov** [490] studied the effect of high concentration sulphide baths in patients with essential hypertension and found a reduction of the blood viscosity which was accompanied by a decrease in the hematocrit (original in Russian).

4.1.62.3 Discussion

There are only a few papers available on the effect of a low-fat and low-salt diet with simultaneous training, the effect of carbon dioxide baths on the flow properties of blood in patients with chronic arterial occlusive disease, the effect of sulphide baths on the flow properties of blood in patients with hypertension, fever therapy in patients with inflammatory arteriopathies, and intra-arterial ozone therapy. The effect of endurance training on the flow properties of blood in patients with intermittent claudication has nonetheless been investigated by five study groups. They all found a decrease, though slight, in the whole blood viscosity, plasma viscosity, and erythrocyte aggregation. Erythrocyte filtrability was improved.

4.1.63 Oxygen Therapy in Cor Pulmonale

4.1.63.1 Introduction

In chronic pulmonary diseases pure oxygen may be given either under normal atmospheric pressure or under hyperbaric pressure (high-pressure chamber).

4.1.63.2 Publications

Gluskowski and co-workers [380] studied the hematocrit values and the whole blood viscosity (Wells-Brookfield viscosimeter at a shear rate of 230 s^{-1}) before and after controlled oxygen respiration for 17 h per day for 6 weeks. In some cases β-adrenergic stimulators and digoxin were given as accompanying medication. The patients were suffering from chronic, obstructive bronchitis which had led to secondary polyglobulism with cor pulmonale. After the 6-week therapy there was a statistically significant drop in the blood viscosity and a statistically significant decrease in the hematocrit value, while the serum protein concentration remained unchanged. Most of the patients had a reduction of the pulmonal arterial pressure as well.

Note: Other rheologic parameters were not studied.

In 1985 **Mathieu** and co-workers [622] studied erythrocyte deformability according to the method of Hanss and Stoltz in 70 patients (no details on the diseases) before and after a 90-min exposure to hyperbaric oxygen at 2 bar. While the hematocrit and the plasma viscosity remained unchanged, there was a decrease in the filtrability index in the short-term experiment as well as in a series of hyperbaric oxygen therapies (abstract, no details).

4.1.63.3 Discussion

Oxygenation in the form of a long-term therapy led to a reduction of the hematocrit values with a corresponding decrease in the whole blood viscosity. A short-term exposure to hyperbaric oxygen for 90 min on the other hand, caused no changes in the hematocrit and the plasma viscosity; there was, however, a decrease in the filtrability index. The two findings are not contradictory since it is absolutely feasible that the stimulation of the erythropoetic system is inhibited within the framework of a long-term hyperbaric oxygen therapy in patients with cor pulmonale and that the whole blood viscosity is decreased with hematocrit values lowered correspondingly.

4.1.64 Extracorporal Ultraviolet Radiation of the Blood

4.1.64.1 Introduction

The extracorporal irradiation of blood with ultraviolet rays is one of the methods used to improve peripheral blood flow. Only a few institutes use this method.

4.1.64.2 Publications

In 1982 **Bäumler** and co-workers [40] studied blood and plasma viscosities in six patients with chronic arterial occlusions. Blood viscosity was measured with a Weissenberg rheogoniometer at shear rates of 0.146–461 s^{-1} and at 37 °C and plasma viscosity was measured with the aid of a capillary viscosimeter at 25 °C. Venous blood (1 ml per kilogram body weight) was removed from the patients, anticoagulated with sodium citrate at a relation of 10:1 and reinfused within 5 min after irradiation with ultraviolet light (wavelength 254 nm). The procedure was carried out in 2-day intervals. As a comparison, six patients were studied after placebo. The blood for the measurement of the rheologic parameters was set to a hematocrit of 45%. There was a decrease in the apparent whole blood viscosity at low shear rates, while the viscosity at shear rates above 46 s^{-1} remained unchanged, as did the plasma viscosity. In the placebo group, there was no influence on the apparent whole blood viscosity at low shear rates. The authors suspected a decrease in erythrocyte aggregation to be the cause of the viscosity behavior, but they pointed out that the small number of cases studied made an interpretation of the results difficult.

Note: No measurement of erythrocyte filtrability and no direct measurement of erythrocyte aggregation. The temperature of the measurements was below body temperature.

Similar results were presented elsewhere by *Lerche* and co-workers from the same study group [567].

4.1.64.3 Discussion

The authors themselves point out difficulties in the interpretation of the results. Further studies are necessary in order to corroborate or deny a hemorheologic effect.

4.1.65 Hemodialysis and Kidney Transplants

4.1.65.1 Publications

In 1983 **Williams** and co-workers [917] studied the effect of hemodialysis in five patients from whom 3.4 l fluid was removed over a period of 3 h. Blood viscosity was measured with a rotational viscosimeter at 450 s^{-1}. There was an increase in blood viscosity, which was dependent on the time after the start of hemodialysis (abstract, no details).

In 1983 **Zerefos** and co-workers [931] found that a successful kidney transplant improved the flow properties of blood which had deteriorated as a result of chronic dialysis. Blood viscosity was measured in 16 patients using a Wells-Brookfield microviscosimeter at two different shear rates (abstract, no details).

In 1985 **Delsignore** and co-workers [165] studied erythrocyte filtrability (Nucleopore filters) as well as the plasma viscosity and the blood viscosity at two different shear rates using a Wells-Brookfield LVT viscosimeter at 37 °C during hemodialysis of patients with uremia. There were no significant changes in these parameters during the course of hemodialysis.

In 1985 **Gueguen** and co-workers [400] reported on the effect of hemodialysis on the flow properties of blood in patients with chronic kidney insufficiency. The blood was examined immediately before and after the dialysis. The whole blood and plasma viscosity were measured with a low shear viscosimeter, and erythrocyte filtrability was measured with two different methods (erythrometer and hemorheometer). The authors could not determine any significant changes resulting from hemodialysis in the hemorheologic paramters measured (abstract, no details).

In 1985 **Konishi** and co-workers [521] reported on the effect of hemodialysis on the flow properties of blood. They found an increase in the whole blood and plasma viscosity as well as in the fibrinogen concentration and the erythrocyte sedimentation rate (abstract, no details).

4.1.65.2 Discussion

While *Williams* and co-workers and *Konishi* and co-workers reported on an increase in the whole blood and plasma viscsosities within the framework of hemodialysis, *Delsignore* and co-workers as well as *Gueguen* and co-workers could not find any such influence. Possibly the changes in the flow properties of blood during dialysis depend very much on the type of procedure and on the question of whether surplus intravascular water is eliminated during hemodialysis or not. If there is an increase in the hematocrit during the procedure, an increase in the whole blood viscosity is to be expected as well. There is one publication available on the whole blood viscosity in a patient with successful kidney transplant [931]; here the viscosity values improved after the operation.

4.1.66 Carnitine

4.1.66.1 Introduction

Carnitine stimulates stomach and pancreas secretion.

4.1.66.2 Publications

In 1984 **Pola** and co-workers [703] studied the effect of carnitine at a dosage of 2 g per day for 60 days on the whole blood viscosity (Haake rotational viscosimeter at 37 °C) and on erythrocyte filtration (according to Reid and co-workers) in patients with chronic arterial vascular disorders. There was a reduction of the whole blood viscosity as well as an improvement in the erythrocyte filtration rate.

Note: No measurements of further hemorheologic parameters.

4.1.67 Benfurodil

4.1.67.1 Introduction

Benfurodil is a vasodilator.

4.1.67.2 Publications

In 1984 **Mozzi** and co-workers [660] studied the effect of benfurodil on the flow properties of blood in patients with chronic arterial occlusive disease. In a double-blind experiment, ten patients were given 50 mg benfurodil intravenously, and ten other patients were given 3 × 150 mg orally for 10 days. The whole blood viscosity was measured with a Wells-Brookfield LVT/CP viscosimeter at various shear rates and at a temperature of 37 °C. In addition, erythrocyte deformability was studied using Nucleopore filters with 5-μm pore diameter. The authors found a decreased whole blood viscosity and an improved filtrability of the blood, while the plasma viscosity did not show a uniform change.

4.1.68 Molsidomine

4.1.68.1 Introduction

Molsidomine is a coronary vasodilator.

4.1.68.2 Publications

In 1984 **Vakaliuk** [970] described the effect of molsidomine on the flow properties of blood in patients with stable angina pectoris and found a decrease in the blood viscosity values (original in Russian).

In 1985 **Furkalo** and co-workers [358] reported on the effect of anti-anginal medications on the whole blood viscosity and plasma viscosity in patients with chronic coronary insufficiency. They found a decrease in the hemorheologic parameters with the use of molsidomine (publication in Russian).

In 1985 **Korabelshchikowa** and co-workers [525] treated six patients with stable angina pectoris and found a reduction of the whole blood viscosity (measured with a rotational viscosimeter) and a decrease in erythrocyte aggregation (determined according to the photo-optic method of O'Brien) after oral application of 4–6 mg molsidomine per day for 8 weeks (original in Russian).

4.1.68.3 Discussion

Vakaliuk, Furkalo and co-workers, as well as *Korabelshchikowa* and co-workers found a reduction of the whole blood viscosity after molsidomine. *Korabelshchikowa* and co-workers also found a reduction of the erythrocyte aggregation.

4.1.69 Trapidyl

4.1.69.1 Introduction

Trapidyl is a coronary vasodilator.

4.1.69.2 Publications

In 1985 **Lerche** and co-workers [568] reported on the effect of trapidyl for a period of 4 weeks at a dosage of 300 mg per day on the flow properties of blood in patients with chronic arterial vascular disorders. The whole blood viscosity was studied at various shear rates; plasma viscosity and erythrocyte aggregation were measured with a photometric method. The authors found a decrease in the blood viscosity at low shear rates, while no changes could be found at high shear rates (abstract, no details).

4.1.70 Glycerine solution

4.1.70.1 Introduction

Among other things, glycerine solutions are used for osmotherapy in acute cerebral ischemias.

4.1.70.2 Publications

In 1985 **Ponari** and co-workers [705] reported on the effect of a 1 h long infusion of 500 ml 10% glycerine solution in ten patients with acute cerebral ischemia. The samples were taken before, 2, and 6 h after the end of treatment and studied hemorheologically. There was a significant decrease of the plasma viscosity; the whole blood viscosity, the hematocrit, and erythrocyte filtrability also decreased, but not statistically significantly (abstract, no details).

4.1.71 Polydeoxynucleotide

4.1.71.1 Introduction

Polydeoxynucleotides are obtained from lung tissue and are said to have a fibrinolytic effect.

4.1.71.2 Publication

In 1985 **Cappelli** and co-workers [115] reported on the effect of defibrotide (polydeoxynucleotide from lung tissue) on the blood and plasma viscosities and erythrocyte filtrability. An infusion of 1800 mg defibrotide in 100 ml saline solution was administered over a period of 30 min. The authors reported on a reduction of the blood viscosity, the plasma viscosity, and the fibrinogen concentration, as well as an improvement in erythrocyte filtrability. They attributed the improvement in the flow properties of blood primarily to a decrease in the fibrinogen concentration (abstract, no details).

4.1.72 l-Thyroxine

4.1.72.1 Introduction

L-Thyroxine is a thyroid hormone.

4.1.72.2 Publications

In 1985 **Larsson** and co-workers [552] described a reduction of blood and plasma viscosities as a result of the administration of thyroxine in patients with hypothyrosis with autoimmune thyroiditis. Blood viscosity was measured with a rotational viscosimeter at 24 °C, plasma viscosity with an Ostwald capillary viscosimeter at 24 °C. The therapy with l-thyroxine led to a decrease in the blood and plasma viscosity at low shear rates. A decreased erythrocyte aggregation was suspected as a result of the decrease in the blood cell sedimentation rate.

Note: No measurements of further hemorheologic parameters.

4.2 Simultaneous Therapy with Hemorheologically Active Substances

4.2.1 Introduction

Simultaneous rheologic therapy means the utilization of two or more rheologically active therapies at the same time (see Sect. 2.3.3). The consideration that prompted **Zederfeldt,** *in 1963, to carry out a venesection within the framework of a hemodilution therapy [930] was probably to let the hematocrit reduction become much more marked than it would have been possible with only one of these methods.*

This attempt at simultaneous therapy later received special significance as the so-called isovolemic hemodilution. With the same goal in mind other authors then combined plasma substitutes with greater amounts of physiologic saline solution, which today must also be considered as a simultaneous rheologic therapy.

In 1968 **Azzena** *and co-workers [36] may have had similar ideas when, for the first time, they treated plasma hyperviscosity due to multiple myeloma both with the rheologic measure of plasmapheresis and with a simultaneous medicinal therapy using steroids and cytostatics. The aim of both therapeutic measures was the reduction of the paraprotein concentration in the plasma.*

The case is somewhat different with the simultaneous therapy suggested in 1976 and 1977 for the treatment of chronic arterial occlusive diseases using hemodilution with plasma substitutes and simultaneous defibrinogenation with ancrod [267, 274]. Here the reasoning was that, in addition to the hematocrit reduction resulting from the decreased whole blood viscosity following the infusion of low molecular weight dextran, a specific decrease in the plasma fibrinogen concentration should decrease particularly the erythrocyte aggregation and reduce plasma viscosity. Thus, in this case the aim of the two measures carried out simultaneously was different from a rheologic point of view: hemodilution reduced the hematocrit and the whole blood viscosity, and defibrinogenation decreased erythrocyte aggregation and plasma viscosity.

This combination of therapeutic measures which influence various hemorheologic parameters of the flow properties of blood was attempted in the following years with a whole range of substances. Thus, hemodilution with plasma substitutes was used in combination with pentoxifylline, which was used in particular to normalize the decreased erythrocyte deformability in patients with vascular diseases. Further combinations of various kinds of rheologically active substances will be discussed in the following section. The order is again chronologic, i.e., the combination of venesection and hemodilution with plasma substitutes, which has been known since 1963 [930], will be treated first.

As mentioned before (see Sect. 2.3.3), the evaluation of rheologic changes in the blood resulting from the simultaneous application of two or more medications or rheologic measures is much more difficult than that of single substances. It remains true that the clinical effect does not have to be better if two different rheologic drugs are combined. The clinical aim of simultaneous therapy and the quantification of the results in the sense of clinical effects and clinical efficacy are in principle the same as with monotherapy with rheologically active substances (see Sect. 2.1).

If one compares the number of publications on simultaneous hemorheologic therapies with the great number of studies on the rheologic effects of single substances, it becomes obvious that the therapeutic possibilities of a simultaneous or combined hemorheologic therapy have neither been completely thought through nor have they been fully utilized clinically. In my opinion, there are still important tasks for the future in this field.

4.2.2 Venesection and Hemodilution with Plasma and Plasma Substitutes

4.2.2.1 Venesection and Dextran Solutions

4.2.2.1.1 Publications

In 1963 **Zederfeldt** [930], from Gelin's team, studied a patient with severe burns whose whole blood viscosity was measured with a Brookfield micro-cone-plate viscosimeter at shear rates of 6–60 s^{-1} and at a temperature of 37 °C. The viscosity values were clearly increased, especially at low shear rates, and were almost normalized by the venesection of 1000 ml blood and the reinfusion of an equivalent volume of low molecular weight dextran (no concentration data) (first isovolemic hemodilution). At the same time, there was a decrease in the hematocrit. On the other hand, blood transfusions increased both the hematocrit and the blood viscosity.

Note: No measurements of further rheologic parameters.

Clinical findings: The patient's anuria disappeared.

A further report on the hemorheologic effect of a combination of venesection and the application of 10% mid-molecular weight dextran followed in 1971 by **Lewis** and co-workers [571]. Blood viscosity was measured in four patients with polycythemia using a Brookfield micro-cone-plate viscosimeter at a shear rate of 6 s^{-1}. Simultaneously with the reduction of the increased hematocrit values, the authors found a decrease in the blood viscosity. In addition, a slightly increased muscular capillary flow rate was found using a double isotope technique (^{133}Xe, ^{51}Cr-EDTA).

Note: No temperature data; no measurement of further rheologic parameters.

In 1976 **Rieger** and co-workers [727, 728] reported on the treatment of ten patients with chronic occlusive arterial disease in stage IV according to Fontaine. Blood viscosity was measured with a rotational viscosimeter (no data on brand name) at various shear rates of 15–78 s^{-1}, and erythrocyte aggregation was determined quantitatively with a photometric method. Within the framework of a 4-week treatment period 500 ml blood were extracted per day until the hematocrit had reached a value of 30%. This volume was substituted each time with an immediate infusion of 300–500 ml 10% low molecular weight dextran solution, and the autologous plasma that was separated by centrifugation was reinfused. At the end of treatment, there was a significant decrease in the whole blood viscosity, especially at low shear rates.

Note: No temperature data; no measurements of plasma viscosity or erythrocyte filtrability.

Clinical findings: All patients felt an improvement in their subjective symptoms, such as, for example, pains at rest. The local clinical evidence did not change uniformly. In four patients the ulcerations healed, in two other patients the results were satisfactory, and in two patients the therapy was not obviously successful.

In 1979 **Köhler** and co-workers [517] treated seven patients with chronic occlusive arterial disease in stage IV according to Fontaine. Blood viscosity was measured at various shear rates of 15–78 s^{-1}. In this acute isovolemic hemodilution the blood was first removed and then autologous plasma in combinated with low molecular weight dextran was reinfused until after 4–7 days, the hematocrit had reached a value of 30%. Simultaneously with the hematocrit reduction there was a decrease in the blood viscosity, which was most marked at low shear rates.

Note: No data on the instrument used and the temperature of the measurements; no measurements of further rheologic parameters.

In 1979 **Rieger** and co-workers [729] studied 25 patients with peripheral arterial occlusive disease in stage IV according to Fontaine, whose clinical symptoms could not be improved to any extent with conventional therapy. For a period of 8–10 weeks they were treated by subacute isovolemic hemodilution. After venesection and the administration of 500 ml 10% dextran solution and the later reinfusion of the plasma separated by means of centrifugation, the hematocrit was set at a value of 31% within 3–4 days. Blood viscosity was measured with a Brookfield micro-cone-plate viscosimeter, (with a Couette chamber and a guard ring to protect against surface artifacts), at various shear rates of 15–74 s^{-1}, and in addition with a Contraves Low Shear 100 coaxial cylinder viscosimeter at various shear rates of 0.05–128 s^{-1}. Plasma viscosity was determined with a Harkness capillary viscosimeter. The quantitative determination of erythrocyte aggregation was performed photometrically. During the course of treatment there was a significant decrease in the blood viscosity at all shear rates, the viscosity drop being most marked at low shear rates. Plasma viscosity did not change significantly. On the other hand, there was a decrease in the erythrocyte aggregation.

In 1981 a publication appeared by **Schmid-Schönbein** and **Rieger** [763] with similar results.

Note: No temperature data for the viscosity measurements; no measurement of erythrocyte deformability.

Clinical findings: Depending on the number and type of occlusions, there was a more or less favorable result. The skin lesions of the upper extremities healed completely, contrary to those of the lower extremities where a favorable result could be gained only in single or double occlusions in a vessel segment.

In 1981 **Angelkort** and co-workers [18] reported on studies of a combined therapy with venesection and the administration of low molecular dextran in 16 patients with chronic arterial occlusive disease in stages II–IV according to Fontaine. There were rheologic changes as a result of the venesections and the daily infusions of 500 ml dextran 40 000 solution for a period of 12–14 days. Blood viscosity, measured with a Wells-Brookfield viscosimeter, showed a more marked decrease in the whole blood viscosity at low shear rates than at higher shear rates. In addition, the authors found that erythrocyte filtrability tended to decrease, as measured through Nucleopore filters of a pore diameter of 5-μm at a pressure of 20 cm H_2O.

Note: No measurement of the erythrocyte aggregation.

Clinical findings: In ten patients in stages III and IV and in three patients in stage IIb there was an improvement in the symptoms; in one patient in stage II and three patients in stage IIb the conditions remained unchanged.

In 1981 **Borgolte** [83] studied the effect of a simultaneous therapy with venesection and an infusion of dextran 40 solution on the flow properties of the blood in patients with chronic obstructive respiratory diseases and hemoglobulin values of at least 18%. In these patients 500 ml blood was removed by venesection and substituted with the same amount of dextran 40 solution (10%). As a comparison, physiologic saline solution was also given. There was a clear reduction of the whole blood viscosity at a shear rate of 4.6 s^{-1} for the dextran group; while these differences were also found in the saline group immediately after the end of infusion, they disappeared in the further course of the experiment. In the dextran group, on the other hand, there was a decrease in the whole blood viscosity even 24 h after the end of the combination therapy.

Note: No measurements of other rheologic parameters; no exact data as to the viscosimeter used.

In 1983 **Wolfe** and co-workers [922] reported on the effect of a combination therapy with venesection and the administration of dextran 70 solution in six patients with stable intermittent claudication. Blood removal and the rheologic measurements were carried out before, 3 days after having reached a hematocrit of 35%, and 1 month later. The authors found a statistically significant decrease in the whole blood viscosity at low and high shear rates, but no decrease in plasma viscosity (abstract, no details).

Clinical findings: The ergometric measurements of the pain-free walking distance showed no improvement in most cases.

The results of these studies were published elsewhere in 1985 [923].

In 1983 **Guerrini** and co-workers [409] studied the effect of normovolemic hemodilution on the flow properties of blood in patients with chronic arterial occlusive disease. After minor venesections, low molecular dextran was infused to achieve a decrease of the hematocrit value to about 37%. Corresponding hemodilutions were carried out daily. The whole blood viscosity was studied with a Wells-Brookfield ¼ LVT viscosimeter at 37 °C. Erythrocyte filtrability was determined according to the method of Reid and co-workers. In a 7-day study in ten patients as well as a long-term therapy in 8 patients who underwent an isovolemic hemodilution once a week, there was a statistically significant decrease in the hematocrit values and in the whole blood viscosity, while erythrocyte filtrability was not influenced.

Kuznetsov and co-workers [531] studied the effect of dextran solutions and albumin solutions within the framework of normovolemic hemodilution in patients with obstructive jaundice and found that some of the hemorheologic abnormalities could be normalized (original in Russian).

In 1984 **Strand** and co-workers [833] studied the effect of a combination of venesection and an infusion of dextran 40 solution on the hemorheologic parameters of patients with acute shock. Fifty patients were subjected to hemodilution, 50 others acted as controls. Hemodilution was started with a venesection of 200 ml, while 150–200 ml 10% dextran 40 solution was infused. Later 300–350 ml dextran solution was administered for the next 2–4 h. On the following days, further venesections were performed and, if needed, dextran 40 infusions in order to reduce the hematocrit from on average 43 to 37%. In seven patients the whole blood viscosity was measured with a Wells-Brookfield viscosimeter. In agreement with the hematocrit decrease, there was already a statistically significant decrease in the whole blood viscosity on the 1st day, the lowest values being achieved after 1 week.

Note: No data on the temperature; no measurements of further rheologic parameters.

Clinical findings: The authors found a marked clinical improvement in the hemodilution group; two patients manifested an anaphylactoid reaction.

Chevreaud and co-workers [123] reported on the effect of normovolemic hemodilution in patients with severe and most severe chronic arterial occlusive disease. The daily venesections of 500 ml were followed by infusions of 200 ml 4% human albumin solution and 250 ml dextran 40 solution until a hematocrit of about 30%–35% was achieved. The treatment was continued over a period of 8 weeks. The authors found an improvement in the hemorheologic parameters measured (whole blood viscosity, erythrocyte aggregation) (abstract, no details).

Clinical findings: The authors found that the pain-free walking distance was doubled in the patients with intermittent claudication, in patients with stage III and IV disease there was an improvement in 50% of the cases.

In 1985 **Danilov** and **Makarevich** [154] studied the effect of isovolemic hemodilution with reopolyglucine on the flow properties of blood in 138 patients with chronic obliterative bronchitis and found a statistically significant decrease in the whole blood viscosity (original in Russian).

Clinical findings: There was an improvement in the clinical symptoms.

In 1985 **Hansen** and co-workers [416] reported on the effect of isovolemic hemodilution on the vision of patients with ischemic and nonischemic occlusions of the retinal veins. After the venesections, plasma and dextran solutions were substituted. The authors found a clinical improvement in 80% of the patients with ischemically induced impaired vision, as opposed to the 20% of the patients with nonischemically induced impaired vision. These findings correlated with the hemorheologic parameters (abstract, no details).

4.2.2.1.2 Discussion

Simultaneous venesection and an infusion with dextran solution leads to a particularly strong decrease in the hematocrit values. This is understandable since with venesection the total number of circulating erythrocytes is reduced and “endogenous” hemodilution occurs as a result of the influx of extravascular fluid. If mostly hyperoncotic colloidal solutions (here dextran solutions) are added as well, this leads to a further hemodilution. This is on the one hand, the result of the exogenous afflux of erythrocyte-free fluid and, on the other hand, again the result of the influx of extravascular fluid thanks to the hyperoncotic properties of the mostly high percentage plasma substitutes. In the case of a combination of venesection and dextran solution, all the investigators unanimously found a reduction of the whole blood viscosity, while the plasma viscosity remained, on the whole, unchanged.

Rieger and co-workers [729] as well as *Chevreaud* and co-workers found a reduction of the erythrocyte aggregation. *Angelkort* and co-workers only found a slight effect on erythrocyte filtrability, while *Guerrini* and co-workers found no effect on this parameter. Accordingly, in combinations of venesection and dextran infusions, there was mainly a hematocrit-dependent reduction of the whole blood viscosity.

4.2.2.2 Venesection and Plasma Protein Solutions

4.2.2.2.1 Publications

In 1967 **Baum** [52] reported on the improvement in the flow properties of blood in newborns with strongly raised hematocrit values and respiratory problems. Blood was removed from the babies, and the equivalent volume of human plasma was reinfused. The blood viscosity measurements were carried out with a rotational viscosimeter of unknown type at a shear rate of 11.5 s^{-1}. There was a decrease in the raised blood viscosity values parallel to a decrease in the hematocrit value.

Note: No temperature data for the viscosity measurements; no measurements of further rheologic parameters.

Clinical findings: The cerebral symptoms (cramps and lethargy) improved. Respiratory problems and cyanotic symptoms were influenced favorably.

In 1967 **Putnam** and co-workers [716] followed with a report on a child with strongly raised hematocrit and pulmonary atresia. A total of 450 ml blood was removed and replaced with 450 ml human plasma. Blood viscosity was measured with a Brookfield micro-cone-plate viscosimeter at 37 °C and at shear rates of 1.15–46 s^{-1}. The authors found a decrease in the blood viscosity.

Note: No measurement of further rheologic parameters.

In 1969 **Mackintosh** [605] studied newborns with raised viscosity and hematocrit values. A total of 20–30 ml blood per kilogram body weight were removed from the babies, and the blood deficit was replaced with autologous plasma. Blood viscosity was measured with a Brookfield LTV micro-cone-plate viscosimeter at shear rates of 1.16–232 s^{-1}. The author found a strong decrease in the blood viscosity and the hematocrit.

Note: No measurements of further rheologic parameters; no temperature data for the blood viscosity measurements.

In 1970 **Kontras** and co-workers [522] treated children with inborn valvular insufficiency and raised viscosity values by means of venesection and simultaneous replacement of the blood removed by autologous plasma. As a result, there was both a reduction of the hematocrit and of the blood viscosity (Brookfield micro-cone-plate viscosimeter, 37 °C, 11.5 s^{-1}). Similar results in newborns with polycythemia were published by **Kontras** [523] in 1972.

Note: No measurements of further rheologic parameters.

In 1970 **Rosenthal** and co-workers [741] reported on the treatment of patients with secondary polycythemia and congenital valvular insufficiency. The blood removed was replaced by a reinfusion of autologous plasma. Simultaneously with a drop in the hematocrit, there was a reduction of the blood viscosity (GDM rotational viscosimeter, 38 °C, shear rates of 0.1–100 s^{-1}) as well as the yield stress.

Note: No measurements of further rheologic parameters.

Clinical findings: The patients reported a decrease in headaches and dyspnea.

In 1972 **Walker** and **Mackintosh** [891] studied newborns with polycythemia and raised blood viscosity values. Isovolemic hemodilution was carried out. A total of 20–30 ml blood per kilogram body weight was removed and replaced by an equivalent volume of autologous plasma. The blood viscosity of the heparinized blood samples was measured with Brookfield micro-cone-plate viscosimeter at 37 °C and at shear rates of 1.16–232 s^{-1}. Simultaneously with the drop in the hematocrit, there was a reduction of the blood viscosity.

Note: No measurements of further rheologic parameters.

Clinical findings: As a result of the treatment, the plethora, cyanosis, cerebral symptoms, and respiratory distress syndrome disappeared.

In 1973 **Gross** and co-workers [393] treated polycythemic newborns by removing 50–60 ml blood, followed by an infusion of autologous plasma. The blood viscosity of the heparinized blood samples was measured with a Brookfield micro-cone-plate viscosimeter at 37 °C and shear rates of 11 and 106 s^{-1}. Erythrocyte deformability was determined according to Gross (no details). At the same time as the reduction of the blood viscosity, there was an improvement in erythrocyte deformability.

Note: No measurements of further rheologic parameters.

Clinical findings: The investigators reported that the cerebral symptomology, plethora, cyanosis, and respiratory distress syndrome disappeared, and that cardiomegaly improved.

In 1974 **Bergqvist** and **Zetterström** [67] treated newborns with raised hematocrit values by means of isovolemic hemodilution. A total of 30 ml blood was removed and replaced by 20% serum albumin solution, 5% glucose solution, and isotonic saline solution. Blood viscosity was measured with a rotational viscosimeter of an unknown type. The authors found a decrease in the blood viscosity and the hematocrit values.

Note: No measurements of further rheologic parameters; no details on the viscosimeter used.

In 1975 **Klövekorn** and co-workers [512] treated anesthetized patients before an operation. A total of 1500–2000 ml blood was removed, and this volume was replaced by a 5% albumin solution in order to avoid the necessity of blood transfusions after the operation. The blood extracted was kept and used later for a necessary homologous blood transfusion. The authors reported that there was a significant drop in the blood viscosity, which was especially marked at low shear rates.

Note: No mention of the type of rotational viscosimeter used; no measurements of further rheologic parameters.

In 1981 **Marazzini** and co-workers [612] studied eight patients with chronic pulmonary insufficiency and high hematocrit values. They measured the whole blood viscosity using an LVT microviscosimeter and erythrocyte filtrability according to the whole blood method of Reid and co-workers. A total of 500 ml blood was removed from the patients at 12-h intervals, and this volume was replaced by plasma and saline solution until a hematocrit value of 45% ± 3% was reached. There was a decrease in the whole blood viscosity at various shear rates during the course of the normovolemic hemodilution as well as an improvement in the filtration values.

Note: No measurement of plasma viscosity and erythrocyte aggregation.

Clinical findings: The authors reported an improvement in the clinical symptoms.

In 1983 and 1984 **Wedzicha** and co-workers [909, 910] studied the effect of isovolemic hemodilution using erythrapheresis in ten patients with chronic obstructive bronchitis whose hematocrit values lay over 45% in men and over 47% in women. A Haemonetics instrument was used and the blood samples were anticoagulated with heparin. The deficit fluid after the removal of blood was compensated for by an infusion of human plasma protein fractions. Blood sampling for the rheologic studies was carried out before and after erythrapheresis. The whole blood viscosity was measured with a Brookfield LVT viscosimeter at 37 °C and at shear rates of 230 and 23 s^{-1}. There was a significant decrease in the whole blood viscosity at both shear rates measured.

Note: No measurements of further rheologic parameters.

Clinical findings: Most of the patients showed an improvement in the clinical symptoms after erythrapheresis.

In 1983 **Wintrich** and co-workers [918] carried out a combined rheologic therapy with venesection (500 ml) and replacement of this volume by albumin solution until a hematocrit of 35% was reached. The hematocrit and the whole blood viscosity showed a therapy-induced reduction of the values (abstract, no details).

Kuznetsov and co-workers [531] studied the effect of dextran solutions and albumin solutions within the framework of normovolemic hemodilution in patients with obstructive jaundice and found that some of the hemorheologic abnormalities could be normalized (original in Russian).

Chevreaud and co-workers [123] reported on the effect of normovolemic hemodilution in patients with severe chronic arterial vascular disease. The daily venesections of 500 ml were followed by infusions of 200 ml 4% human albumin solution and 250 ml dextran 40 solution until a hematocrit of about 30%–35% was achieved. The treatment was continued over a period of 8 weeks. The authors found an improvement of the hemorheologic parameters measured (whole blood viscosity, erythrocyte aggregation) (abstract, no details).

Clinical findings: The authors found that the pain-free walking distance of the patients with intermittent claudication was doubled; in patients with stage III and IV disease there was an improvement in 50% of the cases.

In 1985 **Clivati** and co-workers [137] described the effect of isovolemic hemodilution in patients with secondary polyglobulism due to chronic pulmonary insufficiency. A total of 500 ml blood was removed from the patients and replaced by isotonic saline solution, albumin solution and autologous plasma. The whole blood viscosity and plasma viscosity were measured with a Brookfield LVT/CP viscosimeter; erythrocyte aggregation was determined photometrically and erythrocyte filtrability according to the whole blood method of Reid and co-workers. There was a statistically significant decrease in the hematocrit values (on average from 55.5% to 46.5%), a statistically significant reduction of the whole blood viscosity and an improvement in erythrocyte filtrability, while the plasma viscosity remained unchanged.

In 1985 **Toti** and co-workers [862] reported on the effect of normovolemic hemodilution in patients undergoing surgery. Before the operation the hematocrit was lowered to values of around 30%–35% by means of venesection and the infusion of autologous plasma and isotonic saline-albumin solution. This hemodilution led to a decrease in the blood viscosity erythrocyte deformability, and erythrocyte aggregation. Postoperatively the plasma viscosity and the erythrocyte aggregation rose again (abstract, no details).

In 1985 **Hansen** and co-workers [416] reported on the effect of isovolemic hemodilution on the vision of patients with ischemic and nonischemic occlusions of the retinal veins. After the venesections plasma and dextran solutions were substituted. The authors found a clinical improvement in 80% of cases with ischemically induced impaired vision, contrary to the 20% of cases of nonischemically induced impaired vision. These findings correlated with the hemorheologic parameters (abstract, no details).

4.2.2.2.2 Discussion

If after venesection or erythrapheresis the deficit intravasal volume is replaced by plasma protein solutions (isovolemic hemodilution), there follows similar to the combination of venesection and dextran solution a reduction of the hematocrit values with a consecutive decrease in the whole blood viscosity.

Gross and co-workers, *Marazzini* and co-workers, and *Clivati* and co-workers found an improvement in erythrocyte filtrability. *Toti* and co-workers and *Chevreaud* and co-workers were able to show a decrease in erythrocyte aggregation as well. In addition, the authors unanimously reported clinical improvements as a result of this therapy.

4.2.2.3 Venesection and Hydroxyethyl Starch Solutions

4.2.2.3.1 Publications

In 1978 **Watzek** and co-workers [903] reported on isovolemic hemodilution with hydroxyethyl starch solution before vascular surgery in patients with arterial occlusions. Blood viscosity, relative blood viscosity and plasma viscosity were measured with a Brookfield micro-cone-plate viscosimeter (47–230 s^{-1}). One hour after premedication, approximately 1200 ml blood was removed from the patients and replaced by 1000 ml 6% hydroxyethyl starch solution (MW 200 000) and 500 ml 3.6% human albumin solution. The authors found a significant decrease in the relative blood viscosity with an unchanged plasma viscosity.

Note: No measurement of further rheologic parameters; no temperature data.

In 1981 **Angelkort** and **Frömel** [19] studied the effect of a simultaneous therapy with venesection (450 ml whole blood) and daily infusions of 500 ml 10% hydroxyethyl starch solution (MW 200 000) in ten patients with chronic occlusive arterial disease at different stages. The hemorheologic data were studied before and 12 h after venesection, as well as 14 days after the end of the hemodilution therapy and again 7 days later. Simultaneously with the reduction of the hematocrit values of on average, 46.5% to 36.7%, there was a reduction of erythrocyte aggregation, a slight improvement in erythrocyte filtrability, as well as a slight increase in plasma viscosity (abstract, no details).

In 1984 **Kiesewetter** and co-workers [493] described the effect of a 6-week isovolemic hemodilution therapy with mid-molecular weight hydroxyethyl starch solution in patients with chronic occlusive arterial disease in stages II and III. A total of 300–400 ml whole blood was removed, and the corresponding amount of hydroxyethyl starch (mid-molecular weight 200 000/0.5) was infused. The yield shear stress, erythrocyte aggregation (light transmission method), erythrocyte filtrability (SER), as well as plasma viscosity (capillary viscosimeter) were studied. The pain-free walking distance was also tested on a treadmill ergometer. There was a significant decrease in the flow shear stress and erythrocyte aggregation. Plasma viscosity also decreased, especially in patients who had previously shown raised values.

Clinical findings: The authors found a statistically significant extension of the pain-free walking distance.

In 1986 **Ernst** and co-workers [333] published a double-blind study on the effect of isovolemic hemodilution in patients with intermittent claudication. If the hematocrit of these patients was higher than 45%, then 500 ml blood was removed and replaced by a 10% hydroxyethyl starch solution (mid-molecular weight 200 000). This procedure was carried out three times per week for 3 weeks. After a wash-out period the patients received a placebo infusion. The blood viscosity was determined with a rotational viscosimeter at 37 °C, and the hematocrit with a microhematocrit centrifuge. There was a statistically significant reduction of the whole blood viscosity and the hematocrit.

Clinical findings: The pain-free walking distance increased in the hemodilution group but not in the placebo group.

Preliminary results of this study were also presented elsewhere [329].

In 1986 **Kiesewetter** and co-workers [501] studied the effect of 10% low molecular dextran solution compared to low molecular hydroxyethyl starch (both MW 40 000) during a 14-day hemodilution in combination with physical therapy in patients with chronic arterial occlusive disease. Erythrocyte aggregation, erythrocyte rigidity, and the yield shear stress were measured. After the administration of low molecular hydroxyethyl starch solution, there was a distinct reduction of the erythrocyte aggregation, blood viscosity, and plasma viscosity. Erythrocyte filtrability remained uninfluenced. With the use of dextran there was a marked decrease in erythrocyte rigidity and a slight decrease in erythrocyte aggregation, while the plasma viscosity and yield shear stress remained uninfluenced by dextran.

4.2.2.3.2 Discussion

The combined therapy of venesection and the intravenous infusion of hydroxyethyl starch solution led, as was expected, to a particularly marked decrease in the hematocrit values and consequently to a reduction of the whole blood viscosity. *Watzek* and co-workers found no change in plasma viscosity, while *Angelkort* and *Frömel* found a slight increase in this parameter, and *Kiesewetter* and co-workers found a reduction. The differences may be attributed to the fact that *Watzek* and co-workers carried out a short-term study within a matter of hours, while the studies of *Angelkort* and *Frömel* as well as of *Kiesewetter* and co-workers were extended over a period of two weeks. In addition, *Kiesewetter* and co-workers found a reduction of the erythrocyte aggregation.

4.2.2.4 Venesection and Isotonic Saline Solution

4.2.2.4.1 Publications

In 1985 **Porro** and co-workers [708] carried out isovolemic hemodilution in patients with chronic obstructive pulmonary diseases by substituting sampled blood with the same amount of isotonic saline solution. The authors found a reduction of the whole blood viscosity and a decrease in the hematocrit; erythrocyte deformability, on the other hand, was not influenced by this procedure (abstract, no details).

In 1985 **Ferretti** and co-workers [349] studied the effect of normovolemic hemodilution, which was carried out via venesection and subsequent infusion of isotonic saline solution, in patients with secondary polyglobulism due to chronic pulmonary diseases. The only hemorheologic parameter studied was erythrocyte filtrability (according to Reid and co-workers. There was a reduction of the hematocrit from an average of 53.2% to 45.6% and a slight but statistically significant improvement in the erythrocyte filtration values.

Note: No measurements of further hemorheologic parameters.

4.2.2.4.2 Discussion

Porro and co-workers as well as *Ferretti* and co-workers combined venesection with the infusion of equal amounts of physiologic saline solution and called this isovolemic or normovolemic hemodilution. Strictly this is not really isovolemic hemodilution, since the infused isotonic saline solution remains in the intravasal space for only a short time. According to their hemorheologic effect, these studies should actually be included in Sect. 4.1.2 (Venesection, Erythrapheresis).

4.2.3 Isovolemic Hemodilution and Defibrinogenation

4.2.3.1 Publications

In 1976 and 1977 **Ehrly** and **Saeger-Lorenz** [267, 274] reported on the combined hemodiluting and defibrinogenizing therapy in therapy-refractory patients with multiple occlusions in stages III and IV of chronic arterial occlusive disease. During the first 3 days blood was removed from the patients and replaced with 10% low molecular weight dextran. Following this hemodilution the patients received ancrod. Before the administration of this snake poison enzyme, the blood viscosity and, to a small extent, also the plasma viscosity had decreased as a result of the therapy with low molecular dextran alone. After the administration of ancrod, there was, in addition, a more marked reduction of the plasma viscosity and a clear decrease in erythrocyte aggregation (quantitative photometric method).

Note: No measurements of erythrocyte filtrability; no details on the viscosimeter used.

Clinical findings: The administration of dextran solution with the follow-up administration of ancrod led to a subjective improvement in the pain symptoms.

In 1981 **Böhme** and **Everts** [74] studied ten patients with chronic arterial occlusive disease in stages III and IV. After an isovolemic hemodilution lasting about 14 days (removal of smaller amounts of blood and replacement with low molecular weight dextran solution) the patients underwent an additional defibrinogenizing therapy with ancrod and batroxobin. The whole blood viscosity and the plasma viscosity were determined with a Contraves LS 30 viscosimeter before and after therapy. There was a clear reduction of the blood and plasma viscosities at low shear rates.

Note: No data on the temperature of measurement; no studies of further rheologic parameters.

In 1981 and 1983 **Hossmann** and co-workers [452, 455, 456] reported on a randomized blind study of 30 patients with acute stroke who were treated with either an administration of low molecular dextran and mannitol or, in addition, with the defibrinogenizing enzyme ancrod. The blood and the plasma viscosities were measured with a Couette viscosimeter at various shear rates. The reduction of the blood viscosity, especially at low shear rates, was considerably more marked in the group treated with ancrod than in the control group, which received only the hemodilution treatment.

Note: No measurements of further rheologic parameters.

Clinical findings: There was an improvement in the clinical symptoms following the combined therapy of hemodilution and defibrinogenation.

In 1982 **Hossmann** and **Auel** [454] reported on 12 patients with chronic arterial occlusive disease of the lower extremities in stages II–IV according to Fontaine; they were divided into two groups. In the first group the fibrinogen level was first decreased to 100 mg% with the administration of ancrod, followed by isovolemic hemodilution with venesections and reinfusions of plasma and low molecular dextran solutions. In the second group isovolemic hemodilution was carried out over 14 days, followed by ancrod therapy to reduce the fibrinogen concentration. The apparent whole blood viscosity of the heparinized blood was determined with a Couette rotational viscosimeter; the plasma viscosity was measured with a capillary viscosimeter. In both groups a significant reduction of the whole blood viscosity could be shown at all shear rates; however, it was most marked during the combination of both therapy principles.

Note: No measurements of further rheologic parameters.

Clinical findings: The authors reported that the pains decrease significantly.

In 1985 **Böhme** and **Everts** [75] combined defibrinogenizing substances with the administration of mid-molecular weight hydroxyethyl starch solution (MW 200 000/0.5) in patients with peripheral arterial occlusive diseases in stages II–IV. They found a decrease in the blood viscosity and erythrocyte aggregation.

4.2.3.2 Discussion

If isovolemic hemodilution is carried out with venesection on the one hand and with the infusion of colloidal plasma substitutes on the other hand and if, afterwards or at the same time, the fibrinogen concentration of the plasma is lowered with defibrinogenizing measures such as the administration of ancrod, then there follows not only a decrease in the whole blood viscosity (as with an isovolemic hemodilution alone; see Section 4.2.1), but there is also, as expected, a reduction of plasma viscosity and erythrocyte aggregation. Thus we have the combination of two therapeutic principles: hemodilution (with reduction of the hematocrit) and defibrinogenation (with reduction of the fibrinogen concentration). The three groups, who published studies on this type of simultaneous hemorheologic therapy, found an improvement in the clinical symptoms.

4.2.4 Plasmapheresis and Chemotherapy

4.2.4.1 Publications

In 1968 **Azzena** and co-workers [36] reported on the rheologic effects of plasmapheresis and chemotherapy with steroids and cytostatica in three patients with multiple myeloma. The relative serum viscosity (H_2O = 1) was measured with an Ostwald capillary viscosimeter at 24 °C. Synchronously with the normalization of the pathologic protein pattern, a decrease was observed in the serum viscosity.

Note: No measurements of further rheologic parameters; the viscosity measurements were carried out at 24 °C.

Clinical findings: There was an improvement in the retinopathies, the nosebleeds stopped, and neurologic pains (e.g., back pains) decreased.

In 1968 **Lawson** and co-workers [553] treated three patients with Waldenström's disease. Over a period of 1–2 weeks, 4–6 l plasma per patient were removed by means of plasmapheresis, and the erythrocytes were reinfused. In addition, the patients received medications such as hydrocortisone, cyclophosphamide, and others over a longer period. Serum viscosity was measured relative to water (H_2O = 1). Synchronously with the drop in serum viscosity, there was a decrease in the total proteins as well as in the gamma globulins.

Note: Neither the type of viscosimeter used nor the temperature of the measurements were given.

Clinical findings: Bleeding of the gastric mucosa as well as retinal hemorrhages improved with this therapy.

In 1972 **MacKenzie** and **Fudenberg** [604] carried out an alkalizing therapy with chlorambucil, cyclophosphamide, and melphaline as well as a therapeutic plasmapheresis in 40 patients with Waldenström's disease, 13 of whom also had a so-called hyperviscosity syndrome. The relative serum viscosity (H_2O = 1) was measured at room temperature with an Ostwald capillary viscosimeter. In the case of cryoglobulins, the viscosity measurements were carried out at 37 °C. Without exeption, plasmapheresis, which was performed as an initial therapy in all patients with hyperviscosity, led to a drop in serum viscosity as well as in the IgM globulins.

Note: No measurements of further rheological parameters; the viscosity measurements were carried out at different temperatures (20° and 37 °C).

Clinical findings: The bleeding of the gastric mucosa stopped, and the retinopathies improved. Good results were also achieved in patients in coma.

In 1972 **Sugai** [850] treated a patient with multiple myeloma, cyroglobulins, and hyperviscosity. Within the framework of a 2-week plasmapheresis, 3.5 l plasma were exchanged. In addition, cyclophosphamide and prednisone were administered. The relative serum viscosity (H_2O = 1) was measured with an Ostwald capillary viscosimeter at 20° and 37 °C. As a result of plasmapheresis, the macroglobulin level and the cryoglobulin level were reduced.

Note: No measurements of further rheologic parameters.

Clinical findings: There was a decrease in the tendency to bleeding, as well as an improvement in the visual disturbances and in the neurologic symptoms.

4.2.4.2 Discussion

If, in addition to plasmapheresis, cytostatics and steroids are given as a simultaneous or combined therapy in paraproteinemias, then theoretically an induced decrease in the plasma and/or serum viscosity as a result of plasmapheresis should become even more marked. The results of the four study groups who combined plasmaphereis with a chemotherapeutic therapy also point in this direction. All four studies showed a reduction of the plasma and/or serum viscosity; the clinical symptoms could also be improved.

4.2.5 Streptokinase and Defibrinogenation

4.2.5.1 Publications

In 1971 [224], 1972 [226] and in 1974 [228] **Ehringer** and co-workers reported on a combination of streptokinase and ancrod. The measurements of the blood viscosity (a Brookfield micro-cone-plate viscosimeter at shear rates of 11.5–239 s^{-1} and at 37 °C) resulted in a reduction of the viscosity values as a result of streptokinase, which could be reduced even further by the follow-up therapy with ancrod.

4.2.5.2 Discussion

With the combination of streptokinase on the one hand and ancrod on the other hand *Ehringer* and co-workers combined two hemorheologically active substances which are based on the same principle, i.e., the decrease of the fibrinogen concentration in the plasma. As expected with the combination, there was a particularly marked decrease in the viscosity values, particularly at low shear rates.

4.2.6 Electrolyte Solutions and Pentoxifylline

4.2.6.1 Publications

In 1977 **Schulz** and co-workers [778] treated patients with acute and chronic hearing defects using 500 ml Ringer lactate solution to which 400 mg pentoxifylline was added. From the 2nd day 3 × 2 tablets pentoxifylline of 100 mg each were administered orally. Blood viscosity was measured with a capillary viscosimeter at 37 °C. As a comparison, patients were also treated with a therapy scheme excluding pentoxifylline. The authors found a clear decrease in the blood viscosity as a result of the combined therapy.

Note: No correction for hematocrit; no measurements of further rheologic parameters.

Clinical findings: A clinical improvement of the hearing defects was reported during the course of this therapy.

In 1982 **Schneider** and **Kiesewetter** [768] described initial results of a parental administration of pentoxifylline in ischemic insult. Before the treatment with pentoxifylline the hematocrit of all the patients was set at 40% by means of hemodilution. A total of 1200–1500 mg pentoxifylline per day was infused in an electrolyte solution or a 5% levulose solution. Plasma viscosity was measured with a Harkness viscosimeter; erythrocyte aggregation and erythrocyte deformability were determined according to the methods of Kiesewetter and co-workers. The yield shear stress was also studied. With this treatment there was a significant improvement in the yield shear stress as well as in erythrocyte deformability. Plasma viscosity and erythrocyte aggregation did not change significantly.

Clinical findings. In nine out of ten cases there was an improvement in the clinical symptoms; one patient died of the sequelae of a severe brain stem attack.

These results were also published elsewhere [769].

4.2.6.2 Discussion

With the combination of hemodilution on the one hand and the administeration of pentoxifylline on the other hand, two hemorheologic therapy principles are combined: a reduction of the hematocrit (reduction of the whole blood viscosity) and an improvement in erythrocyte deformability. While with a hemodilution one normally thinks of colloidal plasma substitutes or protein solutions, a fast infusion of 500 ml noncolloidal solution (saline solution, Ringer lactate solution, levulose solution) also leads to a reduction of the hematocrit values with corresponding hemorheologic changes (see Section 4.1.3). Quite often so-called vasoactive substances are applied in greater or lesser volumes of intravenously administered saline solution or Ringer lactate solution; however, the publications available on this subject are rare.

Thus *Schulz* and co-workers found a reduction of the whole blood viscosity after the administration of 500 ml Ringer lactate solution containing 400 mg pentoxifylline. The reduction of the whole blood viscosity may be mainly attributed to hemodilution by the carrier solution. In the publication of *Schneider* and *Kiesewetter* pentoxifylline was not only given in a 5%

levulose solution, but before the administration of the medication the hematocrit was set to a value of 40% by means of an isovolemic hemodilution. Strictly speaking, three therapeutic principles were combined here: venesection, the infusion of cell-free solutions and the administration of a drug which influences erythrocyte deformability.

4.2.7 Hemodilution and Flavon Glycosides

4.2.7.1 Publications

In 1985 **Anadere** and co-workers [14] reported on the effect of a daily administration of 1000 ml 10% low molecular dextran (molecular weight 40 000) with the addition of 100 mg *Ginkgo biloba* extract. Another group just received 1000 ml dextran solution per day. The patients were suffering from acute stroke. The viscoelasticity of the blood was measured with an oscillating capillary rheometer at 23 °C, as was the plasma viscosity. The hematocrit was determined with a microhematocrit centrifuge. The authors found a statistically significant decrease in the viscoelasticity of the blood 5 and 10 days after the begining of therapy, while in the control group showed no statistically significant changes. The plasma viscosity rose statistically significantly in the control group; in the group treated with low molecular dextran plus *Ginkgo biloba* extract there was also an increase, which, however, was not statistically significant.

4.2.8 Urokinase and Human Albumin Solutions

4.2.8.1 Publications

In 1970 **Itoh** and co-workers [468] reported on the effect of a combination of urokinase and serum albumin solution on the flow properties of blood. Patients with diabetic retinopathies and microthrombi of the retinal vessels were given 48 000 IU daily on the 1st 3 days and on the following 3 days 24 000 IU per day. The urokinase was administered intravenously with 200 ml of a serum albumin solution. The authors found a statistically significant decrease in the fibrinogen concentration and in the whole blood viscosity (rotational viscosimeter, 37 °C).

Clinical findings: The retinopathy was improved by the therapy.

4.2.8.2 Discussion

The authors found a decrease in the fibrinogen concentration and in the whole blood viscosity with the simultaneous administration of urokinase and human albumin. The reduction of the fibrinogen concentration can be explained by hemodilution (infusion of human albumin solution) as well as via a fibrinogenolysis. The very low dosages of urokinase indicate that the effect of this substance on the fibrinogen reduction in the studies at hand must have been rather minimal. With normal dosages, on the other hand (see Section 4.1.18), there was a clear effect of the urokinase on the fibrinogen concentration and the corresponding hemorheologic parameters.

4.2.9 Pentoxifylline and Nadolol

4.2.9.1 Publications

In 1983 **Botas** and co-workers [84] reported on the simultaneous administration of nadolol and pentoxifylline in 16 patients with chronic coronary heart disease. The patients received a basic dosage of nadolol for a period of 1 month; then they received either pentoxifylline or placebo. The authors only found an increase in the filtrability of the whole blood in the pentoxifylline group (abstract, no details).

4.2.9.2 Discussion

The simultaneous administration of pentoxifylline on the one hand and of a β-blocker on the other hand could theoretically lead to an addition of hemorheologic effects. The single study available does not suffice to clarify this question.

4.2.10 Pentoxifylline and Piracetam

4.2.10.1 Publications

In 1985 **Ciuffetti** and co-workers [134] described the effect of piracetam on the flow properties of blood in patients with cerebrovascular insufficiency. The whole blood viscosity and the plasma viscosity were measured with a Wells-Brookfield cone-plate viscosimeter at 37 °C; erythrocyte filtrability was determined with a modification of the original whole blood method of Reid. The temperature of the measurements was 37 °C. Piracetam was compared with pentoxifylline, placebo and a combination of piracetam/pentoxifylline (piracetam 3 × 800 mg daily, pentoxifylline 3 × 400 mg daily). The authors found that there was an improvement in erythrocyte deformability which could only be achieved with the combination of both medications.

4.2.10.2 Discussion

The combination of two so-called vasoactive substances in order to influence the flow properties of blood has rarely been studied comparatively. Such a combination would only be reasonable if different hemorheologic goals were in sight.

4.2.11 l-Dopa and Budipine

4.2.11.1 Publications

In 1985 **Ott** and **Lechner** [679] described the effect of a combined therapy with l-dopa (average dosage 300 mg per day) and budipine (0.46–0.92 mg per kilogram body weight) on the whole blood viscosity (measured with a cone-plate viscosimeter at 37 °C and at four different shear rates). Twenty-five patients with Parkinson's disease were treated; there were no significant changes in the whole blood viscosity.

Note: No measurements of other rheologic parameters.

4.2.12 Combinations of Several Hemorheologically Active Drugs

4.2.12.1 Publications

In 1983 **Manrique** [611] reported on 50 patients with acute myocardial infarction who received hydrocortisone, streptokinase, pentoxifylline, low molecular weight dextran, and heparin according to various therapeutic schemes. The hemorheologic changes shown were particularly marked in the patient group in whom hemodilution was combined with defibrinogenation. Especially favorable results were found in the group with simultaneous rheologic therapy (abstract, no details).

In 1983 **Heinen** and co-workers [441] reported on a study in 20 patients with occlusion of the arteria or vena retinalis, who were subjected to various rheologic treatment principles at the same time. The patients received pentoxifylline and prednisolone as a basic therapy: some of the patients were treated with isovolemic hemodilution and the others with hypervolemic hemodilution (500 ml dextran 40 solution). Blood viscosity was studied at various shear rates; erythrocyte deformability was also studied. Depending on the combination of the substances, there were different effects on the flow properties of blood (abstract, no details).

4.2.12.2 Discussion

From time to time a number of therapeutically effective hemorheologic drugs have been used simultaneously. Any conclusions on the results gained can only have limited validity, since nothing can be said as to which of the substances used had an effect and whether these substances perhaps have an additive or even a weakening effect.

4.3 Table of the Rheologic Measures Used by Various Authors with Regard to Diseases or Groups of Diseases

Polycythemia, polyglobulism	327, 871, 786, 523, 857, 858, 815, 460, 120, 656, 890, 547, 899, 88, 740, 38, 55, 804, 428, 430, 651, 463, 649, 380, 52, 716, 605, 522, 741, 891, 393, 67, 612, 909, 910, 137
Heart failure, valvular heart disease	914, 613, 8, 96, 97, 522, 735, 377, 901, 317, 882, 247, 518, 264, 471, 477
Angina pectoris, myocardial infarction, coronary heart disease	667, 173, 659, 745, 698, 739, 629, 544, 545, 189, 546, 505, 843, 543, 189, 243, 60, 61, 62, 86, 63, 64, 856, 888, 27, 598, 841, 611, 129, 130, 177, 507, 358, 633, 795, 854, 358, 911, 802, 896, 897, 898, 525, 777, 668, 161, 162, 794, 525, 870, 84
Cerebral ischemia, cerebral insult	663, 562, 852, 460, 919, 528, 812, 813, 814, 89, 119, 689, 643, 617, 618, 619, 677, 678, 644, 665, 874, 79, 685, 770, 345, 633, 662, 411, 503, 110, 634, 502, 508, 532, 673, 34, 82, 686, 684, 00, 705, 833, 452, 455, 456, 768, 769, 14, 134, 679
Arteriosclerosis, peripheral chronic occlusive arterial disease	578, 5, 519, 12, 462, 463, 121, 519, 793, 427, 658, 351, 133, 80, 112, 111, 81, 13, 676, 421, 423, 426, 692, 25, 539, 540, 501, 505, 92, 892, 893, 580, 356, 301, 235, 42, 255, 228, 229, 883, 759, 223, 253, 875, 238, 244, 224, 245, 246, 249, 248, 227, 251, 256, 263, 254, 252, 245, 879, 877, 860, 203, 205, 269, 260, 776, 206, 202, 207, 465, 586, 929, 591, 506, 115, 597, 904, 155, 767, 156, 620, 835, 22, 421, 422, 424, 257, 70, 445, 878, 818, 819, 881, 475, 444, 425, 262, 266, 880, 514, 775, 279, 283, 598, 751, 15, 16, 609, 18, 17, 554, 30, 132, 194, 403, 407, 834, 691, 694, 837, 869, 838, 655, 839, 20, 473, 195, 811, 330, 331, 24, 365, 816, 21, 627, 696, 842, 201, 536, 7, 72, 866, 755, 166, 185, 905, 906, 405, 187, 1, 2, 272, 509, 33, 515, 323, 184, 186, 160, 744, 406, 697, 118, 664, 535, 873, 110, 817, 752, 494, 482, 483, 293, 32, 836, 718, 448, 209, 791, 621, 693, 701, 563, 196, 661, 742, 747, 354, 107, 453, 299, 300, 82, 876, 404, 58, 927, 391, 163, 894, 77, 322, 624, 789, 628, 170, 363, 695, 328, 40, 567, 703, 660, 568, 727, 728, 517, 729, 763, 922, 923, 123, 409, 918, 333, 493, 267, 274, 74, 454, 75, 226
Raynaud's syndrome	41, 853, 191, 671, 907, 908, 889, 451, 690, 687, 35, 215, 216, 645, 596, 470
Vascular disorders, thromboembolic diseases, thrombosis prophylaxis	376, 76, 276, 495, 42, 255, 475, 307, 497, 352, 584, 585, 47, 859, 760, 560, 69, 595, 510, 511
Shock, trauma, burns	66, 371, 746, 796
Diabetes mellitus (diabetic retinopathy)	657, 122, 771, 44, 364, 835, 415, 555, 655, 805, 806, 807, 113, 73, 457, 458, 904, 48, 49, 59, 362, 737, 873, 769, 140, 139, 559, 801, 822, 218, 219, 681, 478, 479, 743, 480, 481, 169, 558, 6, 884, 706, 592, 867, 868, 848, 583, 680, 562, 533, 561, 750, 556, 496, 416, 468, 441
Hyperlipidemia	372, 348, 623, 653, 654, 505, 843, 190, 804, 700, 782, 26, 799, 515

Essential hypertension	12, 845, 280, 608, 920, 682, 472, 415, 569, 748, 108, 895, 127, 109, 861, 499, 453, 652, 161, 414, 498, 43
Surgery	776, 85, 157, 198, 820, 417, 581, 688, 318, 376, 928, 776, 85, 116, 464, 98, 581, 864, 599, 384, 68, 442, 390, 220, 672, 375, 527, 193, 28, 417, 99, 738, 326, 512, 862
Macroglobulinemia, cryoglobulinemia, multiple myeloma, plasmacytoma	780, 781, 342, 135, 803, 450, 525, 602, 603, 733, 357, 65, 709, 715, 921, 719, 579, 865, 392, 104, 646, 711, 925, 771, 469, 11, 347, 346, 105, 810, 843, 54, 844, 872, 399, 772, 378, 379, 401, 167, 171, 151, 831, 773, 774, 731, 779, 524, 36, 570, 714, 733, 604, 850, 71, 117, 647, 648, 804, 553
Pernicious anemia, sickle cell anemia	526, 486, 487, 429, 788, 138, 651, 474
Pregnancy disorders, eclampsia, contraception	485, 797, 435, 433, 29, 674, 221, 565, 587, 92, 432, 683, 93, 436, 437, 542, 798, 431, 439
Rheumatic diseases	630, 4, 926, 45, 56
Chronic bronchitis, pulmonary hypertension	849, 380, 83, 154, 612, 909, 137, 708, 349
Diseases of the kidney	526, 917, 165, 400, 521, 931

5 Summary

This book was written in order to give a comprehensive critical presentation of publications on therapeutic influences on the flow properties of blood. A short *historical survey* and a glossary of the specific *rheologic terms* is followed by discussions on the possibilities of therapeutic hemorheology, the indications for therapy with rheologically active medications, as well as the pathophysiologic validity of the various subparameters of the flow properties of blood. An enumeration and commentary on in vitro studies with rheologically active medications was avoided on purpose, as was the case with animal experiments and studies on healthy subjects. The significance of the results of rheologic measures in patients is also described critically within the framework of the clinical aims for using such therapeutics; possibilities are shown for objectivizing the effect of rheologic medications.

In the *general part* special emphasis is placed on the fact that the measurement of the rheologic properties of the blood and a corresponding improvement of the parameters as a result of therapeutic measures does not allow a direct inference as to the therapeutic efficacy in the sense of a clinical improvement. Other methods, such as, for example, the measurement of the tissue oxygen pressure, not to mention clinical studies, are absolutely necessary. If, however, an improvement in the flow properties of the blood of patients has been proven with adequate therapeutic measures, then this is a significant indication for a possible mechanism of action of a certain form of therapy, and the often very costly methods of proving the efficacy are then justified. On the other hand, if the rheologic efficacy of a drug cannot be proven, this does not necessarily mean that the substance is clinically ineffective, because perhaps other nonrheologic properties of the substance may be held responsible for a possible therapeutic success.

The publications on the utilization of rheologically effective drugs or measures in patients are presented in *chronologic order.* This means that every substance and every measure for the improvement of the flow properties of blood has a section of its own in which those studies are presented first that were published first. For this reason, the section on the effect of iodine on the flow properties of blood (first publication in 1904) comes before the effect of dextran on the rheology of blood (first publication in 1961). From the first description (e.g., for streptokinase in 1969) to the end of 1986, all the studies found are presented with a commentary and the clinical relevance. At the end of each section on a particular therapeutic measure, all the studies on it are summarized and compared. Thus the reader is provided with a coherent survey on each individual drug and on every measure for the improvement of the flow properties of blood from the first studies on this subject to the usually more precise studies of recent years. Particular emphasis is placed on a critical commentary and a comparative consideration of the individual studies with, in part, discrepant results. Thus a hemorheologic effect can be proven only for some of the measures and medications, and for a range of other substances a supposed or asserted effect on the flow properties of blood is very improbable or still has to be proven.

Since, up to the present day, a great number of publications on more than 50 therapeutic possibilities for the improvement of the flow properties of blood has appeared and is treated in the specific part, it seemed reasonable to present a *survey* of the most important rheologically active substances and therapeutic measures before *the actual specific part with its detailed data.* In this section medicinal and other therapeutic measures for the improvement of the flow properties of blood are described whose effect has been proven, to a great extent, on the basis of scientific results and which are also being increasingly used therapeutically. For some of these measures the clinical efficacy has also been proven or at least been shown to be very probable. In this survey, the significant areas of application for therapeutic hemorheology known to date are also presented.

In the *specific part* of the book several hundred publications on the therapeutic effect of rheologically active drugs in patients are cited and analyzed. There is no doubt that certain therapeutic methods and drugs not only improve the flow properties of blood (therapeutic hemorheology) but are also clinically active (effective hemorheologic therapy). On the other hand, there are a number of publications whose conclusions on hemorheologic effects cannot be fully accepted owing to methodologic or other deficiencies. This is also true of some more recent publications. Hemorheologic parameters must not be looked at in isolation, but must be seen and evaluated as an important piece in the mosaic of the micro- and macrocirculation. In other words: a critical assessment of the validity of therapeutic improvement in the flow properties of blood with the help of nonhemorheologic methods is absolutely necessary. As presented in a section of its own, it may be assumed that in future therapeutic hemorheology will not only strengthen its position in the treatment of various diseases, but will even expand this position. The scientific field of therapeutic hemorheology is comparatively young, and there are valid hopes that in the years and decades to come scientific research will present further interesting possibilities for the therapeutic influence on the flow properties of blood in various diseases.

6 Addendum

When some years ago I decided to examine, compile, and analyze the literature on therapeutic hemorheology I was not aware how extensive this task would be. If I had known how much work the research, the translation, the compilation, and the completion entailed, I would possibly have viewed this plan with some detachment. Now that this book, with its over 900 bibliographic notes, is completed, I must admit that all in all the work was informative and stimulating.

As was to be expected, the individual contributions to the subject therapeutic hemorheology were of varying quality, which made it difficult to give a statement on the actual hemorheologic effect of a drug or other therapeutic measure. An additional difficulty was that some results were published several times. Some of the contributions cited are, according to the inclusion criteria, rightly part of the book, but their clinical relevance must be relativized. On the whole, it would be desirable if some authors would keep to the premise "quality before quantity."

From the publications of the past few years it becomes obvious that, in addition to hemorheologic data, microcirculatory parameters are also described. The attempts to keep clinical hemorheology from separating from the area of clinical microcirculation became clear when, among other things, the German Society for Clinical Hemorheology, founded in 1979, was renamed in 1988 as the German Society for Clinical Microcirculation and Hemorheology. This will be of relevance in the future both for basic clinical research and for clinical practice.

If today I read through my editorial entitled "Why Hemorheology?" in the *Medical Tribune* in the year 1969 and at the same time I am aware that in the year 1989 – 20 years later – the Sixth European Congress on Clinical Hemorheology is taking place in Frankfurt/Main, this really says a lot. The little sapling of those days has grown into a tree which will continue to grow and whose fruits will benefit our patients.

7 References

1 Aarts PAMM, Banga JD, van Houwelingen HC, Heethaar RM, Sixma JJ (1985) Influence of isoxsuprine on red cell deformability in vivo and effects on platelet adherence in vitro (Abstract). Clin Hemorheol 5: 753

2 Aarts PAMM, Banga JD, van Houwelingen HC, Heethaar RM, Sixma JJ (1986) Increased red blood cell deformability due to isoxsuprine administration decreases platelet adherence in a perfusion chamber: A double-blind cross-over study in patients with intermittent claudication. Blood 67: 1474

3 Abel JJ, Rowntree LG, Turner BB (1914) Plasma removal with return of the corpuscles. J Pharmacol Exp Ther 5: 625

4 Abruzzo JL, Heimer R, Giuliano V, Martinez J (1970) The hyperviscosity syndrome, polysynovitis, polymyositis and an unusual 13 S serum IgG component. Am J Med 49: 258

5 Adam H (1909) Zur Viskosität des Blutes. Z Klin Med 68: 177

6 Akiyama M, Isogai Y, Yokose T, Okabe M, Ashikaga M, Ikemoto S, Maeda T, Abe M (1981) Changes in hemorheological factors of diabetics during use of artificial pancreas (Abstract). Clin Hemorheol 1: 300

7 Albanese B (1983) Modificazioni dell'emoreologia indotte dal farmaco α-litico timoxamina. Ric Clin Lab 13 (Suppl 3): 469

8 Albers D (1937) Die Viskosität des Blutes in der kardialen Herzinsuffizienz. Z Kreislaufforsch 29: 914

9 Albers D (1937) Multiple Myelome und Viscosität. Z Klin Med 132: 807

10 Alderman MJ, Ridge A, Morley AA, Ryall RG, Walsh JA (1981) Effect of total leucocyte count on whole blood filterability in patients with peripheral vascular disease. J Clin Pathol 34: 163

11 Allan TL (1981) Plasma exchange in macroglobulinaemia. In: Lowe GDO, Barbenel JC, Forbes CD (eds) Clinical aspects of blood viscosity and cell deformability. Springer, Berlin, p235

12 Alwall N (1939) Jodtherapie bei Hypertonie und Arteriosklerose. Acta Med Scand 101: 83

13 Amtoft A (1973) The effect of Dextran 40 on capillary blood flow and capillary diffusion capacity in the anterior tibial muscle of subjects with intermittent claudication. Scand J Clin Lab Invest 31: 263

14 Anadere I, Chmiel H, Witte S (1985) Hemorheological findings in patients with completed stroke and the influence of a ginkgo biloba extract. Clin Hemorheol 5: 411

15 Angelkort B (1979) Influence of pentoxifylline on erythrocyte deformability in peripheral occlusive arterial disease. Curr Med Res Opin 6: 255

16 Angelkort B, Boateng K, Maurin N (1980) Blood fluidity and coagulation phenomena in chronic arterial occlusive disease. J Int Med Res 8: 242

17 Angelkort B, Kiesewetter H (1981) Influence of risk factors and coagulation phenomena on the fluidity of blood in chronic arterial occlusive disease - Effects of pentoxifylline. Scand J Clin Lab Invest 41 (Suppl 156): 185

18 Angelkort B, Kiesewetter H, Maurin N (1981) Einfluß von Hämodilution und oraler Pentoxifyllin-Medikation auf Muskeldurchblutung, Fließverhalten und Hämostase des Blutes bei chronischer arterieller Verschlußkrankheit. In: Breddin K (ed) Thrombose und Atherogenese. Pathophysiologie und Therapie der arteriellen Verschlußkrankheit. Bein-Beckenvenen-Thrombose. Witzstrock, Baden-Baden, p114

19 Angelkort B, Frömel M (1981) Influence of hemodilution on fluidity and hemostasis in chronic occlusive arterial disease - Study with high molecular hydroxyethyl starch solution (Abstract). Clin Hemorheol 1: 453

20 Angelkort B, Spürk P, Ludwig P (1984) An appraisal of interrelationship between blood fluidity and walking performance in peripheral arterial occlusive disease. IRCS Med Sci 12: 360

21 Angelkort B, Spürk P, Habbaba A, Mähder M (1985) Blood flow properties and walking performance in chronic arterial occlusive disease. Angiology 36: 285

22 Annoni F, Monti D, Mozzi E, Chiurazzi D, Marincola F (1983) Ciclonicate in patients with chronic occlusive peripheral arterial disease: Haemodynamic and rheologic effects. Acta Ther 9: 29

23 4th Annual meeting of Japan Society of Biorheology (1982) (Abstracts). Biorheology 19: 375
24 Antignani PL, Todini AR, Saliceti F, Pacino G, Bartolo M (1985) Clinical, instrumental and hemorheological effects after acute therapy with pentoxifylline (Abstract). Clin Hemorheol 5: 731
25 Apsatarov EA, Goryachev SP, Lyubinsky VL, Timukhina RA, Stakhov OV (1985) Changes of microcirculation and blood rheologic properties in patients with Leriche's syndrome. Khirurgiia (Mosk) 6: 44
26 Arntz HR, Leonhardt H, Dreykluft HR (1979) Influence of clofibrate on blood viscosity in primary hyperlipoproteinemia. Klin Wochenschr 57: 43
27 Arntz HR, Heitz J, Schäfer JH, Zingler G, Schröder R (1985) High-dose intravenous streptokinase (SK) in acute myocardial infarction (AMI): Hemorheological aspects (Abstract). Clin Hemorheol 5: 729
28 Aronson HB, Magora F, London M (1970) The influence of droperidol on blood viscosity in man. Br J Anaesth 42: 1089
29 Aronson HB, Magora F, Schenker JG (1971) Effect of oral contraceptives on blood viscosity. Am J Obstet Gynecol 110: 997
30 Artale F, Sergio G, Di Palma A, Giunti P, Perego MA (1980) Effetto dell'infusione venosa di pentossifillina sulla deformabilità eritrocitaria nell'arteriopatia ostruttiva periferica. Il Policlinico, Sez Medica 87: 338
31 Artale F, Francisci A, Espureo M, Sergio G, Cramarossa L, Di Palma A, Perego MA (1982) Effects of S-adenosyl-L-methionine on hemorheological changes induced by exercise in patients with peripheral occlusive arterial diseases. Curr Ther Res 32: 520
32 Aukland A, Hurlow RA, George AJ, Stuart J (1982) Platelet inhibition with ticlopidine in atherosclerotic intermittent claudication. J Clin Pathol 35: 740
33 Auteri A, Pieragalli D, Blardi P (1984) Azione della CDP-colina sulla sindrome da iperviscosità ematica secondaria delle vasculopatie croniche ischemizzanti. Minerva Cardioangiol 32: 53
34 Avellone G, Pinto A, Mandalà V, Galati D, Di Garbo V, Strano A (1985) Modificazioni emoreologiche indotte dalla somministrazione di buflomedil cloridrato nelle arteriopatie obliteranti degli arti inferiori. Ric Clin Lab 15 (Suppl 1): 577
35 Ayres ML, Jarrett PEM, Browse NL (1981) Blood viscosity, Raynaud's phenomenon and the effect of fibrinolytic enhancement. Br J Surg 68: 51
36 Azzena D, Costa U, Ghigliotti G, Astengo F (1968) Serum hyperviscosity in multiple myeloma and its clinical and neuropsychiatric implications. Confin Neurol 30: 65
37 Bach R, Kiesewetter H, Jung F, Kotitschke G, Witt R, Nüttgens HP, Ladwig KH, Waterloh E, Roebruck P, Gerhards M, Schieffer H, Bette L, Wenzel E (1987) Koronare Herzkrankheit: Prävalenz, Risikofaktoren-Profil und rheologisches Profil. Erste Ergebnisse der Aachen-Studie. In: Strauer BE, Ehrly AM, Leschke M (eds) Fortschritte in der kardiovaskulären Hämorheologie (5. DGKH-Kongreß, Marburg, 1986). Münchner Wissenschaftliche Publikationen, München, p7
38 Bada HS, Korones SB, Kolni HW, Fitch CW, Ford DL, Magill HL, Anderson GD, Wong SP (1986) Partial plasma exchange transfusion improves cerebral hemodynamics in symptomatic neonatal polycythemia. Am J Med Sci 291: 157
39 Baer MR, Stein RS, Dessypris EN (1985) Chronic lymphocytic leukemia with hyperleukocytosis. The hyperviscosity syndrome. Cancer 56: 2865
40 Bäumler H, Lerche D, Scherf HP, Bilsing R, Meier W, Kucera W, Wiesner S (1982) Experimentelle Untersuchungen zum rheologischen Verhalten von Blut bei Ultraviolettbestrahlung (UVB). Z Ges Inn Med 37: 458
41 Balabanova RM, Guseva NG, Apenysheva NP, Oskilko TG, Anikina NV (1983) Effect of reopolyglucin on the microcirculation of patients with sclerodermia systematica. Revmatologiia (Moskva) 3: 26
42 Bapistella B (1972) Untersuchungen über den Einfluß der Streptokinase auf die Blutviskosität und die Aggregation der Erythrozyten. Dissertationsschrift, Frankfurt am Main
43 Barnard RJ, Hall JA, Pritikin N (1985) Effects of diet and exercise on blood pressure and viscosity in hypertensive patients. J Cardiac Rehabil 5: 185
44 Barnes AJ, Oughton J, Willars EJ, Clark PA, Kohner EM (1983) Effect of o-(β-hydroxyethyl)-rutosides (HR) on blood rheology in normals and diabetic patients (Abstract). Clin Hemorheol 3: 287
45 Baron M, Fam AG, Elkan I, Underdown B (1982) Hyperviscosity syndrome in rheumatoid arthritis. J Rheumatol 9: 843
46 Barras JP (1969) The capillary flow of suspensions of human red blood cells in plasma substitutes. Bibl Anat 10: 38
47 Barras JP, Maibach E (1971) Effect du HR sur la viscosité sanguine dans l'insuffisance veineuse chronique. Abstract-Booklet 4th Int Congr Phlebol, Lucerne, p93

48 Barras JP, Graf C (1980) Hyperviscosity in diabetic retinopathy treated with Doxium (calcium dobesilate). Vasa 9: 161
49 Barras JP, Graf C (1981) Effects of medication on blood and plasma viscosity in diabetic subjects with retinopathy. Horm Metab Res 13 (Suppl 11): 120
50 Barras JP, Blattner J (1985) Viscoelastic properties of blood and plasma in diabetes. Vasa 14: 216
51 Bartolo M (1985) Emodiluizione, defibrinogenazione e plasmaferesi in emoreologia. Ric Clin Lab 15 (Suppl 1): 439
52 Baum RS (1967) Hyperviscous blood and perinatal pathology (Abstract). Pediatr Res 1: 288
53 Bayliss LE (1960) The anomalous viscosity of blood. In: Copley AL, Stainsby G (eds) Flow properties of blood and other biological systems. Pergamon, London, p29
54 Beck JR, Quinn BM, Meier FA, Rawnsley HM (1982) Hyperviscosity syndrome in paraproteinemia. Managed by plasma exchange; monitored by serum tests. Transfusion 22: 51
55 Becker B (1976) Untersuchungen über den Einfluß einiger Medikamente auf die Fließeigenschaften des Blutes. Dissertationsschrift, Frankfurt am Main
56 Bedi SS, Eberl R, Million R, Silas AM (1984) Fenclofenac and diclofenac in the treatment of rheumatoid arthritis. Br J Rheumatol 23: 214
57 Begg TB, Hearns JB (1966) Components in blood viscosity. The relative contribution of haematocrit, plasma fibrinogen and other proteins. Clin Sci 31: 87
58 Belch JJF, Lowe GDO, Drummond MM, Forbes CD, Prentice CRM (1981) Prostacyclin reduces red cell deformability (Leserbrief). Thromb Haemost 45: 189
59 Benarroch IS, Brodsky M, Rubinstein A, Viggiano C, Salama EA (1985) Treatment of blood hyperviscosity with calcium dobesilate in patients with diabetic retinopathy. Ophthalmic Res 17: 131
60 Benda L, Redtenbacher L, Spieß A, Steinbach T (1971) Fibrinolytische Behandlung bei schwerer Angina pectoris. Dtsch Med Wochenschr 96: 771
61 Benda L (1972) Streptokinase und Blutviskosität bei akutem Herzinfarkt. In: Sailer S (ed) Aktuelle Probleme der Fibrinolyse-Behandlung. Hollinek, Wien, p27
62 Benda L (1973) Influence of streptokinase on blood rheology and coagulation. Postgrad Med J 49 (Suppl 5): 129
63 Benda L, Klösel W, Redtenbacher L, Spieß A, Zenz W (1976) Fibrinolyse und Antikoagulation bei schwerer Angina pectoris. Wien Med Wochenschr 126: 1
64 Benda L, Redtenbacher L (1977) Zur Dosierung der Streptokinase beim akuten Herzinfarkt. In: Sailer S (ed) Die Fibrinolyse - Behandlung des akuten Myokardinfarktes. Hollinek, Wien, p19
65 Benninger GW, Kreps SI (1971) Aggregation phenomenon in an IgG multiple myeloma resulting in the hyperviscosity syndrome. Am J Med 51: 287
66 Bergentz S, Gelin L, Rudenstam C, Zederfeldt B (1963) The viscosity of whole blood in trauma. Acta Chir Scand 126: 289
67 Bergqvist G, Zetterström R (1974) Blood viscosity and peripheral circulation in newborn infants. Acta Paediatr Scand 63: 865
68 Bernstein EF, Castaneda AR (1968) Studies of shear-dependent blood viscosity in extracorporal circulation. In: Copley AL (ed) Hemorheology. Pergamon, New York, p433
69 Blamey SL, McArdle BM, Burns P, Lowe GDO, Forbes CD, Carter DC, Prentice CRM (1983) Prevention of fibrinolytic shut-down after major surgery by intramuscular stanozolol. Thromb Res 31: 451
70 Blendin R (1975) Untersuchungen über den Einfluß einiger vasoaktiver Pharmaka auf die Fließeigenschaften des Blutes. Dissertationsschrift, Frankfurt am Main
71 Block KJ, Maki DG (1973) Hyperviscosity syndromes associated with immunoglobulin abnormalities. Hematology 10: 113
72 Blume J, Kiesewetter H, Bulling BJ, Gerhards M (1984) Einfluß von Betablockern auf klinische und hämorheologische Parameter bei AVK-Patienten. In: Heilmann L, Kiesewetter H, Ernst E (eds) Klinische Rheologie und Beta-1-Blockade. Zuckschwerdt, München, p153
73 Bode WA, Fonk T, van der Veen EA, van der Meer J (1982) Effects of long-term treatment with isoxsuprine on the blood viscosity profile in maturity-onset diabetics. Clin Hemorheol 2: 189
74 Böhme H, Everts B (1981) Hemodilution combined with defibrinogenation. Bibl Haematol 47: 165
75 Böhme H, Everts B (1985) Hämodilution und Defibrinogenierung in der Behandlung arterieller Durchblutungsstörungen. Int Welt 8: 225
76 Bös W (1976) Wirkungen zweier Diuretika auf die Fließeigenschaften des Blutes bei Patienten mit chronisch venösen Ödemen. Dissertationsschrift, Frankfurt am Main
77 Bogar L, Matrai A, Walker RT, Flute PT, Dormandy JA (1985) Haemorheological effects of a 5-HT_2 receptor antagonist (ketanserin). Clin Hemorheol 5: 115
78 Bogar L, Flute PT, Walker RT, Dormandy JA (1985) Haemorheological effects of a 5-HT_2 receptor antagonist (ketanserin) in patients with leg ischaemia (Abstract). Clin Hemorheol 5: 739

79 Boisseau MR (1985) Changes in blood filtration during acute C.V.A. Effect of treatment (Abstract). Clin Hemorheol 5: 712

80 Boisseau MR, Freyburger G, Louit F, Roudaut MF, Serise JM, Manciet G, Julien J (1985) Vascular pathology and hemorheological disorders: trial of buflomedil, pentosan polysulfate and vinburnin (Abstract). Clin Hemorheol 5: 724

81 Bollinger A, Simon HJ, Köhler R, Lüthy E (1968) Wirkung von niedermolekularem Dextran auf die Blutviskosität und Extremitätendurchblutung. Z Kreislaufforsch 57: 456

82 Bonniot R, Quemada D, Dufaux J, Hazard B, Lamberth M (1980) Traitement des artériopathies par l'acetylsalicylate de lysine - Résultats cliniques chez 68 malades avec 97 études rhéologiques. Arch Méd Ouest 12: 577

83 Borgolte H (1981) Veränderungen der Blutviskosität und -elastizität bei der Gabe von Dextran 40. In: Breddin K (ed) Thrombose und Atherogenese. Pathophysiologie und Therapie der arteriellen Verschlußkrankheit. Bein-Beckenvenen-Thrombose. Witzstrock, Baden-Baden, p 119

84 Botas L, Freitas JP, Teresa Laginha M, Barroca JP, Martins JM, Fátima Silva M, Martins E Silva J (1983) The effects of pentoxyfilline on blood filterability and functional capacity in patients with stable angina pectoris under a beta-blocking trial (Abstract). Clin Hemorheol 3: 328

85 Boyan CP, Underwood PS, Howland WS (1966) The effect of operation, anesthesia and plasma expanders on blood viscosity. Anesthesiology 27: 279

86 Breddin K, Ehrly AM, Fechler L, Frick D, König H, Kraft H, Krause H, Krzywanek HJ, Kutschera J, Lösch HW, Ludwig O, Mikat B, Rausch F, Rosenthal P, Sartory S, Voigt G, Wylicil P (1973) Die Kurzzeitfibrinolyse beim akuten Myokardinfarkt. Dtsch Med Wochenschr 98: 861

87 Brinkmann R, Zijlstra WG, Jansonius NJ (1963) Quantitative evaluation of the rate of rouleaux formation of erythrocytes by measuring light reflection ("Syllectometry"). Proc Kon Ned Akad Wet (Ser C) 66: 236

88 Brown MM, Pearson TC, Marshall J (1981) Effect of changes in plasma viscosity on cerebral blood flow in man (Abstract). Clin Hemorheol 1: 457

89 Brown MM, Marshall J (1982) Effect of plasma exchange on blood viscosity and cerebral blood flow. Br Med J 284: 1733

90 Browning GG, Gatehouse S, Lowe GDO (1986) Blood viscosity as a factor in sensorineural hearing impairment. Lancet 1: 121

91 Brügger W, Imhof P, Müller P, Moser P, Reubi F (1985) Effect of nitroglycerin on blood rheology in healthy subjects. Eur J Clin Pharmacol 29: 331

92 Buchan PC, MacDonald HN (1980) Altered haemorheology in oral-contraceptive users. Br Med J 280: 978

93 Buchan PC (1982) Haemorheological consequences of oestrogen and progestagen therapy. Int J Microcirc Clin Exp 1: 262

94 Buffet E, Adjizian JC, Pignon B, Droulle C, Potron G (1985) Rheological and haemostatic parameters in geriatric population under pentosane polysulfate treatment (Abstract). Clin Hemorheol 5: 738

95 Burch GE, De Pasquale NP (1962) Hematocrit, blood viscosity and myocardial infarction. Am J Med 32: 161

96 Burch GE, De Pasquale NP (1963) Phlebotomy - Use in patients with erythrocytosis and ischemic heart disease. Arch Intern Med 111: 687

97 Burch GE (1965) Hematocrit, viscosity and coronary blood flow. Dis Chest 48: 225

98 Burke AM, Chien S, McMurty JG (1979) Effects of low molecular weight dextran on blood viscosity after craniotomy for intracranial aneurysms. Surg Gynecol Obstet 148: 9

99 Burns P, Small M, Blamey SL, Lowe GDO, Forbes CD (1983) Effects of stanozolol on viscosity in volunteers and in surgical patients (Abstract). Clin Hemorheol 3: 287

100 Burton-Opitz R (1900) Vergleich der Viscosität des normalen Blutes mit der des Oxalatblutes, des defibrinirten Blutes und des Blutserums bei verschiedener Temperatur. Arch Ges Physiol 82: 464

101 Burton-Opitz R (1904) The changes in the viscosity of the blood produced by alcohol. J Physiol 32: 8

102 Burton-Opitz R (1906) Weitere Studien über die Viskosität des Blutes. Arch Ges Physiol 112: 189

103 Burton-Opitz R (1911) The viscosity of the blood. JAMA 57: 353

104 Buskard NA, Galton DA, Goldman JM, Kohner EM, Grindle CF, Newman DL, Twinn KW, Lowenthal RM (1977) Plasma exchange in the long term management of Waldenström's macroglobulinaemia. Can Med Assoc J 117: 135

105 Buskard NA (1981) Blood flow studies in the hyperviscosity syndrome before and after plasma exchange. In: Borberg H, Reuter P (eds) Plasma exchange therapy. Thieme, Stuttgart, p 89

106 Cabannes R, Pistonne M, Faget P, Meite M, Bassembie J (1983) Rheological study of sickle-cells in S.C.D. patients treated with pentoxifylline - A comparative study (Abstract). Clin Hemorheol 3: 232

107 Cäsar K, Wegener B, Hossmann K (1980) Wirkung von Isosorbiddinitrat auf Blutdruck, peripheren Gefäßwiderstand und venöse Kapazität bei Patienten mit arterieller Hypertonie. In: Rudolph W, Schrey A (eds) Nitrate II - Wirkung auf Herz und Kreislauf. Urban und Schwarzenberg, München, p 58
108 Caimi G, Frazzetta F, Capobianco D, Catania A, Francavilla G, Sarno A (1982) Comportamento del pattern emoreologico in soggetti con ipertensione essenziale trattati con beta-bloccanti. Clin Ter 103: 383
109 Caimi G, Catania A, Frazzetta F, Capobianco D, Sarno A (1983) Essential hypertension - The behaviour of the haemorheological determinants and their trends during beta-blocker treatment. Jpn Heart J 24: 723
110 Caimi G, Catania A, Frazzetta F, Capobianco D, Francavilla G, Romano A, Sarno A (1983) Relievi emoreologici in un gruppo di vasculopatici aterosclerotici trattati con suloctidil. Ric Clin Lab 13 (Suppl 3): 465
111 Calabrò A, Rossi A, Piarulli F, Coscetti G, Baiocchi MR, Fellin R, Crepaldi G (1985) Effect of sulodexide administration in a group of patients with peripheral atherosclerosis (Abstract). Clin Hemorheol 5: 723
112 Calabrò A, Rossi A, Baiocchi MR, Coscetti G, Fellin R, Crepaldi G (1985) Effetto del sulodexide su alcuni parametri emoreologici in un gruppo di pazienti con vasculopatia arteriosclerotica periferica. Ric Clin Lab 15 (Suppl 1): 455
113 Calvert GD, Blight L, Franklin J, Oliver J, Wise P, Gallus AS (1980) The effects of clofibrate on plasma glucose, lipoproteins, fibrinogen and other biochemical and haematological variables in patients with mature onset diabetes mellitus. Eur J Clin Pharmacol 17: 355
114 Canovi MC, Spandri A, Lambrughi S, Marcelli D, Piovella C (1985) Confronto in vivo ed in vitro del sangue di soggetti trattati con pentossifillina nei riguardi dell'aggregazione intravasale degli eritrociti. Ric Clin Lab 15 (Suppl 1): 603
115 Cappelli R, Pecchi S, Pieragalli D, Acciavatti A, Galigani C, Messa GL, Vittoria A, Guerrini M, Forconi S, Di Perri T (1985) Effect on rheological and some peripheral haemodynamic parameters of defibrotide (Abstract). Clin Hemorheol 5: 741
116 Carey JS (1972) Colloids and crystalloids for volume replacement during and after surgery. In: Messmer K, Schmid-Schönbein H (eds) Hemodilution - Theoretical basis and clinical application. Karger, Basel, p 229
117 Case DC jr (1982) Combination chemotherapy (M-2) protocol (BCNU, cyclophosphamide, vincristine, melphalan and prednisone) for Waldenström's macroglobulinemia: Preliminary report. Blood 59: 934
118 Catalano M, Sacchi E, Tomasini M, Coazzoli E, Parolini A, Dall'Oglio A, Libretti A (1985) La flunarizina nel trattamento delle arteriopatie degli arti inferiori. Ric Clin Lab 15 (Suppl 1): 475
119 Cavestri R, Agosti R, Clivati A, Pelucchi A, Cherubini P, Mastropasqua D, Longhini E (1985) Effects of isovolemic hemodilution on cerebral blood flow and on hemorheologic parameters in acute ischemic stroke (Abstract). Clin Hemorheol 5: 668
120 Challoner T, Briggs C, Thomas DJ (1981) The effect of reducing haematocrit and viscosity on blood pressure (Abstract). Clin Hemorheol 1: 462
121 Challoner T, Briggs C, Rampling MW, Thomas DJ (1986) A study of the haematological and haemorheological consequences of venesection. Br J Haematol 62: 671
122 Chazan BI, Ferguson BD, Castelli WP, Touborg JNF, Balodimos MC, Rutstein DD (1969) Lipernia retinalis: Microcirculatory changes and lipid studies in family. Metabolism 18: 978
123 Chevreaud C, Voisin P, Burdin D, Guimont C, Laxenaire MC, Stoltz JF (1985) Deliberate normovolemic hemodilution during the medical treatment of arteriopathy of the lower limbs: Rheological effects (Abstract). Clin Hemorheol 5: 722
124 Chien S, Usami S, Taylor HM, Lundberg JL (1966) Effect of hematocrit and plasma proteins on human blood rheology at low rates of shear. J Appl Physiol 21: 81
125 Chien S (1972) Present state of rheology. In: Messmer K, Schmid-Schönbein H (eds) Hemodilution - Theoretical basis and clinical application. Karger, Basel, p 1
126 Chien S, Schmalzer EA, Lee MML, Impelluso T, Skalak R (1983) Role of white cells in filtration of blood cell suspensions. Biorheology 20: 11
127 Chien S, Letcher RL, Pickering TG, Devereaux RB (1983) Blood rheology in hypertension (Abstract). Clin Hemorheol 3: 240
128 Chmiel H (1973) Zur Blutrheologie in Medizin und Technik. Habilitationsschrift, Aachen
129 Chmiel H, Pavlovic Z (1974) Der Einfluß von Carbocromen-HCl (Intensain) auf die Blutviskosität bei starken Rauchern und frischem Herzinfarkt. Z Kardiol 63 (Suppl 1): 53
130 Chmiel H, Pavlovic Z, Effert S (1975) Neue rheologische Methoden zur Differenzierung pathologischer Veränderungen der Blutviskosität. Verh Dtsch Ges Kreislaufforsch 41: 282

131 Chmiel H, Schweizer P, Krause KH, Framing W (1976) New methods of measuring the different rheological parameters in blood (Abstract). Microcirculation 1: 94
132 Chmiel H, Anadere I, Heimburg P (1981) The influence of i.v. application of pentoxifylline on the viscoelasticity of blood in patients with peripheral vascular disorders. Ric Clin Lab 11 (Suppl 1): 189
133 Ciuffetti G, Lupattelli G, Cusco S, Pistola L, Mercuri M, Susta A, Palazzetti D, Mannarino E (1985) Effetti del sulodexide sui parametri emoreologici, lipidici, lipoproteici ed apoproteici di pazienti iperlipemici con arteriopatia obliterante periferica. Ric Clin Lab 15 (Suppl 1): 529
134 Ciuffetti G, Parnetti L, Lupattelli G, Mercuri M, Orecchini G, Senin U (1985) Il pattern emoreologico nel deterioramento mentale senile primitivo: Risultati di uno studio a lungo termine con l'impiego di piracetam e pentossifillina. Ric Clin Lab 15 (Suppl 1): 583
135 Cline MJ, Solomon A, Berlin NI, Fahey JL (1963) Anemia in macroglobulinemia. Am J Med 34: 213
136 Clinical Hemorheology 7 (1987) Nr1 (Special issue)
137 Clivati A, Agosti R, Cherubini P, Ogliari L, Belloni A, Longhini E (1985) Effetti del salasso isovolumetrico sulle proprietà reologiche del sangue in corso di insufficienza respiratoria e policitemia secondaria. Ric Clin Lab 15 (Suppl 1): 611
138 Clivati A, Agosti R, Cherubini P, Spelta P, Bonvini G, Ricciardi G, Pagliano L (1985) Hemorheologic properties in beta-thalassemia (Abstract). Clin Hemorheol 5: 686
139 Coccheri S, Palareti G, Poggi M, Tricarico MG (1982) Improvements in the rheologic properties of blood induced by medium-term treatment with buflomedil in diabetic patients. J Int Med Res 10: 394
140 Coccheri S, Palareti G, Parenti M, Poggi M, Tricarico MG (1982) Buflomedil improves erythrocyte deformability in diabetic patients (Abstract). Haemostasis 12: 119
141 Cokelet GR, Merrill EW, Gilliland ER, Shin H, Britten A, Wells RE (1963) The rheology of human blood. Measurements near and at zero shear rate. Trans Soc Rheol 7: 303
142 Copley AL (1949) Rheological problems in biology. In: Burgers JM et al. (eds) Proceedings of the International Congress on Rheology, Holland, 1948. North-Holland, Amsterdam, I-47; III-8
143 Copley AL (1952) The rheology of blood. A survey. J Colloid Sci 7: 323
144 Copley AL, Stainsby G (eds) (1960) Flow properties of blood and other biological systems. Pergamon, London
145 Copley AL (1962) The endoendothelian fibrin film and fibrinolysis. In: Proceedings of the 8th International Congress on Hematology, Tokyo, 1960, vol3. Pan-Pacific, Tokyo, p1648
146 Copley AL (1968) Hemorheology. Pergamon, New York
147 Copley AL, Seaman GVF (1981) The meaning of the terms rheology, biorheology and hemorheology (Editorial). Clin Hemorheol 1: 117
148 Copley AL (1982) The future of the science of biorheology. Biorheology 19: 47
149 Copley AL (1985) The history of clinical hemorheology. Clin Hemorheol 5: 765
150 Couette M (1890) Études sur le frottement des liquides. Ann Chim Phys 21: 433
151 Crawford J, Cox EB, Cohen HJ (1985) Evaluation of hyperviscosity in monoclonal gammopathies. Am J Med 79: 13
152 Crolle G, Cucca S, Gosparini B, Baso G, Martechini A, Urbani N, Vattolo S (1985) E'possibile migliorare la deformabilità eritrocitaria nell'anziano vasculopatico? Minerva Med 76: 1349
153 Crowell JW, Ford RG, Lewis VM (1959) Oxygen transport in hemorrhagic shock as a function of the hematocrit ratio. Am J Physiol 196: 1033
154 Danilov IP, Makarevich AE (1985) Correction of hemorheological disorders in patients with chronic obstructive bronchitis. Klin Med (Mosk) 63 (7): 49
155 Davis E, Rozov H (1973) The effects of xanthinol nicotinate on the small blood vessels. Bibl Anat 11: 334
156 Davis E, Rozov H (1975) Xanthinol nicotinate in peripheral vascular disease. Practitioner 215: 793
157 Dawidson I, Barrett J, Miller E, Litwin MS (1975) Blood viscosity studies in postoperative patients. Bibl Haematol 41: 76
158 De Clerck F, De Cree J, Brugmans J, Wellens D (1977) Reduction by flunarizine of the increase of blood viscosity after ischemic forearm occlusion. Arch Int Pharmacodyn Thér 230: 321
159 De Clerck F (1985) Rheological effects of flunarizine, a Ca^{2+}-entry blocker: rationale and mechanism (Abstract). Clin Hemorheol 5: 759
160 De Cree J, De Cock W, Geukens H, De Clerck F, Beerens M, Verhaegen H (1979) The rheological effects of cinnarizine and flunarizine in normal and pathologic conditions. Angiology 30: 505
161 De Cree J, Leempoels J, Geukens H, De Cock W, Verhaegen H (1981) RBC deformability in patients with essential hypertension and acute myocardial infarction. Improvement by treatment with a selective 5-HT_2 antagonist (R 41468). Abstract-Booklet 2nd Europ Conf Clin Haemorheol, London, p31

162 De Cree J, Leempoels J, De Cock W, Geukens H, Verhaegen H (1983) Hemorheological effects of ketanserin, a 5-HT_2 receptor antagonist. Abstract-Booklet 1st Europ Meeting Hypertension, Milan, Italy

163 De Cree J, Leempoels J, Geukens H, De Cock W, Verhaegen H (1983) Oral treatment with ketanserin in intermittent claudication. Abstract-Booklet 13th World Congr Int Union Angiol, Rochester, Minnesota, USA

164 Deguchi K, Ito T, Shima H, Uno N, Nakamori I, Ueno H, Umemoto D, Ito U, Kobayashi T, Li S, Tanaka T, Kato M, Konishi M, Kudo M, Kusunose K, Yamada S (1977) Pentoxifylline. Clinical use of pentoxifylline - Concerning blood viscosity and platelet function. Mie Med 21 (3): 375

165 Delsignore R, David S, Francia MT, Bottazzi G, Crotti G (1985) Alterazioni emoreologiche determinate dall'uremia. Effetti dell'emodialisi. Ric Clin Lab 15 (Suppl 1): 335

166 De Quiros JFB, Hess H (1976) Über den Einfluß von Isoxsuprin-HCl auf die Fließfähigkeit des Blutes. Fortschr Med 94: 1661

167 Derksen RHWM, Hene RJ, Gmelig Meyling FHJ, Kater L (1983) Plasma exchange in the treatment of cryoglobulinaemia. Neth J Med 26: 294

168 Determann H (1908) Das Verhalten der Blutviskosität bei Joddarreichung. Dtsch Med Wochenschr 34: 871

169 Diamantopoulos EJ, Raptis S, Karaiskos K, Hadjidakis D, Moulopoulos S (1981) Haemorheological parameters in diabetics under different modes of treatment (Abstract). Clin Hemorheol 1: 465

170 Diehm C, Gallasch G, Wirth A, Schettler G (1984) Hämorheologische Veränderungen nach körperlichem Training. Dtsch Z Sportmed 8: 286

171 Di Lollo F, Fazzini G, Avanzi G (1983) L'emodiluizione e la plasmaferesi nella terapia delle sindromi da iperviscosità. Ric Clin Lab 13 (Suppl 3): 225

172 Dintenfass L (1966) A preliminary outline of the blood high viscosity syndromes. Arch Intern Med 118: 427

173 Dintenfass L, Rozenberg MC (1967) Some observations on the viscosity of blood in various diseases. Effect of intravenous heparin. Angiologica 4: 116

174 Dintenfass L (1968) Fluidity (internal viscosity) of the erythrocyte and its role in physiology and pathology of circulation. Haematologia 2: 19

175 Dintenfass L (1971) Blood microrheology - Viscosity factors in blood flow, ischaemia and thrombosis. Butterworths, London

176 Dintenfass L, Milton GW (1973) Blood viscosity factors and prognosis in malignant melanoma. Effect of ABO blood groups. Med J Aust 1: 1091

177 Dintenfass L, Lake B (1976) Beta blockers and blood viscosity. Lancet 1: 1026

178 Dintenfass L (1976) Rheology of blood in diagnostic and preventive medicine. An introduction to clinical haemorheology. Butterworths, London

179 Dintenfass L, Seaman GVF (1981) Blood viscosity in heart disease and cancer. Pergamon, New York

180 Dintenfass L (1981) Hyperviscosity in hypertension. Pergamon, New York

181 Dintenfass L (1983) Action of drugs on the aggregation and deformability of red cells: effect of ABO blood groups. Ann NY Acad Sci 416: 611

182 Dintenfass L (1983) Effect of drugs and diet on blood viscosity, red cell aggregation and red cell rigidity (Abstract). Clin Hemorheol 3: 288

183 Dintenfass L (1985) Blood viscosity. Hyperviscosity and hyperviscosaemia. MTP Press, Lancaster

184 Di Perri T, Forconi S, Guerrini M, Laghi Pasini F, Del Cipolla R, Rossi C, Agnusdei D (1977) Action of cinnarizine on the hyperviscosity of blood in patients with peripheral obliterative arterial disease. Proc R Soc Med 70 (Suppl 8): 25

185 Di Perri T, Forconi S, Agnusdei D, Guerrini M, Laghi Pasini F (1978) The effects of intravenous isoxsuprine on blood viscosity in patients with occlusive peripheral arterial disease. Br J Clin Pharmacol 5: 255

186 Di Perri T, Forconi S, Guerrini M, Laghi Pasini F, Del Cipolla R, Rossi R, Agnusdei D (1979) Action of cinnarizine on the hyperviscosity of blood in patients with peripheral obliterative disease. Angiology 30: 13

187 Di Perri T, Forconi S, Guerrini M, Pecchi S, Pieragalli D, Cappelli R, Acciavatti A (1981) Influence of non-selective and selective beta adrenoceptor blockade on isoxsuprine-dependent hemodynamic and rheologic changes. Angiology 32: 257

188 Di Perri T, Forconi S, Di Lollo F, Pozza G (1984) Le sindromi da iperviscosità ematica. Fisiologia, patologia e clinica. Pozzi, Roma

189 Ditzel J, Bang HC, Thorsen N (1969) The effect of Dextran 40 on hemorheological factors during the course of acute myocardial infarction. Bibl Anat 10: 132

190 Djurdjevic L, Bräuer H, Rahlfs V (1976) Kontrollierte klinische Prüfung von Chemotherapeutika zur

Senkung der Blutfette Cholesterin und Triglyzeride sowie deren Einfluß auf die Fluidität des Blutes. Therapiewoche 26: 5556
191 Dodds AJ, O'Reilly MJG, Yates CJP, Cotton LT, Flute PT, Dormandy JA (1979) Haemorheological response to plasma exchange in Raynaud's syndrome. Br Med J 2: 1186
192 Dodds AJ (1981) Plasma exchange. In: Lowe GDO, Barbenel JC, Forbes CD (eds) Clinical aspects of blood viscosity and cell deformability. Springer, New York, p 227
193 Doletskii AS, Dallakian NO, Aksel'rov AM, Nazarova NB, Sokolova VA (1985) Correction of disorders of peripheral circulation with trental in children in the postoperative period. Anesteziol Reanimatol 4: 56
194 Donaldson DR, Kester RC, Russell CW, Wiggins PA, Hall TJ (1981) Does oxpentifylline (Trental) have a place in the treatment of intermittent claudication? (Abstract). Clin Hemorheol 1: 469
195 Donaldson DR, Hall TJ, Kester RC, Ramsden CW, Wiggins PA (1984) Does oxpentifylline ("Trental") have a place in the treatment of intermittent claudication? Curr Med Res Opin 9: 35
196 Dorigo B, Raspanti D, Trapani M, Albanese B, Cameli AM, Digiesi V (1985) Studio flussimetrico, termometrico e reologico nelle arteriopatie obliteranti degli arti inferiori trattate con buflomedil. Minerva Med 76: 269
197 Dormandy JA (1970) Clinical significance of blood viscosity. Ann R Coll Surg Engl 47: 211
198 Dormandy JA (1971) Influence of blood viscosity on blood flow and the effect of low molecular weight dextran. Br Med J 4: 716
199 Dormandy JA, Edelman JB (1973) High blood viscosity: An aetiological factor in venous thrombosis. Br J Surg 60: 187
200 Dormandy JA, Hoare E, Khattab AH, Arrowsmith DE, Dormandy TL (1973) Prognostic significance of rheological and biochemical findings in patients with intermittent claudication. Br Med J 4: 581
201 Dormandy JA, Gutteridge JMC, Hoare E, Dormandy TL (1974) Effect of clofibrate on blood viscosity in intermittent claudication. Br Med J 4: 259
202 Dormandy JA, Goyle KB, Postlethwaite J (1976) Improvement in the rate of vascular graft patency with controlled postoperative defibrinogenation (Abstract). Br J Surg 63: 661
203 Dormandy JA, Goyle KB (1976) Clinical improvement in peripheral circulation by controlled defibrinogenation (Abstract). Br J Surg 63: 661
204 Dormandy JA, Hoare E, Postlethwaite H (1976) Blood viscosity of patients with intermittent claudication - Concept of "Rheological Claudication". Biorheology 13: 161
205 Dormandy J, Postlethwaite J, Goyle K (1977) Haemorheological therapy. Bibl Anat 16: 481
206 Dormandy JA, Goyle KB, Reid HL (1977) Treatment of severe intermittent claudication by controlled defibrination. Lancet 1: 625
207 Dormandy JA, Reid HL (1978) Controlled defibrination in the treatment of peripheral vascular disease. Angiology 29: 80
208 Dormandy JA (1981) Drug modification of erythrocyte deformability. In: Lowe GDO, Barbenel JC, Forbes CD (eds) Clinical aspects of blood viscosity and cell deformability. Springer, Berlin, p 251
209 Dormandy JA, Ernst E (1981) Effects of buflomedil on erythrocyte deformability. Angiology 32: 714
210 Dormandy JA, Yates CJP, Berent GA (1981) Clinical relevance of blood viscosity and red cell deformability including newer therapeutic aspects. Angiology 32: 236
211 Dormandy J (ed) (1983) Red cell deformability and filterability. Martinus Nijhoff, Boston
212 Dormandy JA, Matrai A (1983) Assessment of pharmacological agents with a hemorheological action. Ann NY Acad Sci 416: 599
213 Dormandy JA (1984) Effects of anaesthesia and surgery on the flow properties of blood. Microcirculation, Endothelium and Lymphatics 1: 151
214 Dormandy J, Ernst E, Matrai A (1986) Clinical aspects of white cell rheology. Clin Hemorheol 6: 455
215 Dowd PM, Kovacs IB, Bland CJH, Kirby JDT (1981) Effect of prostaglandins I_2 and E_1 on red cell deformability in patients with Raynaud's phenomenon and systemic sclerosis. Br Med J 283: 350
216 Dowd PM, Kovacs IB, Kirby JDT, Turner P (1981) Decreased deformability of red blood cells in patients with systemic sclerosis and the effect of prostaglandins I_2 and E_1 on red blood cell deformability (Abstract). Clin Hemorheol 1: 469
217 Droßbach M (1923) Beziehungen zwischen Blutviskosität und Blutkörperchen, ihre Beeinflussung durch Coffein. Z Gesamte Exp Med 34: 84
218 Drouin P, Rousselle D, Pointel JP, Voisin G, Stoltz JF, Clemens AH, Debry G (1979) Dynamic study of the blood viscosity of insulin-dependent diabetics by means of an artificial pancreas. Horm Metab Res 11 (Suppl 8): 107
219 Drouin P, Rousselle D, Stoltz JF, Guimont C, Gaillard S, Vernhes G, Debry G (1981) Study of blood viscosity and erythrocyte parameters in diabetic patients using an artificial pancreas. Scand J Clin Lab Invest 41 (Suppl 156): 165

220 Drummond AR, Drummond MM, Mackenzie PJ, Lowe GDO, Smith G, Forbes CD, Wishart HY (1980) The effect of general and spinal anaesthesia on red cell deformability and blood viscosity in patients undergoing surgery for fractured neck of femur. In: Stoltz JF, Drouin P (eds) Hemorheology and diseases. Doin, Paris, p445
221 Durocher JR, Weir MS, Lundblad EG, Patow WE, Conrad ME (1975) Effect of oral contraceptives and pregnancy on erythrocyte deformability and surface charge (39037). Proc Soc Exp Biol Med 150: 368
222 Easthope PL, Brooks DE (1980) A comparison of rheological constitutive functions for whole human blood. Biorheology 17: 235
223 Ehringer H, Dudczak R, Kleinberger G, Lechner K, Reiterer W (1971) Arvin: Schlangengift als neue Therapiemöglichkeit bei Durchblutungsstörungen. Wien Klin Wochenschr 83: 411
224 Ehringer H (1971) Effects of the purified fraction of the malayan pit viper (Arvin) alone and in combination with streptokinase in patients with arterial occlusion. Abstract-Booklet Symposium "Problems related to fibrinolysis", Capo Boi, Sardinia
225 Ehringer H, Dudczak R, Lechner K, Widhalm F (1972) Therapeutische Defibrinierung mit Schlangengift (Arwin) bei arteriellen Durchblutungsstörungen. Verh Dtsch Ges Inn Med 78: 624
226 Ehringer H, Dudczak R, Widhalm F, Lechner K (1972) Thrombolytische Therapie bei subakuten und chronischen Gliedmaßenarterienverschlüssen. In: Sailer S (ed) Aktuelle Probleme der Fibrinolyse-Behandlung. Hollinek, Wien, p41
227 Ehringer H, Dudczak R, Lechner K (1973) Therapeutische Defibrinierung mit Ancrod. Dtsch Med Wochenschr 98: 2298
228 Ehringer H, Dudczak R, Widhalm F, Lechner K (1974) Wirkung und Möglichkeiten einer Defibrinierung in der Therapie arterieller Durchblutungsstörungen (Untersuchungen mit Arwin). In: Ehringer H (ed) Fortschritte der konservativen Therapie der peripheren arteriellen Verschlußkrankheiten (Aktuelle Probleme der Angiologie, Bd24). Huber, Stuttgart, p110
229 Ehringer H, Dudczak R, Lechner K (1974) A new approach in the treatment of peripheral arterial occlusions: Defibrination with arvin. Angiology 25: 279
230 Ehrly AM (1966) Wirkung von niedermolekularem Dextran auf Erythrozytenaggregate beim Sludgephänomen. Med Klin 61: 982
231 Ehrly AM (1967) Hämorheologische Probleme bei Venenerkrankungen. Zentralbl Phlebol 6: 338
232 Ehrly AM (1969) Warum Hämorheologie? Med Trib 4; 42a: 13
233 Ehrly AM (1969) The effect of plasma substitutes on erythrocyte aggregation and blood viscosity. Bibl Haematol 33: 302
234 Ehrly AM (1969) Untersuchungen über die Senkung der Strukturviskosität menschlichen Blutes im Hinblick auf angiologische Erkrankungen unter besonderer Berücksichtigung der Blutlipide. Habilitationsschrift, Frankfurt am Main
235 Ehrly AM, Lange B (1969) Senkung der Blutviskosität und Desaggregation von Erythrocytenaggregaten durch Streptokinase. Abstract-Booklet 2nd Int Conf Int Soc Hemorheol, Heidelberg, p22
236 Ehrly AM (1970) Rheologische und mikrozirkulatorische Aspekte bei angiologischen Erkrankungen. Med Trib 5; 10a: 13
237 Ehrly AM (1970) Therapie mit Arwin bei peripheren Durchblutungsstörungen. Arbeitsgespräch bei Knoll AG, Ludwigshafen (siehe: Therapeutische Defibrinierung mit Ancrod. Dtsch Med Wochenschr 99 (1974), 112)
238 Ehrly AM (1971) Hemorheology and thrombosis. Abstract-Booklet 2nd Congr Int Soc Thromb Haemost, Oslo, Norway, p72
239 Ehrly AM (1971) Zur Wirkung von Arvin auf die Fließeigenschaften des Blutes (vorläufige Mitteilung). Herrenalber Angiologisches Gespräch
240 Ehrly AM (1971) Therapie der venösen Stauung durch Verbesserung der Fließeigenschaften des Blutes. Abstract-Booklet 4th Int Congr Phlebol, Lucerne, p92
241 Ehrly AM, Fink MM (1971) Blutkörperchensenkungsgeschwindigkeit (BSG) bei verschiedenen Hämatokritwerten. Med Welt 22: 1960
242 Ehrly AM (1971) Disaggregation of erythrocyte aggregates and decrease of the structural viscosity of human blood by 2-phenyl-benzyl-aminomethyl-imidazolidine (Antazolin). In: Hartert HH, Copley AL (eds) Theoretical and Clinical Hemorheology. Springer, Berlin, p184
243 Ehrly AM, Lange B (1971) Reduction in blood viscosity and disaggregation of erythrocyte aggregate by streptokinase. In: Hartert HH, Copley AL (eds) Theoretical and Clinical Hemorheology. Springer, Berlin, p366
244 Ehrly AM (1972) Rheological changes due to fibrinolytic therapy. In: Messmer K, Schmid-Schönbein H (eds) Hemodilution – Theoretical basis and clinical application. Karger, Basel, p289
245 Ehrly AM (1972) Verbesserung der Fließeigenschaften des Blutes über eine therapeutische Defibrinierung mit Arwin. Gefäßwand und Blutplasma 4: 247

246 Ehrly AM (1972) Influence of arwin on the flow properties of blood (Abstract). Biorheology 9: 151
247 Ehrly AM (1972) Das Verhalten der Blutviskosität bei der Therapie mit Salidiuretika. Verh Dtsch Ges Inn Med 78: 617
248 Ehrly AM, Breddin K (1972) Verbesserung der Fließeigenschaften des Blutes durch Arwin. Verh Dtsch Ges Inn Med 78: 620
249 Ehrly AM (1973) Influence of arwin on the flow properties of blood. Biorheology 10: 453
250 Ehrly AM, Jung G (1973) Circadian rhythm of blood viscosity. Biorheology 10: 577
251 Ehrly AM, Jung HJ (1973) Verbesserung der Fließeigenschaften menschlichen Blutes durch subcutane Applikation von Arwin. Verh Dtsch Ges Inn Med 79: 1397
252 Ehrly AM (1973) Verbesserung der Fließeigenschaften des Blutes: Ein neues Prinzip der medikamentösen Therapie chronischer peripherer arterieller Durchblutungsstörungen. Vasa 2 (Suppl 1)
253 Ehrly AM (1973) Senkung der Blutviskosität durch Arwin. Herz/Kreislauf 5: 133
254 Ehrly AM, Bapistella B (1973) Verbesserung der Fließeigenschaften des Blutes bei der Therapie mit Streptokinase und Arwin. In: Gottstein U (ed) Koronarinsuffizienz, periphere Durchblutungsstörungen. Huber, Bern, p315
255 Ehrly AM (1974) Increased blood flow by improvement of the flow properties of blood: A new concept of the treatment of vascular diseases. In: De Temes Cordeiro GC, Mayal RC (eds) Progress on Angiology - Proceedings of the 1972 8th International Congress of Angiology, Rio de Janeiro. I. Sociedade Brasileira de Angiologia 3: 986
256 Ehrly AM (1974) Therapeutische Aspekte bei Störungen der Mikrozirkulation bei peripheren arteriellen Verschlußkrankheiten. Vasa 3: 312
257 Ehrly AM, Blendin R (1974) Untersuchungen zur Frage des Einflusses von Bencyclan auf die Fließeigenschaften des Blutes. Therapiewoche 24: 2894
258 Ehrly AM, Scheinpflug W (1975) Improved deformability of crenated red cells in hyperosmolar human blood by adenosin. Bibl Anat 13: 13
259 Ehrly AM (1975) The effect of pentoxifylline on the flow properties of hyperosmolar blood. IRCS Med Sci 3: 456
260 Ehrly AM (1975) Dosis-Wirkungs-Beziehung von subcutan appliziertem Arwin® bei Patienten mit chronischen arteriellen Durchblutungsstörungen. Vasa 4: 161
261 Ehrly AM, Köhler HJ, Schroeder W, Müller R (1975) Sauerstoffdruckwerte im ischämischen Muskelgewebe von Patienten mit chronischen peripheren arteriellen Verschlußkrankheiten. Klin Wochenschr 53: 687
262 Ehrly AM (1975) Über die Bedeutung der Erythrozytenverformbarkeit bei angiologischen Erkrankungen. Der Praktische Arzt (Österreich Monatsschr Allg Med), Kongreß "Pentoxifyllin Colloquium", Wien, 1975, p9
263 Ehrly AM (1975) Therapie chronischer peripherer arterieller Verschlußkrankheiten mit dem Schlangengift-Enzym Arwin. Eine Übersicht. Med Welt 26: 446
264 Ehrly AM (1975) Ödemausschwemmung und Thrombose. Forschung und Praxis. Med Welt 26: 1877
265 Ehrly AM (1975) Veränderungen der Fließeigenschaften des Blutes und der Erythrozyten bei arteriellen Verschlußerkrankungen. Med Welt 26: 1971
266 Ehrly AM (1975) Beeinflussung der Verformbarkeit der Erythrozyten durch Pentoxifyllin. Med Welt 26: 2300
267 Ehrly AM, Saeger-Lorenz K (1976) Kombinierte hämodiluierende und defibrinogenierende Therapie chronischer arterieller Verschlußerkrankungen (vorläufige Mitteilung). Abstract-Band Jahrestagung Ges Mikrozirkul, Aachen
268 Ehrly AM (1976) Therapy of occlusive arterial disease with Ancrod. Artery 2: 98
269 Ehrly AM, Köhler HJ (1976) Modifiziertes Dosisschema für die subcutane Anwendung von Arwin bei Patienten mit chronischen arteriellen Durchblutungsstörungen. Vasa 5: 155
270 Ehrly AM, Köhler HJ (1976) Altered deformability of erythrocytes from patients with chronic occlusive arterial disease. Vasa 5: 319
271 Ehrly AM, Köhler HJ (1976) Impaired erythrocyte deformability in patients with chronic arterial occlusive disease. In: Seki K, Mishima Y (eds) Proceedings of the 10th International Congress on Angiology, Tokyo. Seis Printing, Tokyo, p407
272 Ehrly AM, Blendin R (1976) Influence of essential phospholipids on the flow properties of the blood. In: Peeters P (ed) Phosphatidylcholine. Springer, Berlin, p228
273 Ehrly AM (1976) Improvement of the flow properties of blood: A new therapeutical approach in occlusive arterial disease. Angiology 27: 188
274 Ehrly AM, Saeger-Lorenz K (1977) Kombinierte hämodiluierende und defibrinogenierende Therapie chronischer arterieller Verschlußerkrankungen. Verh Dtsch Ges Kreislaufforsch 43: 332

275 Ehrly AM (1977) Hemorheological in vitro findings. Are they transferable to the in vivo situation? Microvasc Res 13: 267
276 Ehrly AM, Bös W (1977) Hämorheologische Untersuchungen bei der saluretischen Therapie chronisch venöser Ödeme mit dehydro sanol tri®. Phlebol Proktol 6: 124
277 Ehrly AM, Schmitt HJ (1977) Verfahren und Vorrichtung zum Messen der Aggregationsgeschwindigkeit von roten Blutkörperchen im Blut. Deutsches Patent Nr 2707962
278 Ehrly AM, Schroeder W, Dannhof S (1977) The effect of pentoxifylline on the oxygen pressure of ischemic muscle tissue in patients with chronic arterial occlusions. IRCS Med Sci 5: 411
279 Ehrly AM (1978) The effect of pentoxifylline on the flow properties of human blood. Curr Med Res Opin 5: 608
280 Ehrly AM, Landgraf H, Saeger-Lorenz K, Wodniok M (1979) Fließeigenschaften des Blutes bei Behandlung der essentiellen Hypertonie mit Xipamid. Dtsch Med Wochenschr 104: 218
281 Ehrly AM, Schroeder W (1979) Sauerstoffdruckmessungen im ischämischen Muskelgewebe als Kontrolle therapeutischer Maßnahmen bei chronischen peripheren Durchblutungsstörungen. In: Hild R, Spaan G (eds) Therapiekontrolle in der Angiologie. Witzstrock, Baden-Baden, p 83
282 Ehrly AM, Schroeder W (1979) Zur Pathophysiologie der chronischen arteriellen Verschlußerkrankung. I. Mikrozirkulatorische Blutverteilungsstörung in der Skelettmuskulatur. Herz/Kreislauf 11: 275
283 Ehrly AM (1979) The effect of pentoxifylline on the deformability of erythrocytes and on the muscular oxygen pressure in patients with chronic arterial disease. J Med 10: 331
284 Ehrly AM (1979) Vergleich der Fließeigenschaften des Blutes venengesunder Probanden und Patienten mit primärer Varikosis unter verschiedenen hydrostatischen Bedingungen. Phlebol Proktol 8: 169
285 Ehrly AM, Köhler HJ, Schroeder W (1979) Zeitliche Beziehungen zwischen den Zunahmen des muskulären Gewebesauerstoffdruckes und der schmerzfreien Gehstrecke bei der Therapie der schweren Claudicatio intermittens mit Ancrod. Vasa 8: 28
286 Ehrly AM (1979) New pathophysiological concept of ischemic diseases. Microcirculatory Maldistribution (MBM). In: Puel P, Boccolon H, Enjalbert A (eds) Hemodynamics of the limbs. Toulouse, p 429
287 Ehrly AM, Landgraf H, Saeger-Lorenz K (1980) Quantification of the effects of rheologically active drugs in patients with vascular diseases. In: Stoltz JF, Drouin P (eds) Hemorheology and diseases. Doin, Paris, p 665
288 Ehrly AM (1980) New pathophysiological concept of ischemic diseases. Microcirculatory Blood Maldistribution (MBM). Bibl Anat 20: 456
289 Ehrly AM (1980) Mikrozirkulatorische Blutverteilungsstörung als pathophysiologisches Prinzip bei chronischen arteriellen Verschlußerkrankungen. In: Müller-Wiefel H (ed) Mikrozirkulation und Blutrheologie. Witzstrock, Baden-Baden, p 59
290 Ehrly AM (1981) Therapeutische Hypofibrinogenämie. In: Vinazzer H (ed) Thrombose und Embolie (Anaesthesiologie und Intensivmedizin, Bd 134). Springer, Berlin, p 307
291 Ehrly AM (1981) Welchen Stellenwert haben Gesamtblutviskosität, Erythrozytenverformbarkeit und Erythrozytenaggregation? Dtsch Med Wochenschr 106: 35
292 Ehrly AM, Landgraf H (1981) Analysis of the specific hemorheological effects of low molecular weight dextran (LMWD)- and saline-infusions in healthy volunteers. Clin Hemorheol 1: 99
293 Ehrly AM, Landgraf H (1981) Influence of vincamine on the flow properties of blood (Abstract). Clin Hemorheol 1: 471
294 Ehrly AM (1982) Are there haemorheological reasons for using drugs acting on red cell deformability? Clin Hemorheol 2: 305
295 Ehrly AM (ed) (1983) Determination of tissue oxygen pressure in patients. Pergamon, Oxford
296 Ehrly AM (ed) (1983) Klinische Hämorheologie: Eine Bestandsaufnahme. Zuckschwerdt, München
297 Ehrly AM (1983) Verbesserungen der nutritiven Durchblutung bei peripheren ischämischen Erkrankungen. Vasa 12 (Suppl 11)
298 Ehrly AM (1983) Hemorheological therapy by fibrinogen-lowering drugs (Abstract). Clin Hemorheol 3: 550
299 Ehrly AM, Landgraf H (1983) Einfluß von Isosorbiddinitrat auf die Fließeigenschaften des Blutes von Patienten mit chronischer arterieller Verschlußkrankheit. Abstract-Band 2. Kongr Dtsch Ges Klin Hämorheol, München
300 Ehrly AM, Landgraf H (1984) Einfluß von Isosorbiddinitrat auf die Fließeigenschaften des Blutes von Patienten mit chronischer arterieller Verschlußkrankheit. In: Ehrly AM (ed) Therapie mit hämorheologisch wirksamen Substanzen. Zuckschwerdt, München, p 145
301 Ehrly AM, Landgraf H (1984) Effects on muscle tissue oxygen tension in claudicants: Infusions of

5% albumin versus 10% hydroxyethyl starch solution (200/0,5). Int J Microcirc Clin Exp 3 (3-4): 521
302 Ehrly AM (ed) (1984) Therapie mit hämorheologisch wirksamen Substanzen. Zuckschwerdt, München
303 Ehrly AM (1984) Evaluation of enzymatic defibrinogenation. Clin Hemorheol 4: 151
304 Ehrly AM (1985) Zum Stellenwert der Erythrozytenaggregation für die gestörte Mikrozirkulation. In: Kiesewetter H, Ehrly AM, Jung F (eds) Hämorheologische Meßmethoden. Münchner Wissenschaftliche Publikationen, München, p 17
305 Ehrly AM (1985) Correlation between tissue oxygen supply and flow properties of blood in patients with ischemic diseases. Acta Cardiol Mediterr 3: 433
306 Ehrly AM (1985) Rheoregulation: Influence of vasoconstriction and vasodilation on the flow properties of blood (Abstract). Clin Hemorheol 5: 617
307 Ehrly AM (1985) Influence of urokinase on the flow properties of blood (Abstract). Clin Hemorheol 5: 728
308 Ehrly AM, Landgraf H, Saeger-Lorenz K (1985) Einfluß einer Infusion von 500 ml 5%iger Humanalbumin-Lösung auf den Muskelgewebesauerstoffdruck von Patienten mit Claudicatio intermittens. In: Ehrly AM, Hauss J, Huch R (eds) Klinische Sauerstoffdruckmessung - Gewebesauerstoffdruck und transkutaner Sauerstoffdruck bei Erwachsenen. Münchner Wissenschaftliche Publikationen, München, p 96
309 Ehrly AM (1986) Rheoregulation: Zeitsynchrone Veränderungen der Fließeigenschaften des Bluts in Abhängigkeit vom Sympathikotonus. In: Tillmann W, Ehrly AM (eds) Hämorheologie und Hämatologie. Münchner Wissenschaftliche Publikationen, München, p 75
310 Ehrly AM (1986) Heißt bessere Durchblutung immer auch bessere Versorgung? Fortschr Med 104: 492
311 Ehrly AM (1986) Diagnostic and prognostic relevance of "extra vivum" rheological results in vascular diseases. Clin Hemorheol 6: 211
312 Ehrly AM (1986) Circadian rhythm of blood viscosity. Clin Hemorheol 6: 279
313 Ehrly AM (1986) Hemorheology and tissue oxygen supply: Relationship to pharmacological treatment. Clin Hemorheol 6: 325
314 Ehrly AM, Brecht HM (1987) Einfluß von Etilefrin auf verschiedene mikrozirkulatorische Parameter von Patienten mit Claudicatio intermittens. In: Strauer BE, Ehrly AM, Leschke M (eds) Fortschritte in der kardiovaskulären Hämorheologie (5. DGKH-Kongreß, Marburg, 1986). Münchner Wissenschaftliche Publikationen, München, p 71 [Abstract: Klin Wochenschr 64 (1986), 1156]
315 Ehrly AM, Hauss J, Huch R (eds) (1987) Clinical oxygen pressure measurement. Tissue oxygen pressure and transcutaneous oxygen pressure. Springer, Berlin
316 Ehrly AM, Landgraf H, Moschner PV, Saeger-Lorenz K (1987) Hydroxyäthylstärke gegen Dextran: Verhalten des Gewebesauerstoffdruckes und der Fließeigenschaften des Blutes bei Patienten mit Claudicatio intermittens. Med Welt 38: 976
317 Eisenberg S (1964) Changes in blood viscosity, hematocrit value and fibrinogen concentration in subjects with congestive heart failure. Circulation 30: 686
318 Erdi A, Thomas DP, Kakkar VV, Lane DA, Dormandy JA (1976) Effects of low-dose subcutaneous heparin on whole-blood viscosity. Lancet 2: 342
319 Ernst E, Dormandy J (1981) The effects of arvin and surgery on red cell filterability. Scand J Clin Lab Invest 41 (Suppl 156): 317
320 Ernst E (1983) Therapie-Ansatz: Fließeigenschaften des Blutes. Münch Med Wochenschr 125: 8
321 Ernst E (1984) Sind "Rheologika" therapeutisch von Nutzen? Z Allg Med 60: 192
322 Ernst E, Magyarosy I, Scherer A, Schmidlechner C (1984) Der Einfluß physikalischer Reize auf die Blutfluidität. Z Phys Med Baln Med Klim 13: 359
323 Ernst E, Marshall M (1984) Verbesserung der reduzierten Erythrozytenflexibilität durch Calciumdobesilat. Ergebnisse einer Pilotstudie. Münch Med Wochenschr 124: 125
324 Ernst E, Marshall M (1984) Acetyl salicylic acid is rheologically inactive and hence an ideal reference medication for clinical trials in hemorheology. Clin Hemorheol 4: 563
325 Ernst E, Matrai A (1984) Increased mortality in hypertensives treated with diuretics - A hemorheological problem? Clin Hemorheol 4: 589
326 Ernst E, Rose M, Lee R (1984) Modification of transoperative changes in blood fluidity by hydroxychloroquine: A possible explanation for the drug's antithrombotic effect. Pharmatherapeutica 4: 48
327 Ernst E (1985) Mit Erfolg gegen den Hirnschlag (Editorial). Fortschr Med 103: 10
328 Ernst E, Matrai A (1985) Exercise and blood fluidity (Abstract). Clin Hemorheol 5: 653
329 Ernst E, Matrai A, Kollar L (1985) A randomized, double blind, placebo-controlled, crossover trial on hemodilution in claudicants - preliminary results (Abstract). Clin Hemorheol 5: 756
330 Ernst E, Matrai A, Weihmayr T, Bushe B, Paulsen F (1985) Behandlung der arteriellen Ver-

schlußkrankheit Stadium II. Hämorheologische und klinische Parameter nach oraler Applikation von Pentoxifyllin (Rentylin®). Münch Med Wochenschr 127: 917

331 Ernst E, Matrai A, Weihmayr T, Paulsen F (1985) Hemorheological and clinical effects of pentoxifylline in claudicants (Abstract). Clin Hemorheol 5: 730

332 Ernst E (1986) Hämorheologie für den Praktiker. Zuckschwerdt, München

333 Ernst E, Matrai A, Kollar L (1986) A randomized, double-blind, placebo-controlled, crossover trial on isovolemic hemodilution in claudicants - Preliminary results. Clin Hemorheol 6: 297

334 Ernst E, Matrai A, Schönhaber H, Paulsen HF, Magyarosy I (1986) Schlaganfallrehabilitation im Spiegel hämorheologischer Meßgrößen. Herz/Kreislauf 18: 30

335 2nd European Conference on Clinical Hemorheology, London (1981) (Abstracts). Clin Hemorheol 1: 451

336 3rd European Conference on Clinical Hemorheology, Baden-Baden (1983) (Abstracts). Clin Hemorheol 3: 221

337 4th European Conference on Clinical Hemorheology, Siena (1985) (Abstracts). Clin Hemorheol 5: 611

338 Everts B, Böhme H (1980) Haemostaseological and rheological changes induced by contrast media after catheter angiography. Abstract-Booklet World Congr Angiol, Athen, 1980, p486

339 Everts B, Böhme H (1981) Kontrastmittelbedingte hämostaseologische und rheologische Veränderungen nach Katheterangiographie. Münch Med Wochenschr 123: 1180

340 Everts B, Böhme H (1983) Haemostaseological and rheological changes by a nonionic contrast medium after lower limb phlebography. Abstract-Booklet 13th World Congr Int Union Angiol, Rochester, Minnesota, USA

341 Ewald CA (1877) Ueber die Transpiration des Blutes. Arch Anat Physiol, p208

342 Fahey JL (1963) Serum protein disorders causing clinical symptoms in malignant neoplastic disease. J Chronic Dis 16: 703

343 Fahey JL, Barth WE, Solomon A (1965) Serum hyperviscosity syndrome. JAMA 192: 464

344 Fahraeus R, Lindqvist T (1931) The viscosity of blood in narrow capillary tubes. Am J Physiol 96: 562

345 Falchi Delitala G, Falchi Delitala N, Casali R, Crescenzi GS, Attorri L, Lombardi R, Martinangeli A, Amici S, De Santis M (1984) Studio a medio termine, in doppio cieco versus placebo, con CDP-colina nella insufficienza cerebrale senile - Aspetti psichici, endocrinologici, emoreologici e biochimico ematologici. Gazz Med Ital Arch Sci Med 143: 789

346 Fazzini GP, Albanese B, Manescalchi PG, Lombardi C, Pasquini G, Bartoli V (1981) Biochemical and haemorheological findings during treatment of plasma cell dyscrasias by plasmapheresis or plasma exchange (Abstract). Clin Hemorheol 1: 473

347 Fazzini GP, Albanese B, Manescalchi PG, Lombardi C, Pasquini G, Bartoli V (1981) Biochemical and haemorheological findings during treatment of plasma cell dyscrasias by plasmapheresis or plasma exchange. Clin Hemorheol 1: 565

348 Fendler K, Horváth L (1983) Effect of oral SP 54 treatment on the primary lipoproteinemias and on the viscosity of blood (Abstract). Clin Hemorheol 3: 293

349 Ferretti PG, Gelmini G, Dall'Asta D, Butturini L, Marchini L, Passeri M (1985) L'emodiluizione normovolemica in pazienti anziani poliglobulici con insufficienza respiratoria cronica. Ric Clin Lab 15 (Suppl 1): 617

350 Folkow B, Mellander S (1964) Veins and venous tone. Am Heart J 68: 397

351 Fontányi S, Baranyi P, Kotányi P, Sikos G (1981) Relative whole blood viscosity measurements in long term heparin treatment of atherosclerotic claudicants. Abstract-Booklet 2nd Europ Conf Clin Haemorheol, London, p94

352 Forbes CD, Barbenell J, Prentice CRM (1976) Treatment of thrombosis by sequential therapy with ancrod followed by streptokinase - Clinical pharmacology and rheology. Haemostasis 5: 348

353 Forconi S (1981) Clinical pharmacology of some so-called rheological drugs. Is a rheological treatment of chronic vascular disease possible? (Abstract). Clin Hemorheol 1: 473

354 Forconi S, Pieragalli D, Acciavatti A, Del Bigo C, Galigani C, Ralli L, Guerrini M, Di Perri T (1984) Positive effect of oral buflomedil on exercise-induced haemorheological damage and on claudication distance in peripheral obliterative arterial disease patients. J Int Med Res 12: 188

355 Forconi S, Pieragalli D, Acciavatti A, Galigani C, Materazzi M, Ralli L, Franchi M, Laghi Pasini F, Messa GL, Bianciardi A, Rigato M, Blardi P, Guerrini M, Di Perri T (1985) Blood hyperviscosity and peripheral ischemia; the secondary hyperviscosity syndrome of ischemizing vascular diseases (Abstract). Clin Hemorheol 5: 708

356 Franke I (1974) Die medikamentöse Beeinflussung der Blutviskosität unter besonderer Berücksichtigung von BL 191. Dissertationsschrift, München

357 Fritz KW, Piehl W, Freyland MD (1970) Zur Frage des Zusammenhanges von renalem hyperfunktionellem Syndrom und Hyperviscosität beim Plasmocytom. Klin Wochenschr 48: 1126
358 Furkalo NK, Ivashchenko TI, Romanenko AI (1985) Therapeutic efficiency of anti-anginal drugs in coronary patients in relation to their effects on blood rheological properties. Kardiologiya 25 (9): 55
359 Gaethgens P (1970) Die Hämodynamik der terminalen Strombahn. Untersuchungen zur Analyse des peripheren Widerstandes im Bereich des mesenterialen Gefäßbettes. Habilitationsschrift, Köln
360 Gaethgens P, Kreutz F, Albrecht KH (1979) Optimal hematocrit for canine skeletal muscle during rhythmic isotonic exercise. Eur J Appl Physiol 41: 27
361 Gaillard S, Schooneman F, Voisin P, Pointel JP, Rousselle D, Drouin P, Stoltz JF (1979) Effets comparés sur la filtrabilité in vivo chez l'homme de deux formes galéniques de la pentoxifylline. Ann Méd Nancy 105: 1099
362 Galigani C, Pieragalli D, Acciavatti A, Guerrini M, Di Perri T (1985) Activity of calcium dobesylate on plasmatic hyperviscosity of diabetic microangiopathy (Abstract). Clin Hemorheol 5: 738
363 Gallasch G, Diehm C, Dörfer C, Schmitt T, Stage A, Mörl H (1985) Einfluß von körperlichem Training auf die Blutfließeigenschaften bei Claudicatio intermittens-Patienten. Klin Wochenschr 63: 554
364 Gallasch G, Dörfer C, Schmitt T, Stage A (1985) Wirksamkeit von Troxerutin auf die Fließeigenschaften des Blutes unter definierten Strömungsbedingungen. Eine Doppelblindstudie an Patienten mit Retinopathia diabetica und Retinopathia arteriosclerotica. Klin Monatsbl Augenheilkd 187: 30
365 Gallus AS, Gleadow F, Dupont P, Walsh J, Morley AA, Wenzel A, Alderman M, Chivers D (1985) Intermittent claudication: A double-blind crossover trial of pentoxifylline. Aust NZ J Med 15: 402
366 Gelin LE (1961) Disturbance of the flow properties of blood and its counteraction in surgery. Acta Chir Scand 122: 287
367 Gelin LE, Ingelman B (1961) Rheomacrodex - a new dextran solution for rheological treatment of impaired capillary flow. Acta Chir Scand 122: 294
368 Gelin LE, Thorén OKA (1961) Influence of low viscous dextran on peripheral circulation in man. A plethysmographic study. Acta Chir Scand 122: 303
369 Gelin LE (1962) Flüssigkeitsersatz im Schock. In: Bock KD (ed) Schock - Pathogenese und Therapie. Springer, Berlin, p372
370 Gelin LE (1962) Rheologic disturbances and the use of low viscosity dextran in surgery. Rev Surg 19: 385
371 Gelin LE (1965) Rheologic disturbances following tissue injury. In: Copley AL (ed) Symposium on Biorheology. Part 4 of the Proceedings of the Fourth International Congress on Rheology, Brown University, Providence, Rhode Island, 1963. Interscience, New York, p299
372 Gelin LE, Kerstell J, Svanborg A (1967) The effect of dietary fat on whole-blood and plasma viscosity in normal and hypercholesterolemic subjects. Acta Med Scand 181: 41
373 Gelin LE (1969) Rheological aspects on shock and the experimental and clinical use of dextrans. In: Ingelman B, Grönwall A, Gelin LE, Eliasson R (eds) Properties and application of dextrans. Almqvist & Wiksell, Stockholm, p43
374 Gelin LE (1971) Rheological effects of dextran. In: Derrick JR, Guest M (eds) Dextrans - Current concepts of basic actions and clinical application. CC Thomas, Springfield/Illinois (USA), p137
375 Gibbs NM, Oh TE, Chester A (1984) Blood viscosity and blood film morphology during prolonged sedation with althesin. Br J Anaesth 56: 1179
376 Girolami A, Cella G (1976) Effect of low-dose subcutaneous heparin on whole-blood viscosity. Lancet 2: 909
377 Girolami A, Patrassi G, Bova G (1978) No change of whole blood or plasma viscosity in cardiac patients after subcutaneous calcium heparin. Folia Haematol 105: 391
378 Glöckner WM, Malotta H, Sieberth HG (1983) Cascade filtration without any substitution as long-term treatment of Waldenström's macroglobulinemia with hyperviscosity syndrome (Abstract). Clin Hemorheol 3: 337
379 Glöckner WM, Sieberth HG, Schmid-Schönbein H (1985) Cascade filtration for the control of hyperviscosity by the removal of IgM-paraprotein in Waldenstrom's macroglobulinaemia. Clin Hemorheol 5: 71
380 Gluskowski J, Jedrzejewska-Makowska M, Hawrylkiewicz I, Vertun B, Zielinski J (1983) Effects of prolonged oxygen therapy on pulmonary hypertension and blood viscosity in patients with advanced cor pulmonale. Respiration 44: 177
381 Goldberg K, Wirth FH, Hathaway WE, Guggenheim MA, Murphy JR, Braithwaite WR, Lubchenco LO (1982) Neonatal hyperviscosity. II. Effect of partial plasma exchange transfusion. Pediatrics 69: 419
382 González de Zárate Apiñaniz J, García Domínguez R, Rodríguez Abela J, Baeza Gil C, Girón Montáñez R (1985) Efectos reológicos de la propanidida-cremophor EL. Rev Española Anest Rean 32: 264

383 Gormsen G (1954) Does rapid dehydration in cardiac decompensation produce thrombembolic complications? Acta Med Scand 148: 61
384 Gottschalk M (1985) Infusionsbedingte Blutviskositätsänderungen im Kindesalter. Kinderarztl Prax 53: 579
385 Gottstein U, Held K (1969) Effekt der Hämodilution nach intravenöser Infusion von niedermolekularen Dextranen auf die Hirnzirkulation des Menschen. Dtsch Med Wochenschr 94: 522
386 Gottstein U, Sedlmeyer I, Schöttler M, Gülk V (1970) Der Effekt von niedermolekularem Dextran auf die Unterschenkeldurchblutung von Gesunden und Kranken mit peripheren arteriellen Zirkulationsstörungen. Dtsch Med Wochenschr 95: 1955
387 Gottstein U (1972) Allgemeine Behandlungsrichtlinien einer medikamentösen Therapie der Hirnarterienverschlußfolgen. Verh Dtsch Ges Inn Med 78: 466
388 Gottstein U (1984) Evaluation of isovolemic hemodilution. Clin Hemorheol 4: 133
389 Gousios A, Shearn MA (1959) Effect of intravenous heparin on human blood viscosity. Circulation 10: 1063
390 Gramstad L, Stovner J (1979) Plasma viscosity and cremophor-containing anaesthetics. Br J Anaesth 51: 1175
391 Grasselli S, Parise P, Berrettini M, De Cunto M, Nenci GG (1985) Red blood cell (RBC) deformability during prostacyclin therapy (Abstract). Clin Hemorheol 5: 750
392 Grindle CFJ, Buskard NA, Newman DL (1976) Hyperviscosity retinopathy - A scientific approach to therapy. Trans Ophthalmol Soc UK 96: 216
393 Gross GP, Hathaway WE, McGaughey HR (1973) Hyperviscosity in the neonate. J Pediatr 82: 1004
394 Groth CG, Thorsén HGR (1965) Effect of Rheomacrodex® on factors of importance for the flow properties of blood. Bibl Anat 7: 267
395 Groth CG, Thorsén G (1965) The effect of Rheomacrodex and Macrodex on factors governing the flow properties of the human blood. Acta Chir Scand 130: 507
396 Groth CG (1966) The effect of infused albumin and Rheomacrodex on factors governing the flow properties of the human blood. Acta Chir Scand 131: 290
397 Groth CG (1968) Plasmaexpanders and the flow properties of blood. In: Copley AL (ed) Hemorheology. Pergamon, New York, p 445
398 Groth CG (1969) Hemorheology, plasma expanders and nutritional blood flow. Int Anesthesiol Clin 7: 239
399 Gueguen M, Durand F, Grosbois B, Godinot B, Leblay R, Genetet B (1983) Haemorheology applies in the diagnosis of monoclonal dysglobulinemia and in the following-up of their plasma exchange therapy (Abstract). Clin Hemorheol 3: 335
400 Gueguen M, Le Pogamp P, Avril Y, Le Goff MC, Durand F, Joyeux V, Genetet B, Chevet D (1985) Effects of haemodialysis on the rheological parameters: preliminary results (Abstract). Clin Hemorheol 5: 675
401 Gülzow K, Euler HH, Löffler H (1984) Therapeutische Apheresen bei Hyperviskositätssyndromen: Klinischer Einsatz einer einfachen Methode der Viskosimetrie. In: Ehrly AM (ed) Therapie mit hämorheologisch wirksamen Substanzen. Zuckschwerdt, München, p 121
402 Guerrini M, Pieragalli D, Messa GL (1981) Changes in blood viscosity after acute administration of l-eburnamonine. Drugs Exp Clin Res 7: 103
403 Guerrini M, Pecchi S, Rossi C, Cappelli R, Pieragalli D, Acciavatti A (1981) Azione della pentossifillina sull'iperviscosità ematica e sulla emodinamica periferica in pazienti affetti da arteriopatia obliterante periferica. Ric Clin Lab 11 (Suppl 1): 265
404 Guerrini M, Acciavatti A, Pecchi S, Cappelli R, Sacchetti S, Forconi S, Di Perri T (1981) Effects of a single i.v. dose of nicergoline on hemorheological and hemodynamic parameters in peripheral vascular disease. A double-blind crossover study. Abstract-Booklet 2nd Europ Conf Clin Haemorheol, London, p 114
405 Guerrini M, Pecchi S, Rossi C, Cappelli R, Pieragalli D, Forconi S, Di Perri T (1981) Possibility of treating vascular diseases from a rheological point of view. Study of the effects of isoxsuprine (Abstract). Clin Hemorheol 1: 475
406 Guerrini M, Rossi C, Acciavatti A, Pieragalli D, Pecchi S, Cappelli R, Forconi S, Di Perri T (1981) Possibility of treating vascular diseases from a rheological point of view. Study of the effects of flunarizine and cinnarizine (Abstract). Clin Hemorheol 1: 476
407 Guerrini M, Rossi C, Acciavatti A, Pieragalli D, Pecchi S, Cappelli R, Forconi S, Di Perri T (1981) Possibility of treating vascular diseases from a rheological point of view. Study of the effects of pentoxifilline (Abstract). Clin Hemorheol 1: 476
408 Guerrini M, Acciavatti A, Materazzi M, Rossi C, Del Bigo C, Forconi S, Di Perri T (1982) Protective effects of buflomedil against exercise-induced reductions in regional erythrocyte deformability of patients with peripheral arterial disease. J Int Med Res 10: 387

409 Guerrini M, Pieragalli D, Acciavatti A, Cappelli R, Pecchi S, Forconi S, Di Perri T (1983) L'emodiluizione normovolemica può realmente modificare il flusso ematico degli arti? Ric Clin Lab 13 (Suppl 3): 439
410 Gustafsson L, Appelgren L, Myrvold HE (1985) Effects of hemodilution on skeletal muscle blood flow and blood viscosity in vivo after splanchnic stasis. Eur Surg Res 17: 366
411 Haass A, Kroemer H, Brenner M, Jäger H, Weiss H, Oest A (1986) Hämorheologische und hämodynamische Effekte von Dextran 40- und HAES-200-Langzeitinfusionen. In: Tillmann W, Ehrly AM (eds) Hämorheologie und Hämatologie. Münchner Wissenschaftliche Publikationen, München, p 130
412 Hagen G (1839) Über die Bewegung des Wassers in engen zylindrischen Röhren. Poggendorff's Annalen der Physik und Chemie 46: 423
413 Hagenbach E (1860) Über die Bestimmung der Zähigkeit einer Flüssigkeit durch Ausfluß aus Röhren. Poggendorff's Annalen der Physik und Chemie 109: 385
414 Hall JA, Barnard RJ, Pritikin N (1983) Effects of a high-complex-carbohydrate, low-fat, low-salt diet and daily exercise on blood viscosity and blood pressure in hypertensive patients (Abstract). Medicine Science Sports Exercise Madison 15: 91
415 Hamazaki T, Hasunuma K, Kobayashi S, Shishido H, Yano S (1985) The effects on lipids, blood viscosity and platelet aggregation of combined use of niceritol (Perycit®) and a low dose of acetylsalicylic acid. Atherosclerosis 55: 107
416 Hansen LL, Wiek J, Arntz HR (1985) Effect of isovolemic hemodilution on visual outcome of patients with ischemic and non-ischemic retinal vein occlusion (Abstract). Clin Hemorheol 5: 691
417 Harke H, Pieper C, Meredig J, Rahman S, Rüssler P (1980) Rheologische und gerinnungsphysiologische Untersuchungen nach Infusion von HÄS 200/0,5 und Dextran 40. Anaesthesist 29: 71
418 Harkness J, Whittington RB (1971) The viscosity of human blood plasma: Its change in disease and on the exhibition of drugs. Rheol Acta 10: 55
419 Haro A (1873) Essai sur la transpirabilité du sang. Gazette Hebdomadaire 11
420 Hartert HH, Copley AL (eds) (1971) Theoretical and Clinical Hemorheology. Springer, Berlin
421 Heidrich H, Ott M (1974) Vasodilatantien und Blutviskosität. Herz/Kreislauf 6: 542
422 Heidrich H, Ott M, Schlichting K (1974) Blutviskosität unter Fludilat. In-vivo- und In-vitro-Untersuchungen. Therapiewoche 24: 2900
423 Heidrich H, Schlichting K, Ott M (1975) Untersuchungen zum Einfluß von Pentoxifyllin auf die Blutviskosität und die Erythrozytenoberfläche. Med Welt 26: 1977
424 Heidrich H (1975) Klinisch-experimentelle Untersuchungen zu neuen Wirkungsmechanismen sogenannter Vasodilatantien und gefäßaktiver Pharmaka. Habilitationsschrift, Berlin
425 Heidrich H, Schlichting K, Ott M (1976) Changes in blood viscosity due to pentoxifylline. IRCS Med Sci 4: 368
426 Heidrich H, Wachta T (1977) Verhalten der Blutviskosität unter 2-, 4- und 6stündiger intravenöser Rheomakrodex- und Kochsalz-Infusion. Verh Dtsch Ges Kreislaufforsch 43: 433
427 Heidrich H, Wachta T (1978) Blutviskosität unter intravenöser Langzeitinfusion von niedermolekularem Dextran (Rheomacrodex 10%). Dtsch Med Wochenschr 103: 298
428 Heilmann E, Zimmermann B, Gruss P (1982) Effect of pentoxifylline on hypercoagulability in patients with polycythemia vera. IRCS Med Sci 10: 594
429 Heilmann E, Laage HM, Zimmermann E, Dorst KG (1982) Medikamentöse Beeinflussung der Erythrozytenflexibilität bei Sichelzellanämie. Innere Med 9: 277
430 Heilmann E (1984) Hämatologische und rheologische Untersuchungen bei Polycythaemia vera. In: Ehrly AM (ed) Therapie mit hämorheologisch wirksamen Substanzen. Zuckschwerdt, München, p 186
431 Heilmann L (1980) Hämorheologische Veränderungen bei Risikoschwangerschaften unter der Therapie mit Flunarizin. Vasa 9: 338
432 Heilmann L, Tauber PF, Siekmann U, Ludwig H (1981) Die Fließfähigkeit des Blutes unter Einnahme von Ovulationshemmern. Zentralbl Gynakol 103: 678
433 Heilmann L (1983) Hypervolemic hemodilution in the treatment of preeclampsia (Abstract). Clin Hemorheol 3: 550
434 Heilmann L, Siekmann U (1983) Hämorheologische Veränderungen unter Beta-Blocker-Therapie. In: Irmer M, Weidinger H (eds) Neuere Aspekte zu: Betablockade und Tokolyse. Beltz, Weinheim, p 91
435 Heilmann L, Siekmann U (1983) Die hypervolämische Hämodilution bei der Präeklampsie. Infusionsther Klin Ernahr 10: 311
436 Heilmann L, Siekmann U (1983) Effect of betamimeticum hexoprenaline on the erythrocyte deformability and sodium/potassium-cotransport system of pregnant women (Abstract). Clin Hemorheol 3: 293

437 Heilmann L (1984) Hämorheologisch orientierte Therapieprinzipien in der Geburtshilfe. In: Ehrly AM (ed) Therapie mit hämorheologisch wirksamen Substanzen. Zuckschwerdt, München, p 88
438 Heilmann L, Kiesewetter H, Ernst E (eds) (1984) Klinische Rheologie und Beta-1-Blockade. Zuckschwerdt, München
439 Heilmann L, Hirche H (1986) Rheologische Veränderungen unter Naftidrofuryl bei der Gestose. Angio 8: 33
440 Heilmann L, Beez M (eds) (1987) Neuere klinische Aspekte zur Hämodilution. Schattauer, Stuttgart
441 Heinen A, Brunner R, Wawer T, Roll K, Konen W, Hossmann V (1983) Randomized double-blind trial of isovolaemic haemodilution (IHD) in retinal vessel occlusion (Abstract). Clin Hemorheol 3: 233
442 Hess H (1970) Neue Überlegungen zur Physiopathologie obliterierender Angiopathien. 1. Zur Regulation der Durchblutung. Fortschr Med 88: 833
443 Hess H (1970) Neue Überlegungen zur Physiopathologie obliterierender Angiopathien. 2. Zur Pathogenese und Therapie. Fortschr Med 88: 923
444 Hess H, Franke J, Jauch M (1973) Medikamentöse Verbesserung der Fließeigenschaften des Blutes. Ein wirksames Prinzip zur Behandlung von arteriellen Durchblutungsstörungen. Fortschr Med 91: 743
445 Hess H, Wagner G (1974) Untersuchungen über den Einfluß von Fludilat auf die Blutviskosität. Therapiewoche 24: 2898
446 Hess W (1907) Die Bestimmung der Viskosität des Blutes. Münch Med Wochenschr 45: 2225
447 Hess WR (1912) Der Strömungswiderstand des Blutes gegenüber kleinen Druckwerten. Arch Anat Physiol, Physiolog Abt, Leipzig, p 197
448 Hildebrandt HG (1980) Einfluß von Buflomedil auf die Fließeigenschaften des Blutes und die Gewebesauerstoffdruckwerte in der Muskulatur bei Patienten mit Claudicatio intermittens. Dissertationsschrift, Frankfurt am Main
449 Hirsch C, Beck C (1901) Studien zur Lehre von der Viscosität (inneren Reibung) des lebenden, menschlichen Blutes. Dtsch Arch Klin Med 69: 503
450 Hoffmann LS, Ritzmann SE (1965) Plasmapheresis - Application of modified technic in patients with monoclonal gammopathies. Tex Rep Biol Med 23: 651
451 Holti G (1979) An experimentally controlled evaluation of the effect of inositol nicotinate upon the digital blood flow in patients with Raynaud's phenomenon. J Int Med Res 7: 473
452 Hossmann V, Bewermeyer H, Heiss WD (1981) A controlled trial of ancrod in the treatment of acute ischaemic stroke. Abstract-Booklet 2nd Europ Conf Clin Haemorheol, London, p 41
453 Hossmann V, Wegener H, Wegener B, Saborowski F, Cäsar K (1982) Hämorheologische und hämodynamische Wirkungen von Isosorbiddinitrat bei essentieller Hypertonie und arterieller Verschlußkrankheit. In: Engel HJ, Schrey A, Lichtlen PR (eds) Nitrate III - Kardiovaskuläre Wirkungen. Springer, Berlin, p 287
454 Hossmann V, Auel H (1983) Haemorheologische Therapie der arteriellen Verschlußkrankheit: Eine kontrollierte, prospektive Studie mit isovolämischer Haemodilution, Arwin und einer Kombination beider Methoden. In: Nobbe F, Rudofsky G (eds) Probleme der Vor- und Nachsorge und der Narkoseführung bei invasiver angiologischer Diagnostik und Therapie. Begutachtung arterieller und venöser Gefäßschäden. Pflaum, München, p 325
455 Hossmann V, Heiss WD, Bewermeyer H, Wiedemann G (1983) Controlled trial of ancrod in ischemic stroke. Arch Neurol 40: 803
456 Hossmann V (1983) Controlled study of the hemorheological effects of ancrod, isovolemic hemodilution and combined treatment in advanced stages of peripheral arterial diseases (Abstract). Clin Hemorheol 3: 309
457 Hudomel J, Németh B, Palfalvy M, Farkas A (1977) Effect of calcium dobesilate (Doxium) on blood hyperviscosity in cases of diabetic retinopathy. Ophthalmic Res 9: 25
458 Hudomel J, Németh B, Palfalvy M, Farkas A (1979) Wirkung von Kalzium-Dobesilat auf die Hyperviskosität des Blutes. Fortschr Med 97: 422
459 Hürthle K (1900) Ueber eine Methode zur Bestimmung der Viscosität des lebenden Blutes und ihre Ergebnisse. Arch Ges Physiol 82: 415
460 Humphrey PRD, Michael J, Pearson TC (1980) Management of relative polycythaemia: Studies of cerebral blood flow and viscosity. Br J Haematol 46: 427
461 Humphrey PRD, Michael J, Pearson TC (1980) Red cell mass, plasma volume and blood volume before and after venesection in relative polycythaemia. Br J Haematol 46: 435
462 Humphrey PRD, Du Boulay GH, Marshall J, Pearson TC, Ross Russell RW, Symon L, Wetherley-Mein G, Zilkha E (1981) Venesection in relative polycythaemia: Measurement of cerebral blood flow, viscosity and blood volume. Abstract-Booklet 2nd Europ Conf Clin Haemorheol, London, p 38

463 Humphrey PRD, Du Boulay GH, Marshall J, Pearson TC, Ross Russell RW, Symon L, Wetherley-Mein G, Zilkha E (1981) Management of relative polycythaemia: Comparative effects of venesection, dextran and fludrocortisone. Abstract-Booklet 2nd Europ Conf Clin Haemorheol, London, p112

464 Humphreys WV, Walker A, Cave FD, Charlesworth D (1976) The effect of an infusion of low molecular weight dextran on peripheral resistance in patients with arteriosclerosis. Br J Surg 63: 691

465 Humphreys WV, Walker A, Charlesworth D (1977) A report on the use of arvin in patients with pregangrene of the lower limb. Br J Surg 64: 31

466 5th International Congress of Biorheology, Baden-Baden (1983) (Abstracts). Biorheology 20: 361

467 International Symposium on filterability and red blood cell deformability, Göteborg, Sweden (1980). Scand J Clin Lab Invest 41 (Suppl 156)

468 Itoh C, Ishii A, Iwafune Y, Yoshimoto H (1982) Fibrinolytic therapy and blood viscosity in simple diabetic retinopathy. Nippon Ganka Kiyo 33: 2469

469 Iwamoto H, Nakagowa S, Matsui N, Yoshiyama N, Shinoda T, Shibamoto T, Takenchi J (1980) An experience of plasma exchange by membrane separator for IgA myeloma. In: Sieberth HG (ed) Plasma exchange. Schattauer, Stuttgart, p377

470 Jacobs MJ, Lemmens HAJ, Slaaf DW, Reneman RS (1983) The effect of ketanserin on microcirculation in patients with ischaemic handsyndromes. Abstract-Booklet 13th World Congr Int Union Angiol, Rochester, Minnesota, USA

471 Jahnsen T, Skovborg F, Hansen F, Larsen J, Nordin H, Storm T (1983) Variations in blood viscosity in patients with acute cardiogenic pulmonary oedema treated with frusemide. Scand J Clin Lab Invest 43: 297

472 Jahnsen T, Skovborg F, Jahnsen F, Storm TL (1985) Blood viscosity in patients receiving continuous treatment with bendroflumethiazide (Centyl® with potassium chloride). Ugeskr Læger 147: 409

473 James DR, Holland BM, Hughes MR, Jones JG, Wardrop CAJ (1984) Oxpentifylline: Effects on red cell deformability and oxygen availability from the blood in intermittent claudication. Clin Hemorheol 4: 525

474 Jan K, Usami S, Smith JA (1982) Effects of transfusion on rheological properties of blood in sickle cell anemia. Transfusion 22: 17

475 Jauch M (1973) Der Einfluß von verschiedenen Medikamenten auf die Viskosität des Blutes. Dissertationsschrift, München

476 Johnsson R (1979) Effect of pentoxifylline on red cell flexibility and cation transport in healthy subjects and patients with hereditary spherocytosis. Scand J Haematol 23: 81

477 Johnsson R, Harjola PT, Siltanen P (1981) Effect of pentoxifylline on red cell flexibility in arteriosclerotic patients and in patients with heart valve prosthesis. Scand J Clin Lab Invest 41 (Suppl 156): 297

478 Juhan I, Buonocore M, Calas-Aillaud MF, Moulin JP, Verdot JJ, Vague P (1980) Effects of insulin on erythrocyte deformability and platelet aggregation in diabetics on an artificial pancreas. Abstract-Booklet Int Symp filterability and red cell deformability, Göteborg, Sweden, p53

479 Juhan I, Vague P, Buonocore M, Moulin JP, Calas MF, Vialettes B, Verdot JJ (1981) Effects of insulin on erythrocyte deformability in diabetics - relationship between erythrocyte deformability and platelet aggregation. Scand J Clin Lab Invest 41 (Suppl 156): 159

480 Juhan I, Vague P, Buonocore M, Muratore R (1981) Déformabilité des hématies chez les diabétiques. Effets de l'insuline. Relations avec l'agrégation plaquettaire. Sem Hôp Paris 57: 466

481 Juhan I, Vague P, Buonocore M, Moulin JP, Jouve R, Vialettes B (1982) Abnormalities of erythrocyte deformability and platelet aggregation in insulin-dependent diabetics corrected by insulin in vivo and in vitro. Lancet 1: 535

482 Jung F, Waldhausen P, Schwab J, Braun B, Molter A, Scheffler P, Kiesewetter H, Wenzel E (1986) Wirkung von Naftidrofuryl auf Mikrozirkulation und Hämorheologie bei peripherer arterieller Verschlußkrankheit im Stadium IIb. In: Tillmann W, Ehrly AM (eds) Hämorheologie und Hämatologie. Münchner Wissenschaftliche Publikationen, München, p163

483 Jung F, Waldhausen P, Schwab J, Molter A, Scheffler P, Kiesewetter H, Wenzel E (1986) Einfluß von Naftidrofuryl auf die Mikrozirkulation bei der peripheren arteriellen Verschlußkrankheit. Med Welt 37: 132

484 Kallio V, Salmivalli M, Brummer P (1967) Blood viscosity changes in patients hospitalized because of acute chest pain. Cardiologia 50: 323

485 Karasawa T, Hayano Y, Imai N (1982) Significance and treatment of abnormal blood viscosities in pre-eclamptic patients. Acta Obstet Gynecol Jpn 34: 599

486 Keller F, Leonhardt H (1979) Amelioration of blood viscosity in sickle cell anemia by pentoxifylline - A case report. J Med 10: 429

487 Keller F, Leonhardt H (1980) Verbesserung der Blutviskosität bei Sichelzellanämie durch Pentoxifyllin. Dtsch Med Wochenschr 105: 898

488 Kennedy MS, Domen RE (1983) Therapeutic apheresis. Applications and future directions. Vox Sang 45: 261

489 Keppeler D (1977) Untersuchungen über den Einfluß von Hydroxyäthylstärke auf die Fließeigenschaften des Blutes. Dissertationsschrift, Frankfurt am Main

490 Kharkharov MA, Tirulov MT (1986) Changes in blood viscosity and central hemodynamics in patients with essential hypertension during treatment with sulfide baths. Vopr Kurortol Fizioter Lech Fiz Kult 1: 13

491 Kiesewetter H, Dauer U, Gerhards M, Gesch M, Schmid-Schönbein H (1981) The single erythrocyte rigidometer (SER) as a reference for RBC-deformability tests. Biorheology 18: 157

492 Kiesewetter H, Radtke H, Schneider R, Mußler K, Scheffler A, Schmid-Schönbein H (1982) Das Mini-Erythrozytenaggregometer: Ein neues Gerät zur schnellen Quantifizierung des Ausmaßes der Erythrozytenaggregation. Biomed Techn 27: 209

493 Kiesewetter H, Blume J, Bulling B, Gerhards M, Jung F, Radtke H, Franke RP (1984) Mittelmolekulare Hydroxyäthylstärke als Volumenersatzmittel. Rheologische Wirkung bei peripherer arterieller Verschlußkrankheit. Dtsch Med Wochenschr 109: 1844

494 Kiesewetter H, Blume J, Jung F, Radtke H, Bulling B, Gerhards M, Prünte C, Reim M (1984) Hämorheologische Aspekte peripherer arterieller Verschlußkrankheiten. Wirksamkeitsuntersuchung von Naftidrofuryl. Med Welt 35: 896

495 Kiesewetter H, Blume J, Radtke H, Gerhards M, Bulling B (1984) Zur Wirksamkeit des Diuretikums Slimin® zur Unterstützung der physikalischen Entstauungstherapie beim venösen Ödem. Therapiewoche 34: 4469

496 Kiesewetter H, Jung F, Körber N, Blume J, Radtke H, Reim M (1984) Stellenwert hämorheologischer Parameter bei der diabetischen Retinopathie. In: Ehrly AM (ed) Therapie mit hämorheologisch wirksamen Substanzen. Zuckschwerdt, München, p 158

497 Kiesewetter H, Jung F, Doenecke P, Hellstern P, Scheffler P, von Blohn G, Waldhausen P, Schwab J, Kiehl R, Wenzel E (1985) Klinische Wirksamkeit der Urokinase durch Fibrinolyse und rheologisch wirkende Fibrinogenolyse. In: Wenzel E, Hellstern P (eds) Abstracts and Handout-Poster. 29. Jahrestagung der DAB/GTH, Saarbrücken, p 131

498 Kiesewetter H, Le Cotonnec JY, Gulati O (1985) Effect of bencianol on blood rheology (Abstract). Clin Hemorheol 5: 741

499 Kiesewetter H, Blume J, Jung F, Bulling B, Gerhards M (1986) Rheologische Wirkung von Metoprolol bei Patienten mit essentieller Hypertonie. Klinikarzt 15: 93

500 Kiesewetter H, Jung F (1986) Rheologische Therapie der peripheren arteriellen Verschlußkrankheit im Stadium IIb. Angio 8: 21

501 Kiesewetter H, Jung F, Blume J, Bulling B, Franke RP (1986) Vergleichende Untersuchung von niedermolekularem Dextran- oder Hydroxyäthylstärkelösungen als Volumenersatzmittel bei Hämodilutionstherapie. Klin Wochenschr 64: 29

502 Kiesewetter H, Schneider R, Jung F (1986) Einsatz von Naftidrofuryl bei Patienten mit Basilarisinsuffizienz. Med Welt 37: 211

503 Kiesewetter H, Schneider R, Jung F, Blume J, Gerhards M (1986) Therapie zerebraler Mikroangiopathien. Untersuchungen mit niedermolekularer Hydroxyäthylstärke mit und ohne Pentoxifyllin. Therapiewoche 36: 50

504 Kiesewetter H, Jung F, Roggenkamp HG, Nüttgens HP, Leipnitz G, Ringelstein EB, Wenzel E, Grillmaier R (1987) Zur diagnostischen Bedeutung rheologischer Parameter in der Klinik. In: Strauer BE, Ehrly AM, Leschke M (eds) Fortschritte in der kardiovaskulären Hämorheologie (5. DGKH-Kongreß, Marburg, 1986). Münchner Wissenschaftliche Publikationen, München, p 185

505 Kilpatrick D, Fleming J, Clyne C, Thompson GR (1979) Reduction of blood viscosity following plasma exchange. Atherosclerosis 32: 301

506 Kindl-Mascher J (1984) Untersuchungen über das Verhalten der Fließeigenschaften des venösen Blutes und des Gewebesauerstoffdruckes in der Unterschenkelmuskulatur von Patienten mit Claudicatio intermittens während einer Therapie mit Batroxobin (Defibrase). Dissertationsschrift, Frankfurt am Main

507 Kirigaya H, Itoh H, Miyahara Y, Inada M, Korenaga M, Kanayama M, Niwa A, Taniguchi K (1982) The effects of drugs on blood viscosity (Abstract). Clin Hemorheol 2: 397

508 Klein M, Szobor A, Bräuer H (1979) Kontrollierte Prüfung von Vincamin. Unter besonderer Berücksichtigung der Fließeigenschaften des Blutes. Med Klin 74 (Sondernummer): 29

509 Klemm J (1976) The treatment of arterial and venous circulatory disturbances with EPL. In: Peeters P (ed) Phosphatidylcholine. Springer, Berlin, p 237

510 Klemm J, Enghofer E (1982) Strömungsgeschwindigkeit von Blut in varikösen Venen der unteren Extremitäten. Einfluß eines Venentherapeutikums (Venostasin®). Münch Med Wochenschr 124: 579
511 Klemm J, Enghofer E (1983) Influence of Venostasin® on blood flow in varicose veins of the lower limbs (Abstract). Clin Hemorheol 3, Nr 150C
512 Klövekorn WP, Pichlmaier H, Ott E, Bauer H, Sunder-Plassmann L, Jesch F, Messmer K (1975) Acute preoperative hemodilution in surgical patients. In: Messmer K, Schmid-Schönbein H (eds) Intentional hemodilution. Karger, Basel, p248
513 Klose HJ, Volger E, Brechtelsbauer H, Heinrich L, Schmid-Schönbein H (1972) Microrheology and light transmission of blood. I. The photometric quantification of red cell aggregation and red cell orientation. Pflüger's Arch 333: 126
514 Knarr W (1975) Beeinflussung der Fließeigenschaften des Blutes durch Pentoxifyllin. Dissertationsschrift, Frankfurt am Main
515 Kobayashi S, Hamazaki T, Hirai A, Terano T, Tamura Y, Kanakubo Y, Yoshida S, Fujita T, Kumagai A (1985) Epidemiological and clinical studies of the effect of eicosapentaenoic acid on blood viscosity. Clin Hemorheol 5: 493
516 Köhler HJ (1971) Untersuchungen über den Einfluß von wasserlöslichen Röntgenkontrastmitteln auf die Fließeigenschaften des Blutes. Dissertationsschrift, Frankfurt am Main
517 Köhler M, Rieger H, Sievers D, Krüpe M (1980) Über die Reaktion des Kreislaufs auf eine gezielte Senkung der Blutviskosität. In: Müller-Wiefel H (ed) Mikrozirkulation und Blutrheologie. Witzstrock, Baden-Baden, p76
518 Köhler S (1973) Untersuchungen über den Einfluß von Diuretika auf die Fließeigenschaften des Blutes. Dissertationsschrift, Frankfurt am Main
519 Koga G (1913) Zur Therapie der Spontangangrän an den Extremitäten. Dtsch Z Chir 11: 371
520 Kollár L, Fendler K, Szekeres P, Kiss T (1985) Evaluation of arvin treatment with crossover double blind study (Abstract). Clin Hemorheol 5: 744
521 Konishi M, Nobuko T, Muneki S, Manubu K, Kojiro Y, Ryu N (1985) Microhemodynamic changes before and after hemodialysis (Abstract). Clin Hemorheol 5: 576
522 Kontras SB, Bodenbender JB, Craenen J, Hosier DM (1970) Hyperviscosity in congenital heart disease. J Pediatr 76: 214
523 Kontras SB (1972) Polycythemia and hyperviscosity syndrome in infants and children. Pediatr Clin North Am 19: 919
524 Kopp WL, Beirne GJ, Burns RO (1967) Hyperviscosity syndrome in multiple myeloma. Am J Med 43: 141
525 Korabelshchikova NI, Bichko MV, Rishko NV (1985) Rheological properties of the blood in stenocardia patients in the dynamics of treatment with obsidan, corvaton, corinfar. Vrach Delo 4: 19
526 Kottmann K (1907) Über die Viskosität des Blutes. Korrespondenzbl Schweizer Ärzte 5: 129
527 Koul BL, Nordhas O, Sonnenfeld T, Ekeström S (1984) The effect of pentoxifylline on impaired red cell deformability following open-heart surgery. Scand J Thorac Cardiovasc Surg 18: 129
528 Krepp HP, Beez M, Bock P (1984) Behandlung des akuten Hirninfarktes durch hypervolämische Hämodilutionstherapie. Klinikarzt 13: 597
529 Krosch H (1959) Rheologische Untersuchungen über die Viskosität des Blutes und ihre Bedeutung für klinische Kreislaufprobleme. Habilitationsschrift, Halle/Saale
530 Kuschinsky G, Lüllmann H (1970) Kurzes Lehrbuch der Pharmakologie und Toxikologie. Thieme, Stuttgart
531 Kuznetsov NA, Filimonov MI, Lopukhin SY, Vasilyev VE (1983) Correction of blood rheologic properties in patients with obstructive jaundice. Khirurgiia (Mosk) 8: 96
532 Kuzuya F, Osawa M, Maruyama S (1985) Effects of vinpocetine on blood viscosity and platelet aggregability. Today's Ther Trends 3: 21
533 Lakomek M, Behrens-Baumann W, Schröter W, Tillmann W (1983) Absence of increased aggregation of erythrocytes in children with diabetes mellitus (Abstract). Clin Hemorheol 3: 270
534 Landgraf H, Ruppel C, Saeger-Lorenz K, Vogel C, Ehrly AM (1982) Verbesserung der Fließeigenschaften des Blutes von Patienten durch niedermolekulare Hydroxyäthylstärke (Expafusin®). Infusionsther Klin Ernahr 9: 202
535 Landgraf H, Ehrly AM (1984) Hämorheologische Wirkungen hypervolämischer Hämodilution mit Dextran 40 und HAES 200 bei Patienten mit arterieller Verschlußkrankheit. In: Ehrly AM (ed) Therapie mit hämorheologisch wirksamen Substanzen. Zuckschwerdt, München, p138
536 Landgraf H, Ehrly AM (1984) Wirkungen einer Clofibrat-induzierten Senkung der Fibrinogenkonzentration im Blut bei peripheren arteriellen Durchblutungsstörungen. In: Mahler F, Nachbur B (eds) Zerebrale Ischämie. Huber, Bern, p277
537 Landgraf H (1984) Untersuchungen über die Wirkungen von Plasmaersatzstoffen auf die Fließ-

eigenschaften des Blutes und die Sauerstoffversorgung der Skelettmuskulatur bei chronischer peripherer arterieller Verschlußkrankheit. Habilitationsschrift, Frankfurt am Main

538 Landgraf H, Ehrly AM (1985) Bedeutung und Möglichkeit der Quantifizierung der Plasmaviskosität. In: Kiesewetter H, Ehrly AM, Jung F (eds) Hämorheologische Meßmethoden. Münchner Wissenschaftliche Publikationen, München, p35

539 Landgraf H, Moschner P, Ehrly AM (1985) Hypervolemic hemodilution (HHD) in claudicants: Hemorheological changes due to infusion of plasma substitutes (Abstract). Clin Hemorheol 5: 723

540 Landgraf H, Ehrly AM (1986) Verhalten des Gewebesauerstoffdruckes bei peripherer arterieller Verschlußkrankheit nach hypervolämischer Hämodilution mit verschiedenen Plasmaersatzstoffen. In: Häring R (ed) DGA - Deutsche Gesellschaft für Angiologie. Referate. 5. Gemeinsame Jahrestagung der Angiologischen Gesellschaften der Bundesrepublik Deutschland, Österreichs und der Schweiz, Berlin 1985. Demeter, Gräfelfing, p304

541 Landgraf H, Ehrly AM (1987) Verhalten hämorrheologischer Parameter und des Gewebesauerstoffdruckes im Verlauf einer hypervolämischen Hämodilution bei Patienten mit arterieller Verschlußkrankheit. In: Heilmann L, Beez M (eds) Neuere klinische Aspekte zur Hämodilution (Symp Essen, 1985). Schattauer, Stuttgart, p115

542 Lang GD, Lowe GDO, Walker JJ, Forbes CD, Prentice CRM, Calder AA (1984) Blood rheology in pre-eclampsia and intrauterine growth retardation: effects of blood pressure reduction with labetalol. Br J Obstet Gynaecol 91: 438

543 Langsjoen PH (1966) Blood viscosity changes in acute myocardial infarction. Postgrad Med 39: A-42

544 Langsjoen PH, Immon TW (1967) Rheologic changes in myocardial infarction. Am Heart J 73: 430

545 Langsjoen PH, Immon TW (1968) Hemorheologic observations in acute myocardial infarction. Angiology 19: 247

546 Langsjoen PH (1973) The value of reducing blood viscosity in acute myocardial infarction. Bibl Anat 11: 180

547 Lanzer G, Tilz GP (1986) Die therapeutische Blutkomponentenseparation in der klinischen Notfallmedizin. Wien Med Wochenschr 136: 134

548 Larcan A, Stoltz JF (1970) Microcirculation et hémorhéologie. Masson, Paris

549 Larcan A, Stoltz JF (1979) Médicaments à action hémorhéologique directe ou indirecte. Méthodes d'étude et essais de classement. Thérapie 34: 5

550 Larcan A, Stoltz JF (1983) Classification et physiopathologie du syndrome d'hyperviscosité. Journ Annu Diabetol Hotel Dieu: 179 (73 ref)

551 Larcan A, Stoltz JF (1983) Syndromes d'hyperviscosité sanguine - classification et compréhension physiopathologique - déductions thérapeutiques. Ann Med Interne 134: 395

552 Larsson H, Valdemarsson S, Hedner P, Odeberg H (1985) Reversal of increased whole blood and plasma viscosity after treatment of hypothyreoidism in man. Acta Med Scand 217: 67

553 Lawson NS, Nosanchuk JS, Oberman HA, Meyers MC (1968) Therapeutic plasmapheresis in treatment of patients with Waldenström's macroglobulinemia. Transfusion 8: 174

554 Le Devehat C, Lemoine A, Cirette B, Ramet M, Vimeux M (1980) Kinetics of variations of erythrocyte filterability and 2-3 diphosphoglycerate in diabetic patients with arteriopathy of lower limbs treated with pentoxifylline. In: Stoltz JF, Drouin P (eds) Hemorheology and diseases. Doin, Paris, p683

555 Le Devehat C, Lemoine A, Cirette B, Ramet M, Vimeux M (1980) Filterability and metabolism of erythrocytes under bencyclane in diabetic patients with macroangiopathy. Folia Angiol 28 (Suppl 7): 41

556 Le Devehat C, Cirette B, Lemoine A, Roux E, Ramet M (1981) Activité rhéologique du cervoxan. Étude de la filtrabilité et de l'état métabolique du globule rouge chez le diabétique. Angéiologie 33: 189

557 Le Devehat C, Lemoine A, Cirette B, Ramet M (1981) Pharmacological influences of pentoxifylline on red cell filterability and 2-3 diphosphoglycerate. Scand J Clin Lab Invest 41 (Suppl 156): 301

558 Le Devehat C, Lemoine A, Ramet M, Cirette B, Roux E, Vimeux M (1981) Study of filterability and metabolic state of red cells in diabetics on artificial pancreas. Effects on insulin activity. Abstract-Booklet 2nd Europ Conf Clin Haemorheol, London, p12

559 Le Devehat C, Lemoine A, Cirette B (1983) Activité hémorhéologique du buflomédil i.v. et per os dans l'angiopathie diabétique. Angéiologie 35: 243

560 Le Devehat C, Lemoine A (1985) Haemorheological changes induced by Troxerutine in patients affected by venous insufficiency of lower legs (Abstract). Clin Hemorheol 5: 740

561 Le Devehat C, Lemoine A, Bertrand A (1985) Effect of good control glucose by insulin with pump on haemorheological and microcirculation parameters in diabetic patients (Abstract). Clin Hemorheol 5: 673

562 Leiper JM, Lowe GDO, Anderson J, Burns P, Cohen HN, Manderson WG, Forbes CD, Barbenel JC, MacCuish AC (1984) Effects of diabetic control and biosynthetic human insulin on blood rheology in established diabetics. Diab Res 1: 27
563 Leonardo G, Arpaia MR, Onza C, Del Guercio R (1984) Regarding the action of buflomedil: Haemorheological and blood chemistry aspects. Curr Ther Res 36: 1016
564 Leonhardt H, Grigoleit HG, Reinhardt I (1977) Changes and relationships of erythrocyte flexibility and whole blood viscosity of normo-osmolar and hyperosmolar erythrocyte suspensions influenced by 3,7-dimethyl-1-(5-oxo-hexyl)-xanthine. Vasa 6: 316
565 Leonhardt H, Grigoleit HG, Reinhardt I (1978) Auswirkungen von Zigarettenrauch und oralen Kontrazeptiva auf die Plasma- und Vollblutviskosität. Med Welt 29: 880
566 Leonhardt H (1978) Mikrorheologische Untersuchungen an Gesunden mit Risikofaktoren. Habilitationsschrift, Berlin
567 Lerche D, Bäumler H, Pöhlmann G, Scherf HP, Bilsing R (1983) The treatment of peripheral vascular diseases by ultraviolet blood irradiation. A hemorheological study (Abstract). Clin Hemorheol 3: 283
568 Lerche D, Bäumler H, Winter G, Tourowski A, Meier W (1985) A hemorheological characterization of Rocornal® in case of peripheral circulatory diseases (Abstract). Clin Hemorheol 5: 726
569 Letcher RL, Chien S, Laragh JH (1979) Changes in blood viscosity accompanying the response to prazosin in patients with essential hypertension. J Cardiovasc Pharmacol 1 (Suppl 1): 8
570 Levine MA, Hammack WJ, Frommeyer WB (1968) Treatment of macroglobulinemia with penicillamine. Clin Res 8: 54
571 Lewis DH, Sandegård J, Kutti J, Gelin J (1972) Normovolemic hemodilution in polycythemia. Observations on capillary flow and capillary transport in skeletal muscle. In: Messmer K, Schmid-Schönbein H (eds) Hemodilution. Theoretical basis and clinical application. Karger, Basel, p 133
572 Lewy B (1898) Die Reibung des Blutes. Arch Ges Physiol 65: 447
573 Lichtman MA, Rowe JM (1982) Hyperleukocytic leukemias: Rheological, clinical and therapeutic considerations. Blood 60: 279
574 Liewendahl K, Johnsson R, Majuri H (1981) Effect of anticonvulsant drugs on red cell flexibility, adenosine triphosphatase activities and cation transport. Scand J Clin Lab Invest 41 (Suppl 156): 279
575 Lim C jr, Kostrzewska E, Bergentz SE, Gelin LE (1969) Rheology of human blood following treatment with Dextran-40 and Dextran-70. Bibl Anat 10: 9
576 Linderkamp O, Meiselman HJ, Wu PYK, Miller FC (1981) Blood and plasma viscosity and optimal hematocrit in the normal newborn infant. Clin Hemorheol 1: 575
577 Linderkamp O, Hersch S, Ozanne P, Meiselman HJ (1986) Vergleich von Methoden zur Bestimmung der Erythrozytenaggregation. In: Tillmann W, Ehrly AM (eds) Hämorheologie und Hämatologie. Münchner Wissenschaftliche Publikationen, München, p 140
578 Lindmann K (1908) Klinische und experimentelle Beiträge zur pharmakologischen Beeinflussung der Blutviskosität. Dissertationsschrift, Marburg
579 Lindsley H, Teller D, Noonan B, Petersen M, Mannik M (1973) Hyperviscosity syndrome in multiple myeloma. Am J Med 54: 682
580 Litmanovich KY, Melnikova VN, Fregatova LM, Kargin VD, Egorova LV, Kotovshchikova MA (1985) Methods of correction of hemorheological disturbances in patients with thrombobliterating diseases of the peripheral arteries of the extremities. Vest Khir 135 (7): 96
581 Litwin MS, Dawidson I, Barrett J, Miller E (1981) Physiologic effect in postoperative patients of intravascular cellular aggregate dissolution caused by Dextran-40 infusion. Abstract-Booklet 2nd Europ Conf Clin Haemorheol, London, p 66
582 Lommel F (1904) Über die Viskosität des menschlichen Blutes bei Schwitzprozeduren. Dtsch Arch Klin Med 80: 308
583 Louis J, Voisin P, Kolopp M, Vaillant G, Gaillard S, Debry G, Stoltz JF, Drouin P (1983) Lack of acute effects of insulin in vivo on hemorheological parameters in hyperglycemic diabetic patients (Abstract). Clin Hemorheol 3: 294
584 Lowe GDO, Campbell AF, Meek DR, Forbes CD, Prentice CRM (1978) Subcutaneous ancrod in prevention of deep-vein thrombosis after operation for fractured neck of femur. Lancet 2: 698
585 Lowe GD, Morrice JJ, Fulton A, Forbes CD, Prentice CR, Barbenel JC (1978) Subcutaneous ancrod after operation for fractured hip - A dose-ranging and feasibility study. Thromb Haemost 40: 134
586 Lowe GD, Morris JJ, Forbes CD, Prentice CR, Fulton AJ, Barbenel JC (1979) Subcutaneous ancrod therapy in peripheral arterial disease: Improvement in blood viscosity and nutritional blood flow. Angiology 30: 594
587 Lowe GDO, Drummond MM, Forbes CD, Barbenel JC (1980) Increased blood viscosity in young women using oral contraceptives. Am J Obstet Gynecol 137: 840

588 Lowe GDO, Barbenel JC, Forbes CD (eds) (1981) Clinical aspects of blood viscosity and cell deformability. Springer, Berlin
589 Lowe GDO, Barbenel JC, Forbes CD (eds) (1981) Clinical aspects of blood viscosity and cell deformability. Section E: Therapeutic aspects of blood rheology. Springer, Berlin, p211
590 Lowe GDO, Drummond MM, Forbes CD, Barbenel JC (1981) Occlusive arterial disease and blood rheology. In: Lowe GDO, Barbenel JC, Forbes CD (eds) Clinical aspects of blood viscosity and cell deformability. Springer, Berlin, p133
591 Lowe GDO, Dunlop DJ, Lawson DH, Pollock JG, Watt JK, Forbes CD, Prentice CRM, Drummond MM (1982) Double-blind controlled clinical trial of ancrod for ischemic rest pain of the leg. Angiology 33: 46
592 Lowe GDO, Leiper JM, Forbes CD, Manderson WG, MacCuish AC (1982) Effects of diabetic control and biosynthetic human insulin on blood rheology in established diabetes (Abstract). Diabetologia 23: 471
593 Lowe GDO (1984) Evaluation of rheological therapy by orally administered drugs. Clin Hemorheol 4: 159
594 Lowe GDO (1984) Blood rheology and venous thrombosis. Clin Hemorheol 4: 571
595 Lowe GDO, Trope GE, Small M, Alvarez E, Foulds WS, Forbes CD (1985) Treatment of retinal vein occlusion by viscosity reduction with stanozolol (Abstract). Clin Hemorheol 5: 691
596 Lucas GS, Simms MH, Caldwell NM, Alexander SJC, Stuart J (1984) Haemorrheological effects of prostaglandin E_1 infusion in Raynaud's syndrome. J Clin Pathol 37: 870
597 Lukjan H, Chyzy R, Krupinski K, Bielawiec M (1985) Red cell deformability in patients with arteriosclerosis obliterans and influence of some drugs. Abstract-Booklet 4th Europ Conf Clin Haemorheol, Siena, p70
598 Lusov VA, Belousov JB, Sidelmann AA, Savenkov MP, Cholodova OE, Dudaev VA (1978) The treatment of peripheral atherosclerosis with pentoxifylline. Pharmatherapeutica 2 (Suppl 1): 123
599 Luther B (1983) Klinisch-experimentelle Untersuchungen zur Blutviskosität. Z Exp Chir Transplant Künstliche Organe 16: 256
600 Lutz TW, Barras JP (1983) Rheological behaviour of human blood and plasma at steady flow and oscillatory flow. Vasa 12: 121
601 Lyons BH (1952) Cerebral thrombosis following mercurial diuresis. Can Med Assoc J 66: 545
602 MacKenzie MR, Brown E (1968) Hypervolemia in hyperviscosity syndrome of macroglobulinemia (Abstract). Clin Res 16: 122
603 MacKenzie MR, Brown E, Fudenberg HH (1968) Hypervolemia in the hyperviscosity syndrome of macroglobulinemia (Abstract). J Clin Invest 47: 64a
604 MacKenzie MR, Fudenberg HH (1972) Macroglobulinemia: An analysis for forty patients. Blood 39: 874
605 Mackintosh T (1969) Blood viscosity in newborn infant and diagnosis and treatment of hyperviscous states. Arch Dis Child 44: 781
606 Magora F, Hershko C, Aronson HB (1970) The effect of electrical sleep on the viscosity of blood. Br J Anaesth 42: 1085
607 Manescalchi PG, Morini R, Pasquini G, Albanese B (1983) Sulla non correlazione fra parametri della filtrazione eritrocitaria ed altri parametri emoreologici. Ric Clin Lab 13 (Suppl 3): 337
608 Mann G (1980) Untersuchungen über den Einfluß verschiedener Saluretika auf die Fließeigenschaften des Blutes bei Patienten mit essentieller Hypertonie. Dissertationsschrift, Frankfurt am Main
609 Manrique RV (1979) Avaliaçao clinico-laboratorial da modificaçao da flexibilidade eritrocitária induzida pela pentoxifilina. Rev Bras Clín Terap 8: 572
610 Manrique RV, Müller R (eds) (1981) Disorders of blood flow. New therapeutic aspects. Excerpta Medica, Amsterdam
611 Manrique R (1983) Blood coagulation, rheology and coronary reocclusion after local thrombolysis in acute myocardial infarction. Abstract-Booklet 13th World Congr Int Union Angiol, Rochester, Minnesota, USA
612 Marazzini L, Soroldoni M, Clivati A, Simonati O, Ogliari L, Brasca F, Agosti R, Longhini E (1981) Effects of isovolumetric venesection on blood viscosity, on red cell deformability and on arterial flow velocity in chronic respiratory failure with secondary polycythemia. Respiration 42: 168
613 Markson A (1936) Blood viscosity in congestive heart failure. Glasgow Med J 7: 201
614 Marshall M, Reiser R (1983) Changes of blood viscosity by indirect anticoagulation (Abstract). Clin Hemorheol 3: 244
615 Marshall M, Reiser R (1983) Beeinflussung der Blutviskosität durch indirekte Antikoagulation. In: Nobbe F, Rudofsky G (eds) Probleme der Vor- und Nachsorge und der Narkoseführung bei invasiver angiologischer Diagnostik und Therapie. Begutachtung arterieller und venöser Gefäßschäden. Pflaum, München, p493

616 Marshall M (1984) Beeinflussung der Hämorheologie durch oral zu verabreichende Pharmaka. In: Heilmann L, Kiesewetter H, Ernst E (eds) Klinische Rheologie und Beta-1-Blockade. Zuckschwerdt, München, p 110

617 Martin P (1979) Déformabilité des globules rouges et accidents vasculaires cérébraux. Intérêt de la pentoxifylline. Gaz Méd France 86: 2585

618 Martin P, Vives P (1980) Red blood cell filtration in stroke. Effect of pentoxifylline. In: Stoltz JF, Drouin P (eds) Hemorheology and diseases. Doin, Paris, p 693

619 Martin P, Vives P (1980) The effect of pentoxifylline on red cell deformability in cerebrovascular accidents. Curr Med Res Opin 6: 518

620 Mashiah A, Patel P, Schraibman IG, Charlesworth D (1978) Drug therapy in intermittent claudication: an objective assessment of the effects of three drugs on patients with intermittent claudication. Br J Surg 65: 342

621 Materazzi M, Guerrini M, Rossi C, Cappelli R, Pecchi S, Acciavatti A, Pieragalli D, Forconi S, Di Perri T (1981) Action of buflomedil on peripheral obliterative arterial disease. A rheological and metabolic study (Abstract). Clin Hemorheol 1: 488

622 Mathieu D, Coget J, Vinckier L, Devulder B, Wattel F (1985) Red blood cell deformability and hyperbaric oxygen therapy (Abstract). Clin Hemorheol 5: 625

623 Mátrai A, Fendler K, Horváth L (1981) Haemorheological changes after anticoagulant therapy with a heparinoid (SP 54). Abstract-Booklet 2nd Europ Conf Clin Haemorheol, London, p 84

624 Matrai A, Ernst E, Eck M (1984) Hämorheologie und CO_2-Bäder (Abstract). Z Phys Med Baln Med Klim 13 (Sonderheft 2): 42

625 Matrai A, Ernst E, Flute PT, Dormandy JA (1984) Blood filterability in peripheral vascular disease - red cell deformability or cell sticking? Clin Hemorheol 4: 311

626 Matrai A, Ernst E, Dormandy JA (1985) Erythrozytenfiltrabilität: Unterscheidung zwischen Filterokklusion und Erythrozyten-Passagezeit. In: Kiesewetter H, Ehrly AM, Jung F (eds) Hämorheologische Meßmethoden. Münchner Wissenschaftliche Publikationen, München, p 117

627 Matrai A, Ernst E (1985) Pentoxifylline improves white cell rheology in claudicants. Clin Hemorheol 5: 483

628 Mattassi R, Cardillo B, D'Angelo F (1985) Haemorheological and clinical effects of ozone in arterial occlusive disease (Abstract). Clin Hemorheol 5: 644

629 Mayer GA (1976) Blood viscosity and oral anticoagulant therapy. Am J Clin Pathol 65: 402

630 Mayesima T (1912) Klinische und experimentelle Untersuchungen über die Viskosität des Blutes. Mittlg Grenzgeb Med Chir 24: 413

631 McCaskey GW (1908) The viscosity of the blood; its value in clinical medicine. JAMA 51: 1653

632 Mehta AB, Goldman JM, Kohner E (1984) Hyperleucocytic retinopathy in chronic granulocytic leukaemia: The role of intensive leucapheresis. Br J Haematol 56: 661

633 Merchan R, Doná G (1984) Microcirculation and haemorheology studies of polyunsaturated phosphatidylcholine (EPL; Lipostabil®) in the aged. Clin Trials J 21: 517

634 Merchan R, Zurita IE, Martinez JF (1985) Suloctidil: Long-term trial in the aged. Study of microcirculation, viscosity, red cell deformability, fibrinolysis and lipid fractions. Drug Res 35: 855

635 Mercke CE, Kjellen E (1981) Erythrocyte deformability in patients with malignant tumours - Potential role of prostaglandins. Abstract-Booklet 2nd Europ Conf Clin Haemorheol, London, p 56

636 Merrill EW, Cokelet GC, Britten A, Wells RE jr (1963) Non-Newtonian rheology of human blood - Effect of fibrinogen deduced by "subtraction". Circ Res 13: 48

637 Merrill EW, Gilliland ER, Lee TS, Salzman EW (1966) Blood rheology: Effect of fibrinogen deduced by addition. Circ Res 18: 437

638 Messmer K, Schmid-Schönbein H (eds) (1972) Hemodilution - Theoretical basis and clinical application. Karger, Basel

639 Messmer K, Sunder-Plassmann L (1974) Hemodilution. Prog Surg 13: 208

640 Messmer K, Schmid-Schönbein H (eds) (1975) Intentional hemodilution. Bibl Haematol 41

641 Messmer K (1975) Hemodilution. Surg Clin North Am 55: 659

642 Messmer K, Hammersen F (eds) (1985) Entzündung und Rheologie der Leukozyten. Karger, Basel

643 Metoki H, Kanazawa T, Shibutani K, Izawa M, Soma Y, Onodera K (1979) A study on blood viscosity in CVA - The relationships between blood viscosity and duration after stroke and the effects of pentoxyphyllin. Acta Neurol Scand 60 (Suppl 72): 622

644 Metoki H, Kudo R, Shimanaka Y, Matsui T, Kanazawa T, Mori H, Izawa K, Shibutani K, Muraoka H, Onodera K (1983) A study on blood viscosity in cerebrovascular disease. The changes of blood viscosity during the time after stroke and the effect of pentoxifyllin on the viscosity. Hirosaki Med J 35: 1

645 Mikita J, Khan O, Dormandy JA, Cotton LT (1985) Does prostacyclin influence haemorrheology in Raynaud's phenomenon? (Abstract). Clin Hemorheol 5: 751

646 Milan R, Duric D, Trajkovic B, Radojicic B, Pekic B, Tomasevic R, Duji A (1977) Syndrome of the myeloma serum hyperviscosity treated by plasmapheresis. Med Pregl 30: 471
647 Miller B, Heilmann L (1985) Haemorheological parameters during chemotherapy for breast cancer (Abstract). Clin Hemorheol 5: 625
648 Miller B, Heilmann L, Callies R (1985) Hemorheological studies in mammary carcinoma. Arch Gynecol 238: 559
649 Milligan DW, Davies JA (1982) The effect of iron therapy on whole blood viscosity and calf blood flow in polycythaemia (Abstract). Br J Haematol 52: 137
650 Milligan DW, Tooke JE, Davies JA (1982) Effect of venesection on calf blood flow in polycythaemia. Br Med J 284: 619
651 Molaro G, De Angelis V, Vettore L (1985) Effetto del dipiridamolo sui parametri emoreologici nelle sindromi sclerocitemiche e policitemiche. Ric Clin Lab 15 (Suppl 1): 597
652 Montagnani M, Civitelli R, Nami R, Agnusdei D, Falsini G, Conti F, Bianchini C, Liberatori P, Gennari C (1985) Effects of a calcium blocking agent (nifedipine) on red blood cell deformability in essential hypertension (Abstract). Clin Hemorheol 5: 713
653 Montefusco S, Ciampitiello G, Gnasso A, Postiglione A (1984) Migliorati parametri emoreologici in pazienti iperlipidemici trattati con mucopolisaccaridi. Ric Clin Lab 14 (Suppl 2): 144
654 Montefusco S, Ciampittiello G, Gnasso A, Cicerano U, Postiglione A (1985) Miglioramento dei parametri emoreologici dopo terapia con mucopolisaccaridi. Ric Clin Lab 15 (Suppl 1): 543
655 Moratti F, Cortiana M, Taglioretti D (1983) Applicazione della filtrazione eritrocitaria nello studio della vasculopatia diabetica. Efficacia terapeutica della pentossifillina. Clin Ter 106: 19
656 Moscatelli B, Marmaggi S, Traietti P, Colloca A, Fabbri M, Bologna E (1984) Blood viscosity and circulating platelets activation in polycythemic COLD patients: Effect of erythroapheresis (Abstract). Respiration 46 (Suppl 1): 159
657 Moser M, Klieber M, Winkler R (1985) Vergleichende Studien über den Einfluß von Diät, Bewegungs- und Jodbalneotherapie auf blutrheologische Parameter bei Diabetikern im Rahmen eines 4wöchigen Kuraufenthalts in Bad Hall. Wien Klin Wochenschr 97: 327
658 Motta G, Ratto GB (1980) Osservazioni cliniche sull'azione reologica ed emodinamica di taluni farmaci attivi sul microcircolo. Minerva Med 71: 1367
659 Moulopoulos SD, Zereg N (1968) La viscosité kinétique du sang. Arch Mal Coeur 61: 1156
660 Mozzi E, Annoni F, Chiurazzi D, Rossi R, Pacini F, Martini A (1984) Azione del benfurodil emisuccinato (Eucilat®) su alcuni parametri emoreologici in pazienti portatori di arteriopatie ostruttive croniche periferiche. Minerva Cardioangiol 32: 141
661 Mozzi E, Chiurazzi D, Spinola A, Annoni F, Germiniani R (1985) Effects of buflomedil in a group of patients with chronic peripheral arterial diseases. Clinical and hemorheological evaluation. J Int Med Res 13: 317
662 Müller K, Haaß A, Jäger H, Wagner EM, Hellstern P (1983) Dextran und HAES, vergleichbare und unterschiedliche hämorheologische Parameter. Psycho 9: 381
663 Müller O, Inada R (1904) Jodwirkung bei Arteriosklerose. Dtsch Med Wochenschr 30: 1751
664 Müller-Bühl U, Comberg HU, Diehm C, Allenberg J, Mörl H (1982) Hämodilutions-Therapie der arteriellen Verschlußkrankheit mit Hydroxyäthylstärke 200/0,5. Münch Med Wochenschr 10: 241
665 Muggeo M, Calabrò A, Businaro V, Patussi L, Volpe A, Signorini GP, Crepaldi G (1983) Blood clotting, fibrinolytic and haemorheological parameters in ischaemic vascular disease: the effects of pentoxifylline in the treatment of acute cerebral vascular disease. Pharmatherapeutica 3 (Suppl 1): 74
666 Murashko VV, Yusipova TA, Strutynskii AV, Dzhanashiya PK, Shenshina AM, Sklyarova MA (1984) Effect of sodium nitrite on blood rheology, microcirculation and blood coagulation in patients with circulatory failure. Ter Arkh 56 (4): 108
667 Nalefski LA, Gibert NC, Sheedy JA (1950) Viscosity studies on blood following the administration of anticoagulants. J Lab Clin Med 36: 970
668 Nenci GG (1985) Long-term tolerability and activity on platelet aggregation and erythrocyte deformability of indobufen in cardiovascular patients. Abstract-Booklet 4th Europ Conf Clin Haemorheol, Siena, p 184
669 Neumann V, Cove DH, Shapiro LM, George AJ, Kenny MW, Meakin M, Stuart J (1983) Effect of ticlopidine on platelet function and blood rheology in diabetes mellitus. Clin Hemorheol 3: 13
670 Nusbacher J (1982) Therapeutic hemapheresis. Clin Lab Med 2: 87
671 O'Reilly MJG, Dodds AJ, Tolpos G, Hamilton A, Cotton LT (1979) Plasmapheresis in Raynaud's phenomenon. In: Heidrich H (ed) Raynaud's phenomenon. TM, Bad Oeynhausen, p 291
672 Orr JE, Davidson RJL, Russell IF, Robertson GS (1982) A haemorheological study of althesin. Br J Anaesth 54: 1003
673 Osawa M, Maruyama S (1985) Über den Einfluß von "TCV-3B" (Vinpocetin) auf die Blutviskosität in ischämischen zerebrovaskulären Erkrankungen. Ther Hung 33: 7

674 Oski FA, Lubin B, Buchert EB (1972) Reduced red cell filterability with oral contraceptive agents. Ann Intern Med 77: 417

675 Ostwald W (1924) Zur Viskosimetrie kolloidaler Lösungen. Z Physikal Chem 111: 62

676 Ott M (1973) Untersuchungen über den Einfluß von Vasodilatantien auf die Blutviskosität. Dissertationsschrift, Berlin

677 Ott E, Lechner H (1981) Hemorheologic conditions in patients with cerebrovascular disease altered by treatment with pentoxifylline. Ric Clin Lab 11 (Suppl 1): 253

678 Ott E, Lechner H (1983) Changes of flow properties of the blood in cerebrovascular disease and their medical treatment with pentoxifylline. J Cereb Blood Flow Metab 3 (Suppl 1): 530

679 Ott E, Lechner H (1985) The effect of budipine on the rheology of the blood. In: Gerstenbrand F, Poewe W, Stern G (eds) Clinical experiences with budipine in Parkinson therapy. Springer, Berlin, p 182

680 Oughton J, Lawson P, Kohner EM, Barnes AJ (1983) The effect of improved metabolic control on diabetic complications and on blood rheology (Abstract). Clin Hemorheol 3: 270

681 Paisey RB, Harkness J, Hartog M, Chadwick T (1980) The effect of improvement in diabetic control on plasma and whole blood viscosity. Diabetologia 19: 345

682 Palareti G, De Fabritiis A, Poggi M, Tricarico MG, Borgatti E, Coccheri S (1981) Haemorheological parameters in essential hypertension (EH) and early changes during antihypertensive treatments (Abstract). Clin Hemorheol 1: 495

683 Palareti G, Poggi M, Grauso F, Tricarico MG, Rodorigo G, Coccheri S (1982) Haemorheologic changes in women during oral contraception (Abstract). Haemostasis 12: 145

684 Palareti G, Poggi M, Limoni P, Andreoli A, Tricarico MG, Coccheri S (1983) Changes in the rheologic properties of blood in acute cerebrovascular disease: Effects of treatment with l-eburnamonine. Int Angiol 2: 179

685 Parnetti L, Ciuffetti G, Mercuri M, Senin U (1985) Haemorheological pattern in initial mental deterioration: results of a long-term study using piracetam and pentoxifylline. Arch Gerontol Geriatr 4: 141

686 Passero S, Bonelli G, Guerrini M, Acciavatti A, Nardini M, Battistini N (1985) Effetti del trattamento cronico con farmaci antiaggreganti sulla viscosità ematica in pazienti con malattie cerebrovascolari. Ric Clin Lab 15 (Suppl 1): 593

687 Pau P, Mereu S, Montisci R, Falchi S, Petruzzo P (1983) No changes in red cells deformability in Raynaud phenomenon induced by beta adrenoceptors stimulation (Abstract). Biorheology 20: 406

688 Pavlovskii DP (1965) Effect of anticoagulants on blood viscosity, circulation time and the capillaroscopic picture. Ter Arkh 37: 367

689 Pecker J, Vallée B, Haranger JM, Gueguen M (1978) Torental et accidents ischémiques du système nerveux central. L'Ouest Méd 31: 1571

690 Pennarola R, De Martino F, Barletta R, Mazzarella S, Sacco M (1983) Hemorheologic aspects among 25 cases of microangiopathy by vibrating tools treated with pentoxifylline (Abstract). Clin Hemorheol 3: 289

691 Perego MA, Sergio G, Artale F (1981) Aspetti emoreologici in rapporto alla fisiopatologia e alla clinica dell'arteriopatia ostruttiva periferica. Ric Clin Lab 11 (Suppl 1): 281

692 Perego MA, Sergio G, Artale F, Cramarossa L, Espureo M (1982) Traitement hémorhéologique des artériopathies obstructives périphériques. Angéiologie 34: 33

693 Perego MA, Sergio G, Espureo M, Francisci A, Artale F (1982) Haemodynamic and haemorheological effects of buflomedil in patients with peripheral occlusive arterial disease. Curr Med Res Opin 8: 178

694 Perego MA, Sergio G, Artale F (1983) Haemorheological aspects of the pathophysiology and clinical features of peripheral occlusive arterial disease. Pharmatherapeutica 3 (Suppl 1): 91

695 Perego MA, Sergio G, Artale F, Giunti P, Danese C (1985) Effects of physical training on regional blood filtrability and platelet β-thromboglobulin in patients with peripheral arteriopathy (Abstract). Clin Hemorheol 5: 652

696 Perego MA, Sergio G, Artale F, Giunti P, Danese C (1986) Haemorrheological improvement by pentoxifylline in patients with peripheral arterial occlusive disease. Curr Med Res Opin 10: 135

697 Perhoniemi V, Salmenkivi K, Sundberg S, Johnsson R, Gordin A (1984) Effects of flunarizine and pentoxifylline on walking distance and blood rheology in claudication. Angiology 35: 366

698 Pernigotti L, Neirotti M, Gagliardo S, Musso M, Sclavo R, Bulla A (1985) Effetti immediati della somministrazione di eparina e di eparani (mesoglicano) su alcuni parametri emoreologici e dell'emostasi. Ric Clin Lab 15 (Suppl 1): 537

699 Petralito A, Miano MF, Messina S, Grimaldi LT, Malatino LS (1985) Effects of pentoxifylline on erythrocyte aggregation (Abstract). Clin Hemorheol 5: 733

700 Pfeiffer M, Tilsner V (1978) Einfluß von Etofibrat auf die Plasmaviskosität bei Hyperlipoproteinämien. Med Klin 73: 60

701 Pieragalli D, Guerrini M, Materazzi M, Ralli L, Forconi S, Di Perri T (1983) Protective effect of buflomedil against the exercise-induced haemorheological impairment of vascular patients (Abstract). Clin Hemorheol 3: 292

702 Poiseuille JML (1842) Recherches expérimentales sur le mouvement des liquides dans les tubes de très petits diamètres. Compt rend 15: 1167

703 Pola P, Flore R, Tondi P, Serricchio M, Dal Lago A, Fioroni G (1984) Terapia dell'iperviscosità e della ridotta deformabilità eritrocitaria nell'arteriopatico cronico: necessità dell'impiego di più farmaci. Giorn Ital Angiol 4: 31

704 Pollock S, Challoner T, Thomas D, Harrison MJG (1981) The effect of venesection of red cell deformability (Abstract). Clin Hemorheol 1: 496

705 Ponari O, Catamo A, Molina E, Mombelloni A, Tonelli C, Tardio S, Squeri M (1985) Effects of glycerol on some blood rheology parameters in patients with acute cerebral ischemia. Preliminary report (Abstract). Clin Hemorheol 5: 705

706 Poon PYW, Dornan TL, Orde-Peckar C, Mullins R, Bron AJ, Turner RC (1982) Blood viscosity, glycaemic control and retinopathy in insulin-dependent diabetes. Clin Sci 63: 211

707 Porro F, Pastori P, Lanzoni M, Curti L, Fanelli M (1985) Hemorheology in acute ischemic stroke: Influence of isovolemic hemodilution (Abstract). Clin Hemorheol 5: 669

708 Porro F, Pastori M, Traverso G, Lanzoni M, Pastori P (1985) Isovolemic hemodilution in advanced chronic obstructive lung disease. A preliminary hemorheologic study (Abstract). Clin Hemorheol 5: 727

709 Powles R, Smith C, Kohn J, Hamilton Fairley G (1971) Method of removing abnormal protein rapidly from patients with malignant paraproteinaemias. Br Med J 3 : 664

710 Pre'Denning du A, Watson JH (1906) The viscosity of blood. Proc R Soc B 78: 328

711 Preston FE, Cooke KB, Foster ME, Winfield DA, Lee D (1978) Myelomatosis and the hyperviscosity syndrome. Br J Haematol 38: 517

712 Preston FE, Sokol RJ, Lilleyman JS, Winfield DA, Blackburn EK (1970) Cellular hyperviscosity as a cause of neurological symptoms in leukaemia. Br Med J 1: 476

713 Pretolani E, Battistini G, Iosa G, Tonti D, Zoli I (1983) Il ciclonicato nella terapia medica delle arteriopatie periferiche. Giorn Ital Angiol 3: 268

714 Pronk EAS, Klein F, Elkerbout F, Radema H, Cleton FJ (1969) Therapy of macroglobulinemia. Acta Med Scand 186: 273

715 Pruzanski W, Watt JG (1972) Serum viscosity and hyperviscosity syndrome in IgG multiple myeloma. Ann Intern Med 77: 853

716 Putnam TC, Kevy SV, Replogle RL (1967) Factors influencing the viscosity of blood. Surg Gynecol Obstet 124: 547

717 Rakita L, Gillespie DG, Sancetta SM (1965) The acute and chronic effects of phlebotomy on general hemodynamics and pulmonary functions of patients with secondary polycythemia associated with pulmonary emphysema. Am Heart J 70: 466

718 Randi ML, Fabris F, Crociani ME, Battocchio F, Girolami A (1985) Effects of ticlopidine on blood fibrinogen and blood viscosity in peripheral atherosclerotic disease. Drug Res 35: 1847

719 Reich L, Feizi T, Winchester R, Wechsler B, Walzer P, Wright P (1973) Effects of large scale plasmapheresis performed on patients with macroglobulinemia. Proc Am Assoc Cancer Res 14: 68

720 Reid HA, Chan KE, Thean PC (1963) Prolonged coagulation defect (defibrination syndrome) in malayan viper bite. Lancet 1: 621

721 Reid HL, Barnes AJ, Lock PJ, Dormandy JA, Dormandy TL (1976) A simple method for measuring erythrocyte deformability. J Clin Pathol 29: 855

722 Reid HL, Dormandy JA, Barnes AJ, Lock PJ, Dormandy TL (1976) Impaired red cell deformability in peripheral vascular disease. Lancet 1: 666

723 Replogle RL, Meiselman HJ, Merrill EW (1967) Clinical implications of blood rheology studies. Circulation 36: 148

724 Reynolds O (1883) An experimental investigation of the circumstances which determine whether the motion of water shall be direct or sinuous and the law of resistance in parallel channels. Philos Trans R Soc 174: 935

725 La Ricerca in Clinica e in Laboratorio 15 (1985) (Suppl 1)

726 Rieger H (1976) Zur Physiologie und Pathophysiologie der Blutplättchen unter rheologischen Aspekten. Habilitationsschrift, Aachen

727 Rieger H, Leyhe A, Schmid-Schönbein H, Schoop W, Schneider R, Malotta H (1977) Isovolämische Hämodilution bei peripherer arterieller Verschlußkrankheit. Konzepte, Methoden und vorläufige Ergebnisse. In: Alexander K, Cachovan M (eds) Diabetische Angiopathien. Witzstrock, Baden-Baden, p354

728 Rieger H, Schoop W, Schmid-Schönbein H, Schneider R, Malotta H (1977) Isovolemic hemodilution (IHD) in peripheral vascular occlusive disease (PVOD). Concepts, methods and preliminary results. Microvasc Res 13: 267

729 Rieger H, Köhler M, Schoop W, Schmid-Schönbein H, Roth FJ, Leyhe A (1979) Hemodilution (HD) in patients with ischemic skin ulcers. Klin Wochenschr 57: 1153

730 Rieger H (1982) Neues in der konservativen Therapie arterieller Durchblutungsstörungen. Med Welt 33: 576

731 Ritzmann SE, Coleman SL, Levin WC (1960) The effect of some mercaptanes upon a macrocryogelglobulin; modifications induced by cysteamine, penicillamine and penicillin. J Clin Invest 39: 1320

732 Ritzmann SE, Thurm RH, Truax WE, Levin WC (1960) The syndrome of macroglobulinemia: Review of the literature and a report of two cases of macrocryogelglobulinemia. Arch Intern Med 105: 939

733 Ritzmann SE, Daniels JC, Levin WC (1970) Paralymphomatous disease - Syndrome of macroglobulinemia. In: Leukemia-Lymphoma. Year Book Medical, Chicago, p 169

734 Roba J, Roncucci R, Lambelin G (1977) Pharmacological properties of suloctidil. Acta Clin Belg 32: 3

735 Rogen SA (1940) Viscosity in cardiac failure. Modification by administration of calcium gluconate. Lancet 1: 780

736 Romberg E (1904) Über Arteriosklerose. Verh Kongr Inn Med 21: 60

737 Roncucci R, De Hertogh R, Dormandy JA, Doumont J, Gurevich V, Lambelin G, Lansen J, Rottiers R, Souridis E, Van Stalle F, Versee L (1979) Effects of long-term treatment with suloctidil on blood viscosity, erythrocyte deformability and total fibrinogen plasma levels in diabetic patients. Drug Res 29: 682

738 Rose MS, Ernst E, Lee R, Hill P, Rainsbury R, Kirkham JS (1981) Effects of hydroxychloroquine on transoperative haemorheological events (Abstract). Clin Hemorheol 1: 501

739 Rosenblatt G, Stokes J, Bassett DR (1965) Whole blood viscosity, hematocrit and serum lipid levels in normal subjects and patients with coronary heart disease. J Lab Clin Med 65: 202

740 Rosenkrantz TS, Oh W (1982) Cerebral blood flow velocity in infants with polycythemia and hyperviscosity: Effects of partial plasma exchange transfusion with Plasmanate. J Pediatr 101: 94

741 Rosenthal A, Nathan DG, Marty AT, Button LN, Miettinen OS, Nadas AS (1970) Acute hemodynamic effects of red cell volume reduction in polycythemia of cyanotic congenital heart disease. Circulation 42: 297

742 Rossi C, Pastorelli M, Romeo R, Acciavatti A, Pieragalli D, Galigani C, Pecchi S, Cappelli R, Guerrini M, Forconi S, Di Perri T (1985) Aspects of microcirculatory activity of buflomedil (Abstract). Clin Hemorheol 5: 736

743 Rousselle D, Gaillard S, Voisin P, Pointel JP, Debry G, Drouin P, Stoltz JF (1980) Blood viscosity and erythrocyte filterability in diabetic patients controlled via an artificial pancreas. In: Stoltz JF, Drouin P (eds) Hemorheology and diseases. Doin, Paris, p 411

744 Rudofsky G, Brock FE, Ulrich M, Nobbe F (1979) Clinical evaluation of flunarizine: Walking distance, ergometric performance and hemodynamic and biochemical effects. Angiology 30: 470

745 Ruggiero HA, Castellanos H, Caprissi LF, Caprissi ES (1982) Heparin effect on blood viscosity. Clin Cardiol 5: 215

746 Ryabov GA, Semenov VN, Kosyrev AB, Bunin VM, Pospelov VV, Knyazev IV, Shafransky YA, Sundukov AN (1983) The influence of haemosorption on rheology of the blood. Anesteziol Reanimatol 3: 17

747 Saitta A, Castaldo M, Bonaiuto M, Di Cesare E, Barbera N, Trifiletti A, Pizzoleo MA (1983) Aspetti emoreologici ed emocoagulativi in corso di trattamento con buflomedil cloridrato. Rass Med Int 4: 549

748 Sarno A, Ferrara F, Catania A, Frazzetta F, Capobianco D, Caimi G (1982) Comportamento del pattern emoreologico in soggetti con ipertensione essenziale trattati con beta-bloccanti. Clin Ter 103: 267

749 Sarno A, Caimi G (1985) L'approccio farmacologico in emoreologia. Ric Clin Lab 15 (Suppl 1): 479

750 Sarno A, Galluzzo A, Romano A, Giordano C, Sinagra D, Botta RM, Capobianco D, Caimi G (1985) Diabetici di I tipo trattati con insulina umana: Valutazione emoreologica durante monitoraggio con biostator. Ric Clin Lab 15 (Suppl 1): 607

751 Satewachin II, Iljin VN, Schestakov VA, Alexandrova NP, Rodionov SV, Danilova LM, Iljin SG (1978) The use of pentoxifylline in a combined regimen for the prevention of re-thrombosis in major blood vessels. Pharmatherapeutica 2 (Suppl 1): 109

752 Savi L, Dal Lago A, Serricchio M, Pola P (1979) Action du naftidrofuryl sur les viscosités sanguine et plasmatique et sur la fibrinogénémie. La Vie Méd 4: 285

753 Schäbitz J, Brema C, Teichmann W (1982) Zur Einflußnahme von Röntgenkontrastmitteln auf die Blutviskosität. Dtsch Gesundheitswesen 37: 2006

754 Schäbitz J (1983) Influence of roentgenologic contrast media on blood rheology (Abstract). Clin Hemorheol 3: 306

755 Schlichting K, Heidrich H (1976) Effect of isoxsuprine on blood viscosity. Vasa 5: 51

756 Schmid-Schönbein GW, Engler RL (1987) No-reflow in the microcirculation. The granulocyte tragedy. In: Hartmann A, Kuschinsky W (eds) Cerebral ischemia and hemorheology. Springer, Berlin, p 464

757 Schmid-Schönbein H (1970) Rheologische Eigenschaften von Erythrozyten. Habilitationsschrift, München

758 Schmid-Schönbein H, Gallasch G, von Gosen J, Volger E, Klose HJ (1976) Red cell aggregation in blood flow. I. New methods of quantification. Klin Wochenschr 54: 149

759 Schmid-Schönbein H, Rieger H, Hess H (1977) Quantitative Erfassung der Effekte fibrinolytischer Therapie auf das Fließverhalten des Blutes. Klin Wochenschr 55: 111

760 Schmid-Schönbein H (1978) Die Erhaltung der Fließfähigkeit des Blutes durch O-(β-Hydroxyethyl)-rutoside. In: Voelter W, Jung G (eds) O-(β-Hydroxyethyl)-rutoside - Experimentelle und klinische Ergebnisse. Springer, Berlin, p 129

761 Schmid-Schönbein H (1981) Was ist eine Mikrozirkulationsstörung bei chronisch-degenerativer Gefäßerkrankung? Arzneimittelforschung 31: 1999

762 Schmid-Schönbein H, Messmer K, Rieger H (eds) (1981) Hemodilution and flow improvement. Karger, Basel

763 Schmid-Schönbein H, Rieger H (1981) Isovolaemic haemodilution. In: Lowe GDO, Barbenel JC, Forbes CD (eds) Clinical aspects of blood viscosity and cell deformability. Springer, Berlin, p 211

764 Schmid-Schönbein H, Malotta H, Driessen G, Kiesewetter H, Teitel P (1983) A test profile for routine laboratories in clinical haemorheology (Abstract). Clin Hemorheol 3: 555

765 Schmid-Schönbein H (1985) Antithixotrope Therapie bei arteriellen Verschlußerkrankungen: Was leisten „rheologische Maßnahmen" bei unvermindert bestehender Obliteration? Wien Med Wochenschr 135: 368

766 Schmitt HJ (1977) Entwicklung und Darstellung einer Methode zur Messung der Erythrozytenaggregation. Dissertationsschrift, Frankfurt am Main

767 Schmittner I (1974) Einfluß von 6α-Methylprednisolon und Xantinolnicotinat auf die Fließeigenschaften des Blutes. Dissertationsschrift, Frankfurt am Main

768 Schneider R, Kiesewetter H (1982) Parenterale Pentoxifyllin-Applikation bei ischämischem Insult. Dtsch Med Wochenschr 107: 1674

769 Schneider R, Schmid-Schönbein H, Kiesewetter H (1983) The rheological efficiency of parenteral pentoxifylline (Trental) in patients with ischemic brain lesions. Preliminary results. Eur Neurol 22 (Suppl 1): 98

770 Schneider R, Körber N, Brockmann M, Kiesewetter H (1986) The haemorheological treatment of lacunar strokes: Studies with pentoxifylline (Trental 400). In: Trübestein G (ed) Konservative Therapie arterieller Durchblutungsstörungen. Thieme, Stuttgart, p 321

771 Schooneman F, Pointel JP, Streiff F, Gaillard S, Drouin P, Stoltz JF (1980) Study of rheological properties of blood in various pathological situations following therapeutical plasma exchange (Preliminary study). In: Stoltz JF, Drouin P (eds) Hemorheology and diseases. Doin, Paris, p 607

772 Schooneman F, Gaillard S, Paulus F, Streiff F, Stoltz JF (1983) Macro- and microrheological variations observed during plasma exchange in cases of monoclonal disglobulinemia (Abstract). Clin Hemorheol 3: 335

773 Schooneman F, Paulus F, Gaillard S, Streiff F, Stoltz JF (1985) Macro- and microrheological variations observed during plasma exchange in cases of monoclonal disglobulinemia. Clin Hemorheol 5: 51

774 Schooneman F, Gaillard S, Paulus F, Stoltz JF, Streiff F (1985) Variations hémorhéologiques au cours de 50 dysglobulinémies monoclonales traitées par échanges plasmatiques. Rev Méd Interne 6: 487

775 Schubotz R, Mühlfellner O (1977) The effect of pentoxifylline on erythrocyte deformability and on phosphatide fatty acid distribution in the erythrocyte membrane. Curr Med Res Opin 4: 609

776 Schülke K, Hartel W (1966) Vergleichende Untersuchungen über das Verhalten der Plasmaviscosität und des Blutvolumens nach Gabe von hoch- und niedermolekularen Dextranen. Langenbecks Arch Klin Chir 316: 623

777 Schürmann W, Eberhardt R, Pfafferott C, Volger E (1986) Zusammenhang zwischen der antianginösen Wirkung und der Erythrozytenfiltrabilität und -aggregation von Nifedipin (Abstract). Z Kardiol 75 (Suppl 1): 80

778 Schulz CEJ, Richter W, Naujoks J (1977) Blutviskosität und Innenohrschwerhörigkeit. Laryngol Rhinol Otol (Stuttg) 56: 328

779 Schwab PJ, Fahey JL (1960) Treatment of Waldenström's macroglobulinemia by plasmapheresis. N Engl J Med 60: 574
780 Schwab PJ, Fahey JL (1960) Beneficial effects of serum viscosity reduction in two patients with Waldenström's macroglobulinemia. Clin Res 8: 55
781 Schwab PJ, Okun E, Fahey JL (1960) Reversal of retinopathy in Waldenström's macroglobulinemia by plasmapheresis. Arch Ophthalmol 64: 515
782 Schwartzkopff W, Roetz R (1978) Verträglichkeit und Wirkungen von Sinteroid®. Therapiewoche 28: 2437
783 Schwedoff T (1890) Recherches expérimentales sur la cohésion des liquides. J Physique Théorétique Appliquée 9: 34
784 Scott Blair GW (1974) An introduction to biorheology. Elsevier Scientific, Amsterdam
785 Seaman GVF, Engel R, Swank R, Hissen W (1965) Circadian periodicity in some physicochemical parameters of circulating blood. Nature 207: 833
786 Segel N, Bishop JM (1966) The circulation in patients with chronic bronchitis and emphysema at rest and during exercise, with special reference to the influence of changes in blood viscosity and blood volume on the pulmonary circulation. J Clin Invest 45: 1555
787 Seiffge D (1980) Methods for the investigation of drug-influence on aggregation and deformability of red blood cells (RBC). In: Stoltz JF, Drouin P (eds) Hemorheology and diseases. Doin, Paris, p657
788 Seiffge D, Berthold R, Berthold F (1983) Effect of pentoxifylline on sickle cell thalassaemia: Haemorheological and clinical results. Klin Wochenschr 61: 1159
789 Selvetella L, De Caterina M, Caso R, Amato B, Bracale G (1985) Variazioni emoreologiche ed emocoagulative dopo piretoterapia in pazienti con arteriopatia ad impronta flogistica. Ric Clin Lab 15 (Suppl 1): 571
790 Sengespeik HC, Güntner M, Raz D, Linderkamp O (1984) Hämorheologische Effekte von Intralipid. In: Ehrly AM (ed) Therapie mit hämorheologisch wirksamen Substanzen. Zuckschwerdt, München, p190
791 Sergio G, Artale F, Espureo M, Francisci A, Astorre P, Perego MA (1981) Haemorheological improvement by pentoxifylline and buflomedil in patients with peripheral occlusive arterial disease (Abstract). Clin Hemorheol 1: 504
792 Sgries B, Haedicke C, Altmann S, Stangel W, van den Berg E, Hartmann F (1980) Blood viscosity and peripheral blood flow in patients with polycythemia vera - The influence of therapeutic erythrocytopheresis. In: Stoltz JF, Drouin P (eds) Hemorheology and diseases. Doin, Paris, p405
793 Shabanov AH, Gudynskaja CJ, Komov DV (1967) Changes in blood viscosity in patients with endarteriitis obliterans. Lab Delo 12: 713
794 Shanghai Cooperative Group for the study of tanshinone IIA: Therapeutic effect of sodium tanshinone IIA sulfonate in patients with coronary heart disease. A double blind study. J Tradit Chin Med 4: 21
795 Shigehiro S, Kakubari Y, Muramatsu J, Kato Y, Kawabata K, Kikawada R (1979) Casson viscosity of blood in ischemic heart disease - Decrease of casson viscosity by ticlopidine (Abstract). Biorheology 16: 494
796 Shirinova MN, Goldfarb YS, Burykina IA, Makarenko ON (1983) Non specific detoxication mechanisms in the use of hemosorption for the treatment of burn toxemia and sepsis. Gematol Transfuziol 28: 32
797 Siekmann U, Heilmann L, Ludwig H (1983) The use of low molecular dextran in cases with microcirculatory disorders in pregnancy (Abstract). Clin Hemorheol 3: 273
798 Siekmann U, Heilmann L (1984) Der Einsatz von Beta-Rezeptoren-Antagonisten in der Geburtshilfe: klinische und rheologische Ergebnisse mit Metoprolol. In: Heilmann L, Kiesewetter H, Ernst E (eds) Klinische Rheologie und Beta-1-Blockade. Zuckschwerdt, München, p159
799 Simons LA, Hickie JB, Balasubramaniam S (1985) On the effects of dietary n-3 fatty acids (Maxepa) on plasma lipids and lipoproteins in patients with hyperlipidaemia. Atherosclerosis 54: 75
800 Sirs JA, Rampling MW, Dupont PA (1979) The effect of infusions of arvin into rabbits on blood rheology (Abstract). Microvasc Res 18: 282
801 Skovborg F, Nielsen AV (1969) Variation in whole-blood and plasma viscosity in acidotic diabetics. Bibl Anat 10: 145
802 Slonim A, Cristal N, Erez R, Shainkin-Kestenbaum R (1981) The effect of nifedipine, a calcium antagonist on red blood cell filterability (Abstract). Clin Hemorheol 1: 506
803 Smith E, Kochwa S, Wassermann L (1965) Aggregation of IgG globulin in vivo. Am J Med 39: 35
804 Smud R, Sermukslis B, Kartin D (1976) Changes in blood viscosity induced by pentoxifylline. Pharmatherapeutica 1: 229
805 Solerte SB, Carnevale Schianca GP, Adamo S, Viola C, Baldi S, Sgarro V, Ferrari E (1983) Diabetic

microangiopathy: preventive therapeutical approach with pentoxifylline (Abstract). Clin Hemorheol 3: 290
806 Solerte SB, Adamo S, Viola C, Bozzetti A, Di Santi M, Ferrari E (1985) Pentoxifylline reduces microalbuminuria in microproteinuric diabetic patients (Abstract). Clin Hemorheol 5: 734
807 Solerte SB, Ferrari E (1985) Diabetic retinal vascular complications and erythrocyte filtrability: results of a 2-year follow-up study with pentoxifylline. Pharmatherapeutica 4: 341
808 Somer T (1966) The viscosity of blood, plasma and serum in dys- and paraproteinemias. Acta Med Scand 180 (Suppl 456)
809 Somer T (1975) Hyperviscosity syndrome in plasma cell dyscrasias. Adv Microcirc 6: 1
810 Somer T, Ditzel J (1981) Clinical and rheological studies in a patient with hyperviscosity syndrome due to Waldenström's macroglobulinemia. Bibl Haematol 47: 242
811 Spürk P, Habbaba A, Angelkort B (1985) Improvements of blood-fluidity and walking-distance in patients under i.v.-therapy with pentoxifylline (Abstract). Clin Hemorheol 5: 731
812 Staedt U, Genth KR, Gaa J, Ritz R (1986) Hämodilution bei Patienten mit akuter zerebraler Durchblutungstörung mit niedermolekularem Dextran (Dex 40) und mittelmolekularer Hydroxyethylstärke (HES 200). Therapiewoche 36: 2485
813 Staedt U, Schwarz M, Bayerl JR, Tornow K, Heene DL (1986) Hämodilution bei akuter zerebraler Ischämie. Neurologischer Score und hämorheologische Parameter vor und nach 10tägiger Behandlung mit 10% Dex 40 und 10% HES 200/0,5 (II). Med Welt 37: 695
814 Staedt U, Schwarz M, Heene DL (1986) Hämodilution bei akuter zerebraler Ischämie. Hämorheologische Parameter vor und während Infusion von 10% Dex 40 und 10% HES 200/0,5 (I). Med Welt 37: 660
815 Stangel W, Haedicke C, Sgries B, Altmann S, van den Berg E, Hartmann F (1979) Influence of erythrocytapheresis on blood viscosity and peripheral blood flow in patients with polycythemia vera (Abstract). Symp Blood Cells Hemorheol 16: 502
816 Sternitzky R, Seige K (1985) Clinical investigation of the effects of pentoxifylline in patients with severe peripheral occlusive vascular disease. Curr Med Res Opin 9: 602
817 Stodter KH (1978) Einfluß von Adrevil auf die Fließeigenschaften menschlichen Blutes. Dissertationsschrift, Frankfurt am Main
818 Störmer B, Kleinschmidt K, Loose D, Kremer K (1977) Rheological changes in the blood of patients with chronic arterial occlusive disease after the administration of vasoactive drugs. Curr Med Res Opin 4: 588
819 Störmer B, Kleinschmidt K, Loose D, Kremer K (1978) Rheologische Veränderungen des Blutes bei chronisch arteriellen Verschlußkrankheiten durch Vasodilatantien. Dtsch Med Wochenschr 103: 694
820 Störmer B, Störmer K, Kremer K (1978) Die Veränderung der Blutviskosität während der aortofemoralen Bypassoperation unter besonderer Berücksichtigung niedriger Schergeschwindigkeiten. Biorheology 15: 37
821 Störmer K, Störmer B, Loose D, Engels G, Kremer K (1976) Einfluß von Kontrastmittel auf die Blutviskosität nach Katheterangiographie. Vasa 5: 239
822 Stoltz JF, Drouin P, Gaillard S, Rousselle D, Vera JC, Pointel JP (1979) Variations de la viscosité sanguine de sujets diabétiques insulino-dépendants équilibrés par pancréas artificiel. Ann Biol Clin 37: 191
823 Stoltz JF, Drouin P (eds) (1980) Hemorheology and diseases. Doin, Paris
824 Stoltz JF, Larcan A (1980) Drugs affecting hemorheological parameters. In: Stoltz JF, Drouin P (eds) Hemorheology and diseases. Doin, Paris, p651
825 Stoltz JF, Schooneman F, Streiff F (1981) Hemorheological parameters and plasma exchanges. Abstract-Booklet 2nd Europ Conf Clin Haemorheol, London, p107
826 Stoltz JF (1981) Drugs affecting blood rheology - a review. Scand J Clin Lab Invest 41 (Suppl 156): 287
827 Stoltz JF (1982) Main determinants of red cell deformability. Clinical and pharmacological applications. Clin Hemorheol 2: 163
828 Stoltz JF (1983) Les déterminants de la rhéologie sanguine: implications thérapeutiques. Actua Chim Thér 10: 29
829 Stoltz JF (1984) Blood rheology: etiology of hyperviscosity syndromes. Int Angiol 3: 13
830 Stoltz JF (1985) Données fondamentales en hémorhéologie. II. Les syndromes d'hyperviscosité. Biomed Pharmacother 39: 282
831 Stoltz JF, Schooneman F, Paulus F, Streiff F (1985) Micro- and macrorheological variations observed in 50 patients suffering from monoclonal dysglobulinemia and treated with plasma exchange procedures (Abstract). Clin Hemorheol 5: 728
832 Stoltz JF (ed) (1986) Hémorhéologie et agrégation érythrocytaire. Médicales Internationales, Paris
833 Strand T, Asplund K, Eriksson S, Hägg E, Lithner F, Wester PO (1984) A randomized controlled trial of hemodilution therapy in acute ischemic stroke. Stroke 15: 980

834 Strano A, Davì G, Novo S, Avellone G, Pinto A (1981) Valutazione clinica degli effetti della pentossifillina in pazienti effetti da disturbi circolatori periferici cronici. Ric Clin Lab 11 (Suppl 1): 303
835 Strano A, Novo S, Davì G, Avellone G, Mandalà V (1982) Effects of sorbinicate and nicotinic acid on blood viscosity, red cell deformation and platelet function. Pharmacol Res Commun 14: 639
836 Strano A, Avellone G, Davi G, Mandala V, Raneli G (1983) Ticlopidine induced haemorheological changes in patients suffering from atherosclerosis obliterans of the lower limbs (Abstract). Clin Hemorheol 3: 290
837 Strano A, Avellone G, Pinto A, Mandala V, Adamo L (1983) Pentoxiphylline induced haemorheological changes in patients suffering from atherosclerosis obliterans (Abstract). Clin Hemorheol 3: 291
838 Strano A, Daví G, Novo S, Avellone G, Pinto A (1983) Clinical evaluation of the effects of pentoxifylline in patients with chronic peripheral circulatory disorders. Pharmatherapeutica 3 (Suppl 1): 117
839 Strano A, Davi G, Avellone G, Novo S, Pinto A (1984) Double-blind, crossover study of the clinical efficacy and the hemorheological effects of pentoxifylline in patients with occlusive arterial disease of the lower limbs. Angiology 35: 459
840 Strano A (1985) The rationale for haemorheological treatment of peripheral occlusive arterial disease (POAD) (Abstract). Clin Hemorheol 5: 755
841 Strano A, Avellone G, Novo S, Fasulo S, Di Garbo V (1985) Modificazioni di alcuni parametri emoreologici indotte dal trattamento con pentossifillina in pazienti affetti da pregresso infarto del miocardio. Ric Clin Lab 15 (Suppl 1): 621
842 Strano A, Davì G, Avellone G, Novo S, Pinto A (1986) Double-blind, crossover study of the clinical efficacy and the haemorheological effects of pentoxifylline in patients with occlusive arterial disease of the lower limbs. In: Trübestein G (ed) Konservative Therapie arterieller Durchblutungsstörungen. Thieme, Stuttgart, p 113
843 Strauer BE, Fateh-Moghadam A, Kment A, Samtleben W, Volger E (1981) Use of plasmapheresis and immunosuppressive therapy in coronary microangiopathies. Bibl Haematol 47: 213
844 Strauer BE (1983) Koronare Durchblutungsstörungen aus rheologischer Sicht. Therapiewoche 33: 841
845 Strauß W (1975) Hämorheologische Wirkungen einer längerfristigen Gabe des Diuretikums Furosemid im Rahmen einer antihypertensiven Therapie. Dissertationsschrift, Frankfurt am Main
846 Stuart J, Kenny MW (1981) Sickle-cell disease and vascular occlusion. In: Lowe GDO, Barbenel JC, Forbes CD (eds) Clinical aspects of blood viscosity and cell deformability. Springer, Berlin, p 109
847 Stuart J, Stone PCW, Bareford D, Caldwell NM, Davies JE, Baar S (1985) Evaluation of leucocyte removal methods for studies of erythrocyte deformability. Clin Hemorheol 5: 137
848 Stufano N, Sabbà C, Altomare E, Conese A, Palasciano G, Albano O (1983) Erythrocyte filtration and glycemic control in diabetes mellitus (Abstract). Clin Hemorheol 3: 322
849 Sturani C, Palareti G, Poggi M, Schiavina M, Torricelli P, Papiris S (1985) Effects of pentoxifylline (Px) on cardiopulmonary hemodynamics and haemorheologic parameters at rest and during exercise in chronic cor pulmonale (Abstract). Clin Hemorheol 5: 732
850 Sugai S (1972) IgA pyroglobulin, hyperviscosity syndrome and coagulation abnormality in a patient with multiple myeloma. Blood 39: 224
851 Sunder-Plassmann L, Klövekorn WP, Holper K, Hase U, Messmer K (1971) The physiological significance of acutely induced hemodilution. Proceedings of the 6th European Conference on Microcirculation, Aalborg (1970). Karger, Basel, p 23
852 Swank RL (1951) Changes in blood produced by fat meal and by intravenous heparin. Am J Physiol 164: 798
853 Talpos G, White JM, Horrocks M, Cotton LT (1978) Plasmapheresis in Raynaud's disease. Lancet 1: 416
854 Tekeres M, Sarosi I, Juricskay I, Losoncsy H, Nagy I (1982) Effect of longterm ticlopidine therapy on the rheological properties of blood (Abstract). Haemostasis 12: 120
855 Terekhov NT, Starikov AV, Kushko OV, Karpenko VV (1982) Effect of hemosorption on the rheological properties of the blood and some indices of intoxication in patients with diffuse peritonitis. Klin Khir 3: 31
856 Theiss W, Volger E, Wirtzfeld A, Keisel I, Blömer H (1980) Gerinnungsbefunde und rheologische Messungen bei Streptokinasebehandlung des akuten Myokardinfarktes. Klin Wochenschr 58: 607
857 Thomas DJ, Du Boulay GH, Marshall J, Pearson TC, Ross Russell RW, Symon L, Wetherley-Mein G, Zilkha E (1977) Cerebral blood-flow in polycythaemia. Lancet 2: 161
858 Thomas DJ, Marshall J, Ross Russell RW, Wetherley-Mein G, Du Boulay GH, Pearson TC, Symon L, Zilkha E (1977) Effect of haematocrit on cerebral blood-flow in man. Lancet 2: 941
859 Thulesius O, Gjöres JE (1972) Studies on blood viscosity in chronic venous insufficiency, with regard to treatment with flavonoids. Angiologica 9: 390

860 Tilsner V, Greul W (1975) Erfahrungen mit der defibrinierenden Therapie mit Arwin®. Folia Angiol 23: 407
861 Toikawa H, Tsushima N, Sakakura M, Kaba M, Konishi M, Nakayama R, Yasunaga K (1985) Effect of prazosin on microcirculation system (Abstract). Clin Hemorheol 5: 577
862 Toti G, Agosti R, Cherubini P, Farini P, Clivati A, Longhini E (1985) Normovolemic hemodilution in surgery (Abstract). Clin Hemorheol 5: 688
863 Trevan JW (1918) The viscosity of blood. Biochem J 12: 60
864 Tsyganyi AA, Karpenko VV, Krishtof AN, Pritula SY (1982) The influence of plasma replacers on the rheological properties of the blood, suspension stability of the erythrocytes and microcirculation in patients operated for mitral stenosis. Anesteziol Reanimatol 4: 38
865 Tuddenham EGD, Whittaker JA, Bradley J, Lilleyman JS, James DR (1974) Hyperviscosity syndrome in IgA multiple myeloma. Br J Haematol 27: 65
866 Turczynski B, Znamirowska D (1985) Effect of adrenergic beta receptor blockaders on blood viscosity after their short-term use in patients with arteriosclerosis. Polski Tygodnik Lekarski (Warszawa) 40 (17): 481
867 Tymms DJ (1982) Improvement in blood viscosity following continuous subcutaneous insulin infusion in type 2 (non-insulin-dependent) diabetic patients (Abstract). Diabetologia 23: 207
868 Uchimura I (1983) Effect of treatment on retinal circulation, platelet aggregation and blood viscosity in insulin-dependent and noninsulin-dependent diabetics. Bull Tokyo Med Dent Univ 30: 17
869 Urai L, Makláry E, Kolonics I, Novák L (1983) Effect of pentoxifylline (Trental) on haemorheological factors and the circulation in peripheral occlusive arterial disease. Abstract-Booklet 13th World Congr Int Union Angiol, Rochester, Minnesota, USA
870 Vakaliuk IP (1984) Efficacy of corvaton in non-stable stenocardia. Vrach Delo 11: 18
871 Van den Berg E, Haedicke C, Sgries B, Altmann S, Stangel W, Hartmann F (1980) Einfluß der Erythrozytopherese auf das Hyperviskositätssyndrom bei Polycythaemia vera - Klinik, Viskosität und arterielle Durchblutung. In: Müller-Wiefel H, Barras JP, Ehringer H, Krüger M (eds) Gefäßersatz. Witzstrock, Baden-Baden, p 119
872 Van den Berg E, Haedicke C, Sgries B, Stangel W, Altmann S, Hartmann F (1983) Klinik und Therapie des Hyperviskositätssyndroms. Vasa 12 (Suppl 10): 1
873 Van Stalle F, Lambelin G (1981) Suloctidil: review of its effects on platelet survival time, platelet activation and blood viscosity (Abstract). Clin Hemorheol 1: 509
874 Vaya A (1985) RBC deformability in patients with cerebrovascular disease. Influence of pentoxifylline and dipyridamole (Abstract). Clin Hemorheol 5: 734
875 Vinazzer H (1971) Zur Wirkung von Arvin auf die Blutgerinnung. Wien Z Inn Med 52: 378
876 Vittoria A, Guerrini M, Pieragalli D, Del Bigo C, Martelli G, Franchi M, Messa G, Blardi P, Galigani C, Di Perri T (1983) Azione dell'acetilsalicilato di lisina sulle alterazioni emoreologiche indotte dallo sforzo ischemico in pazienti vasculopatici. Ric Clin Lab 13 (Suppl 3): 315
877 Völker D, Martin M (1973) Verhalten der Blutviscosität unter Defibrase und der kombinierten Therapie mit Defibrase und Streptokinase. Verh Dtsch Ges Inn Med 79: 1338
878 Völker D (1974) Rheologische Veränderungen unter oraler Langzeittherapie mit Bencyclan. Therapiewoche 24: 2901
879 Völker D (1975) Rheologische Veränderungen unter der Therapie mit Defibrase. In: Martin M, Schoop W (eds) Aktuelle Probleme in der Angiologie, Bd 26; p 171
880 Völker D (1976) Haemorheological investigations in patients with vascular disease undergoing treatment with pentoxifylline. Pharmatherapeutica 1: 154
881 Völker D (1979) Orale Therapie mit Bencyclan. Z Allg Med 55: 182
882 Völker R, Eichler K (1965) Nebenwirkungen von Diuretika auf Rheodynamik und Blutgerinnung. Dtsch Med Wochenschr 90: 2150
883 Vogeler C (1973) Untersuchungen über die medikamentöse Beeinflussung der Filtrabilität des Blutes. Dissertationsschrift, Frankfurt am Main
884 Voisin P, Kolopp M, Rousselle D, Gaillard S, Pointel JP, Stoltz JF, Debry G, Drouin P (1982) Influence du contrôle métabolique sur la viscosité sanguine et l'activité plaquettaire chez les diabétiques insulino-dépendants. Nouv Rev Fr Hématol 24: 187
885 Vojnikovic B (1984) Hyperviscosity in whole blood, plasma and aqueous humor decreased by Doxium (calcium dobesilate) in diabetics with retinopathy and glaucoma: A double-blind controlled study. Ophthalmic Res 16: 150
886 Volger E (1980) Experimentelle und klinische Untersuchungen über die Rheologie des Blutes bei kardiovaskulären Erkrankungen und deren Risikofaktoren. Habilitationsschrift, München
887 Volger E (1983) Klinische Hämorheologie und Diabetes mellitus. In: Ehrly AM (ed) Klinische Hämorheologie: Eine Bestandsaufnahme. Zuckschwerdt, München, p 62
888 Volger E, Theiss W (1983) Rheological consequences of systemic streptokinase therapy in acute myo-

cardial infarction. In: Trübestein G, Etzel F (eds) Fibrinolytische Therapie. Schattauer, Stuttgart, p455
889 Von Rhede van der Kloot EJH, Jacobs MJHM, Weber H, Lemmens HAJ (1985) Plasma filtration in patients with Raynaud's phenomenon. Clin Hemorheol 5: 79
890 Wade JPH (1983) Transport of oxygen to the brain in patients with elevated haematocrit values before and after venesection. Brain 106: 513
891 Walker CHM, Mackintosh TF (1972) The treatment of hyperviscosity syndromes of the newborn with hemodilution. In: Messmer K, Schmid-Schönbein H (eds) Hemodilution - Theoretical basis and clinical application. Karger, Basel, p271
892 Walker RT, Matrai A, Dormandy JA, Flute PT (1983) Haemorheological and functional changes in intermittent claudication following plasmapheresis with Haemocell (Abstract). Clin Hemorheol 3: 338
893 Walker RT, Dormandy J (1984) Plasma exchange in peripheral vascular disease. Arch Emerg Med 1 (Suppl): 23
894 Walker RT, Matrai A, Bogar L, Dormandy JA (1985) Serotonin and the flow properties of blood. J Cardiovasc Pharmacol 7 (Suppl 7): 35
895 Waller DG (1982) Whole blood viscosity following cardioselective beta-adrenoceptor blockade in hypertension. Clin Hemorheol 2: 243
896 Waller D, Nicholson H, Roath S (1983) The acute effects of nifedipine on red cell deformability in angina pectoris (Abstract). Clin Hemorheol 3: 294
897 Waller DG, Nicholson HP, Roath S (1984) The acute effects of nifedipine on red cell deformability in angina pectoris. Br J Clin Pharmacol 17: 133
898 Waller DG, Roath OS (1986) Nifedipine and red cell deformability in angina. Clin Hemorheol 6: 275
899 Wallis PJW, Skehan JD, Newland AC, Wedzicha JA, Mills PG, Empey DW (1986) Effects of erythrapheresis on pulmonary haemodynamics and oxygen transport in patients with secondary polycythaemia and cor pulmonale. Clin Sci 70: 91
900 Warstat T (1974) Untersuchungen über den Einfluß von niedermolekularem Dextran auf die Filtrabilität des Blutes in 8 μm-Filtern. Dissertationsschrift, Frankfurt am Main
901 Wasilewski A (1962) Zur Pathophysiologie und Therapie der kardialen Dekompensation unter Berücksichtigung der Viskositätsänderung des Blutes. Med Klin 57: 599
902 Watson WC (1957) Lipaemia, heparin and blood viscosity. Lancet 2: 366
903 Watzek C, Wagner O, Draxler V, Gilly H, Schwarz S, Sporn P, Steinbereithner K, Zekert F (1978) Der Einfluß normovolämischer Hämodilution unter Verwendung von Hydroxyäthylstärke auf Kreislauf und Organfunktionen bei gefäßchirurgischen Eingriffen. Wien Klin Wochenschr 90: 224
904 Watzlaw P (1978) Untersuchungen über den Einfluß von Actovegin und Doxium auf die Fließeigenschaften des Blutes. Dissertationsschrift, Frankfurt am Main
905 Weber G (1979) Kontrolle therapeutischer Effekte im Bereich der Mikrozirkulation. In: Hild R, Spaan G (eds) Therapiekontrolle in der Angiologie. Witzstrock, Baden-Baden, p336
906 Weber G, Kreisel T, Peter S, Künzel J (1980) A double-blind placebo controlled crossover study in patients with peripheral vascular diseases, using a new capillary viscometer. Angiology 31: 1
907 Weber H, von Rhede van der Kloot E, Jacobs MJ, Schmid-Schönbein H, Lemmens HAJ (1983) Plasma filtration therapy in patients with Raynaud's phenomenon - 2. Haemorheological aspects (Abstract). Clin Hemorheol 3: 338
908 Weber H, Schmid-Schönbein H, Lemmens HAJ (1985) Plasmapheresis as a treatment of Raynaud-attacks: Microrheological differential diagnosis and evaluation of efficacy. Clin Hemorheol 5: 85
909 Wedzicha JA, Rudd RM, Apps MCP, Cotter FE, Newland AC, Empey DW (1983) Erythrapheresis in patients with polycythaemia secondary to hypoxic lung disease. Br Med J 286: 511
910 Wedzicha JA, Cotter FE, Rudd RM, Apps MCP, Newland AC, Empey DW (1984) Erythrapheresis compared with placebo apheresis in patients with polycythaemia secondary to hypoxic lung disease. Eur J Respir Dis 65: 579
911 Weinberger I, Fuchs J, Rappaport M, Rotenberg Z, Joshua H, Agmon J (1985) Effect of sublingual isosorbide dinitrate and nifedipine on plasma viscosity, fibrinogen and hematocrit in patients with myocardial infarction. Abstract-Booklet Int Symp Cardiovasc Pharmacother, Geneva, Switzerland, Abstract 340
912 Wells R (1970) Syndromes of hyperviscosity. N Engl J Med 283: 183
913 Wells R (ed) (1973) The microcirculation in clinical medicine. Academic Press, New York
914 Welsh WH (1911) Viscosity of the blood. Heart 3: 118
915 West N (1984) Hyperviscosity syndrome in IgE myeloma. Br Med J 289: 1539
916 Whittacker SRF, Winton FR (1933) The apparent viscosity of the blood flowing in the isolated hindlimb of the dog and its variation with corpuscular concentration. J Physiol (Lond) 78: 339

917 Williams C, Rodriguez M, Llan de Rosos O, Pederson J, Rankin L, Llach F (1983) Blood viscosity (V), systemic vascular resistance (SVR) and blood pressure (BP) changes during hemodialysis (HD) (Abstract). Kidney Int 23: 165

918 Wintrich H, Rudofsky G, Maier J (1983) The influence of IHD on hemodynamic and rheological parameters in patients with arterial occlusive disease (Abstract). Clin Hemorheol 3: 310

919 Witte S, Anadere I, Chmiel H (1983) The influence of a gingko biloba extract on the increased viscoelasticity of blood by complete stroke (Abstract). Clin Hemorheol 3: 291

920 Wodniok M (1981) Untersuchungen über die Wirkung zweier Diuretika auf die Fließeigenschaften des Blutes bei Patienten mit essentieller Hypertonie. Dissertationsschrift, Frankfurt am Main

921 Wolf RE, Alperin JB, Ritzmann SE, Levin WC (1972) IgG-K multiple myeloma with hyperviscosity syndrome - Response to plasmapheresis. Arch Intern Med 129: 114

922 Wolfe JHN, Blackford HN, Waller D, Chapman MB, Prout WG (1983) Does haemodilution benefit patients with intermittent claudication? (Abstract). Clin Hemorheol 3: 310

923 Wolfe JHN, Waller DG, Chapman MB, Blackford HN, Prout WG (1985) The effect of hemodilution upon patients with intermittent claudication. Surg Gynecol Obstet 160: 347

924 Woodcock BE, Smith E, Lambert WH, Morris Jones W, Galloway JH, Greaves M, Preston FE (1984) Beneficial effect of fish oil on blood viscosity in peripheral vascular disease. Br Med J 288: 592

925 Wrabetz-Wölke A, Altmann S, Deicher H, Stangel W (1980) Changes in whole blood and plasma viscosity after plasmapheresis in patients with hyperviscosity syndrome (Abstract). In: Stoltz JF, Drouin P (eds) Hemorheology and diseases. Doin, Paris, p605

926 Yamagata J, Shiozawa K, Shiokawa J (1980) Therapeutic plasma exchange for rheumatic disease. In: Sieberth HG (ed) Plasma exchange. Schattauer, Stuttgart, p266

927 Yamaguchi H (1984) Therapeutic study of reduced red cell deformability in peripheral vascular disease (Abstract). Angio Arch 7: 51

928 Yao ST, Shoemaker WC (1966) Plasma and whole blood viscosity changes in shock and after dextran infusion. Ann Surg 164: 973

929 Yasunaga K, Kumada K, Matsuda K (1979) Coagulation studies on the patients treated with Defibrase®, a snake venom batroxobin. Jpn Arch Int Med 26: 465

930 Zederfeldt B (1965) Rheological disturbances and their treatment in clinical surgery. In: Copley AL (ed) Symposium on Biorheology. Part 4 of the Proceedings of the Fourth International Congress on Rheology, Brown University, Providence, Rhode Island, 1963. Interscience, New York, p397

931 Zerefos NS, Boletis J, Stathakis C, Alexopoulos J, Xefteri E, Katsilabros N, Daikos G (1983) Successful kidneys transplantation corrects the impaired blood viscosity of chronically hemodialyzed patients (Abstract). Kidney Int 23: 298

932 Zijlstra WG (1960) Syllektometrie. In: Kramer K (ed) Oxymetrie - Theorie und klinische Anwendung. Thieme, Stuttgart, p117

8 Index to the General Part

9 Comparison of Generic Names (INN) and Trade Names of a Selection of Drugs Named by the Authors in the Initial Description of a Hemorheologic Effect (Chronologic Order)

Dextran solution, MW 40 000 and 75 000 (Rheomacrodex, Macrodex; AB Pharmacia, Uppsala, Sweden)	[367]
Streptokinase (Streptase; Behringwerke, Marburg/Lahn, FRG)	[235]
Ancrod (Arwin; Knoll AG, Ludwigshafen, FRG)	[253]
Ethinyl estradiol + northisteron (Gynovular; Schering AG, Berlin, FRG)	[29]
O-(β-hydroxyethyl)-rutoside (Venoruton; Zyma AG, Nyon, Switzerland)	[47]
Xanthinol nicotinate (Complamin; Beecham-Wülfing, Neuss, FRG)	[155]
Bencyclane (Fludilat; Thiemann GmbH, Waltrop, FRG)	[676]
Pentoxifylline (Trental; Albert-Roussel, Wiesbaden, FRG)	[444]
Naftidrofuryl (Dusodril; Lipha GmbH, Essen, FRG)	[676]
Carbocromen (Intensain; Cassella-Riedel Pharma GmbH, Frankfurt, FRG)	[129]
Isoxsuprine (Duvadilan; Karl Thomae GmbH, Biberach/Riss, FRG)	[755]
Calcium dobesilate (Dexium; Delalande Arzneimittel GmbH, Köln, FRG)	[457]
Flunarizine (Sibelium; Janssen GmbH, Beerse/Belgium; Neuss, FRG)	[744]
Hydroxyethyl starch solution (Plasmasteril; Fresenius AG, Bad Homburg, FRG)	[489]
Buflomedil (Bufedil; Deutsche Abbott GmbH, Wiesbaden, FRG)	[448]
Sodium pentosan polysulphate (Pentosan polysulphate SP 54; bene-Arzneimittel, München, FRG)	[623]

MIX
Papier aus verantwortungsvollen Quellen
Paper from responsible sources
FSC® C105338

If you have any concerns about our products,
you can contact us on
ProductSafety@springernature.com

In case Publisher is established outside the EU,
the EU authorized representative is:
Springer Nature Customer Service Center GmbH
Europaplatz 3, 69115 Heidelberg, Germany

Printed by Libri Plureos GmbH
in Hamburg, Germany